Non-Destructive Testing of Materials and Parts: Techniques, Case Studies and Practical Applications

Non-Destructive Testing of Materials and Parts: Techniques, Case Studies and Practical Applications

Guest Editors

Luís Miguel Pereira Durão
Nuno Calçada Loureiro

MDPI

Basel • Beijing • Wuhan • Barcelona • Belgrade • Novi Sad • Cluj • Manchester

Guest Editors

Luís Miguel Pereira Durão
Department of Mechanical Engineering
ISEP, Polytechnic of Porto
Porto
Portugal

Nuno Calçada Loureiro
Department of Industrial Production Engineering
Instituto Superior de entre Douro e Vouga
Santa Maria da Feira
Portugal

Editorial Office
MDPI AG
Grosspeteranlage 5
4052 Basel, Switzerland

This is a reprint of the Special Issue, published open access by the journal *Materials* (ISSN 1996-1944), freely accessible at: https://www.mdpi.com/journal/materials/special_issues/1IK320160B.

For citation purposes, cite each article independently as indicated on the article page online and as indicated below:

Lastname, A.A.; Lastname, B.B. Article Title. *Journal Name* **Year**, *Volume Number*, Page Range.

ISBN 978-3-7258-5109-6 (Hbk)
ISBN 978-3-7258-5110-2 (PDF)
https://doi.org/10.3390/books978-3-7258-5110-2

Contents

About the Editors

Luís Miguel Pereira Durão

Luís Miguel Pereira Durão has a PhD in Mechanical Engineering. From 2007 to 2015, he coordinated CIDEM, a Research Center in Mechanical Engineering at ISEP. Presently, he is a Coordinator Professor at the Mechanical Engineering Dept. of ISEP, Polytechnic of Porto, and develops research activities at INEGI and CIDEM, Porto, Portugal. He is a member of the Mechanical Engineering College of the Ordem dos Engenheiros (Portuguese Engineers Association).

Nuno Calçada Loureiro

Nuno Calçada Loureiro has a PhD in Leaders for Technical Industries. Presently, he is a Coordinator Professor at ISVOUGA, Dept. of Industrial Production Engineering, Santa Maria da Feira, Portugal, and develops a research activity at INEGI, Porto, Portugal. He is also a member of the Specialization Commission in Teaching and Research of the Ordem dos Engenheiros (Portuguese Engineers Association).

materials

MDPI

Editorial

Special Issue: Non-Destructive Testing of Materials and Parts—Techniques, Case Studies and Practical Applications

Luis M. P. Durão [1,2,*] and Nuno C. Loureiro [2,3]

1 ISEP, Polytechnic of Porto, rua Dr. António Bernardino de Almeida, 4249-015 Porto, Portugal
2 INEGI Inst. de Ciência e Inovação em Eng. Mecânica e Eng. Industrial, R. Dr. Roberto Frias 400, 4200-465 Porto, Portugal; n.loureiro@doc.isvouga.pt
3 ISVOUGA, Instituto Superior de Entre Douro e Vouga, R. António de Castro Corte Real, Apart. 132, 4520-181 Santa Maria da Feira, Portugal
* Correspondence: lmd@isep.ipp.pt

The simplest definition of Non-Destructive Testing (NDT) is to "Inspect or measure without doing harm". From simple visual inspection to the use of lasers, NDT methods cover a broad range of uses in diverse environments. For example, in daily life, we practice visual inspection when we check the thread of a screw; our luggage is inspected by radiography when we go to an airport; and many medical diagnoses rely on echography, an ultrasound method. Thus, NDT provides security, reliability and sustainability in our technological society.

In fact, due to its ability to evaluate the properties of a material, component or system without causing damage, NDT ensures safety in industries like aerospace, nuclear, oil and gas, and construction, where structural integrity is vital; preserves assets by allowing for regular inspections of infrastructures; and supports quality control. Furthermore, NDT is cost-effective, as components can be tested without being destroyed, avoiding costly repairs, replacements or catastrophic failures, and contributing to sustainability by minimizing waste and conserving resources. Despite its key role in many industries, NDT faces several challenges that can impact the accuracy, efficiency and overall effectiveness of inspections.

Current challenges faced by NDT include the detection of tiny or subsurface flaws, especially in complex geometries or hard-to-reach areas; the problems posed by high material complexity, like composites or highly attenuative/anisotropic materials; low Signal-to-Noise ratios, which make it hard to differentiate actual defect signals from background noise; and subjectivity issues, as the interpretation of NDT results still relies on technicians' experience.

This Special Issue, titled Non-Destructive Testing of Materials and Parts: Techniques, Case Studies and Practical Applications, aims to serve as a forum for presenting the latest research and developments in this field, present new ideas that could help to close knowledge gaps, and highlight future research directions to foster future developments on this interesting topic.

As Guest Editors of this Special Issue, following a rigorous peer-review process of manuscripts received over 16 months, we are pleased to announce the publication of 10 manuscripts. The accomplishment of this Special Issue was only possible due to the interest from proficient researchers from all over the world. The Guest Editors would like to thank all of the authors of the published works and also express their gratitude to the reviewers for their time, and for their valuable comments and suggestions, which improved the quality and value of this Special Issue. The success of this Special Issue would not have been possible without the support of the Section Managing Editor, Ms. Serena Shi, who we thank for her dedication and commitment. The Guest Editors also thank the Editors-in-Chief of *Materials* for this opportunity for collaboration, and congratulate them on their stewardship of such a globally respected journal. The diligence, creativity, and dynamic cooperation of all those mentioned above contributed to the success of this Special Issue.

Several recently published papers focus on the contributions and applications of current Non-Destructive Testing techniques, including Machine Learning, which can be easily combined with other NDT techniques, like acoustic emission (AE) [1] or AE plus radiography, for further data analysis [2]. The monitoring of structures remains an essential issue in NDT, independently of technical fields, from oil and gas [3] to aerospace [4]. Complex materials, like composites, are a recurring theme in NDT studies [2,5], and are also the subject of several thorough reviews of sensing technologies [6], showing the importance of these techniques in detailed defect detection.

We invite the reader to peruse the entire book to become acquainted with the themes of focus in the 10 published papers. One study discusses the use of modern methods, such as acoustic emission combined with machine learning, to conduct effective structural health monitoring, while another presents a method that provides a more effective approach to early fatigue-damage detection by capturing nonlinear Lamb waves, and another investigates the use of laser ultrasonic wave techniques as innovative visualization methods for damage detection. In addition, one study investigates how high-resolution ultrasound can be used to quantify sub-surface wrinkles, for instance, in a CFRP laminate, while another discusses the possibilities presented by combining NDT methods for damage detection in FRP-reinforced elements.

However, there is also some room for studies on the development of electromagnetic tests to study mechanical and thermal features in ferrous alloys, or advances in the use of nanoindentation and sclerometry to evaluate the effects of machining processes on the mechanical characteristics of metallic parts.

Finally, the use of NDT in carbon fiber-reinforced polymers should not be disregarded, as these materials are becoming more important in terms of their applications; both thermal methods and enhanced radiography or image processing have been used for studying machining outcomes like damage extension or other features related to material removal processes primarily developed for metal machining. These recent improvements could contribute to improving the reliability of structural parts made from carbon fiber-reinforced polymers.

As Guest Editors, we hope this Special Issue provides readers with some new insights on the use of Non-Destructive methods like acoustic emission, ultrasound, electromagnetic waves, thermal imaging, nanoindentation, thermography and enhanced radiography.

Author Contributions: Conceptualization, L.M.P.D. and N.C.L.; writing—original draft preparation, L.M.P.D.; writing—review and editing, N.C.L. All authors have read and agreed to the published version of the manuscript.

Funding: This research received no external funding.

Acknowledgments: The Guest Editors wish to acknowledge all of the authors for their vital contributions to this Special Issue, and to the editorial staff of *Materials* for their invaluable support.

Conflicts of Interest: The authors declare no conflicts of interest.

References

1. Almeida, R.S.M.; Magalhães, M.D.; Karim, M.N.; Kamen Tushtev, K.; Kurosch Rezwan, K. Identifying damage mechanisms of composites by acoustic emission and supervised machine learning. *Mater. Des.* **2023**, *227*, 111745. [CrossRef]
2. Osa-uwagboe, N.; Udu, A.G.; Ghalati, M.K.; Silberschmidt, V.V.; Aremu, A.; Dong, H.; Demirci, E. A machine learning-enabled prediction of damage properties for fiber-reinforced polymer composites under out-of-plane loading. *Eng. Struct.* **2024**, *308*, 117970. [CrossRef]
3. Trujillo, M.A.; Dios, J.R.M.; Martín, C.; Viguria, A.; Ollero, A. Novel Aerial Manipulator for Accurate and Robust Industrial NDT Contact Inspection: A New Tool for the Oil and Gas Inspection Industry. *Sensors* **2019**, *19*, 1305. [CrossRef] [PubMed]
4. Deane, S.; Avdelidis, N.P.; Ibarra-Castanedo, C.; Zhang, H.; Nezhad, H.Y.; Williamson, A.A.; Mackley, T.; Davis, M.J.; Maldague, X.; Tsourdos, A. Application of NDT thermographic imaging of aerospace structures. *Infrared Phys. Technol.* **2019**, *97*, 456–466. [CrossRef]
5. Lee, Y.; Roh, H.D.; Park, H.W.; Park, Y.B. Advanced non-destructive evaluation of impact damage growth in carbon-fiber-reinforced plastic by electromechanical analysis and machine learning clustering. *Compos. Sci. Technol.* **2022**, *218*, 109094. [CrossRef]
6. Gupta, R.; Mitchell, D.; Blanche, J.; Harper, S.; Tang, W.; Pancholi, K.; Baines, L.; Bucknall, D.G.; Flynn, D. A Review of Sensing Technologies for Non-Destructive Evaluation of Structural Composite Materials. *J. Compos. Sci.* **2021**, *5*, 319. [CrossRef]

materials

MDPI

Article

Effects of Seawater on Mechanical Performance of Composite Sandwich Structures: A Machine Learning Framework

Norman Osa-uwagboe [1,2], Amadi Gabriel Udu [2,3], Vadim V. Silberschmidt [1,*], Konstantinos P. Baxevanakis [1] and Emrah Demirci [1]

[1] Wolfson School of Mechanical, Electrical, and Manufacturing Engineering, Loughborough University, Loughborough LE11 3TU, UK; n.osa-uwagboe@lboro.ac.uk (N.O.-u.); k.baxevanakis@lboro.ac.uk (K.P.B.); e.demirci@lboro.ac.uk (E.D.)

[2] Air Force Research and Development Centre, Nigerian Air Force Base, Kaduna PMB 2104, Nigeria

[3] School of Engineering, University of Leicester, Leicester LE1 7RH, UK

* Correspondence: v.silberschmidt@lboro.ac.uk; Tel.: +44-(0)-1509-227504

Abstract: Sandwich structures made with fibre-reinforced plastics are commonly used in maritime vessels thanks to their high strength-to-weight ratios, corrosion resistance, and buoyancy. Understanding their mechanical performance after moisture uptake and the implications of moisture uptake for their structural integrity and safety within out-of-plane loading regimes is vital for material optimisation. The use of modern methods such as acoustic emission (AE) and machine learning (ML) could provide effective techniques for the assessment of mechanical behaviour and structural health monitoring. In this study, the AE features obtained from quasi-static indentation tests on sandwich structures made from E-glass fibre face sheets with polyvinyl chloride foam cores were employed. Time- and frequency-domain features were then used to capture the relevant information and patterns within the AE data. A *k*-means++ algorithm was utilized for clustering analysis, providing insights into the principal damage modes of the studied structures. Three ensemble learning algorithms were employed to develop a damage-prediction model for samples exposed and unexposed to seawater and were loaded with indenters of different geometries. The developed models effectively identified all damage modes for the various indenter geometries under different loading conditions with accuracy scores between 86.4 and 95.9%. This illustrates the significant potential of ML for the prediction of damage evolution in composite structures for marine applications.

Keywords: composite sandwich; machine learning; acoustic emission; damage prediction; seawater exposure

1. Introduction

Fibre-reinforced plastic sandwich structures (FRPSS) are a common form of composite materials with growing application in various industrial sectors thanks to their superior mechanical properties, particularly their stiffness-to-weight ratio, which surpasses that of metals [1–3]. In the past few decades, structures made from FRPSS have been employed in the automotive, aerospace, and marine industries. Specifically, significant parts of underwater vessels, boats, and wind turbines are fabricated using FRPSS thanks to their low weight, high strength, the ease of manufacturing complete shapes, cost effectiveness, corrosion resistance, and buoyancy [4–6]. Composite sandwich structures typically comprise two (top and bottom) face sheets, joined together with a lightweight core. This design ensures the fabrication of a structure with relatively low weight yet sufficient stiffness and strength to withstand the in-service loads when compared to its constituent materials [7]. The overall mechanical performance of FRPSS depends on the properties and thickness of the face sheets, core, and the effectiveness of the bonds between the constituents [8,9]. It has been established that glass-fibre reinforcements are generally hydrophobic (do not absorb moisture); however, water ingress can affect the polymetric matrix resulting in plasticisation, swelling, and weakening of the interfacial bonding of the structure, thereby causing degradation of the structure's mechanical properties [10–12]. Consequently, common damage modes for FRPSS subjected to seawater exposure are matrix cracking, face sheet buckling, delamination, and disbonding between the face sheet and the core [13,14]. Historically, investigations into these damage modes mostly relied on experiments and numerical simulations, which can be both time consuming and resource intensive. Non-destructive techniques such as acoustic emission (AE) have been employed in several studies to characterize the damage-resistance properties of composite sandwich structures. The results from these studies showed that AE could identify the major damage modes of the structures and correlate them with the mechanical properties of the samples [15–17].

In recent years, the use of machine learning (ML) algorithms to evaluate the mechanical properties and performances of composite structures under diverse loading and environmental conditions has attracted increasing attention. For instance, transfer learning, a subset of ML, shows the potential to adapt to environmental and operational variabilities, thereby accurately identifying these properties [18–20]. This trend is attributed to the accuracy of the models and their robustness across several applications. For example, four distinct ML methods—namely, decision tree, support vector regression (SVG), Gaussian process regression (GPR), and an ensemble method—were used to predict the damage behaviour of carbon- and glass-fibre-reinforced composites under tensile loading regimes [21]. The developed models were assessed using performance matrices such as mean absolute error (MAE), mean squared error (MSE), root mean square error (RMSE), and the coefficient of determination (R^2). The findings indicated that GPR outperformed the other models with the highest R^2 of 0.98 and the lowest error values [22]. Also, a convolutional neural network based on SqueezeNet was employed, using wavelet scalograms to characterize the mechanical properties of FRP composite structures [23]. The obtained results yielded a performance of more than 85% for three of the four clustered data. A deep learning approach based on the Inception Time model was proposed in [24] for damage classification of AE time- and frequency-domain features to identify fibre breakage, matrix cracking, and delamination damage modes under tension. Additionally, damage prediction models for FRP composite structures under compression [25,26], impact [27,28], and fatigue [29] were also developed using ML approaches. While demonstrations of the application of ML techniques in damage prediction and structural health monitoring abound in the literature, there is a need to further develop them for sandwich structures experiencing a combination of quasi-static localized loads and environmental conditions. Such advancements would offer valuable possibilities for material optimization and significantly enhance their practicality, particularly for offshore applications. Moreover, ML approaches have the potential to offer designers a suitable methodology with a robust dataset that can be tailored for industry-specific applications. Consequently, such models could, therefore, be beneficial for a wide range of applications.

In this study, the damage evolution in sandwich structures exposed to seawater conditions was investigated under quasi-static indentation (QSI) using the AE technique for damage characterization. The primary objective of this research is to enhance our understanding of damage in FRPSS by establishing a predictive methodology. This methodology is based on multiple ML classification algorithms, with the identification of the best-performing models determined through assessments of accuracy, recall, precision, and F1-score. In the study, time- and frequency-domain features are extracted, while the *k*-means ML algorithm is employed to identify various damage conditions. Thereafter, ensemble-based algorithms are used to predict the damage conditions of control and seawater-exposed samples. The damage modes of these two types of samples, that have different cores and are subjected to localized loading from three indenter geometries, are chosen as response variables, while the AE time- and frequency-domain features based on a rigorous feature selection process are selected as predictor variables. This is based on the direct correlation between the damage to the material, its susceptibility to moisture ingress, and the variation in the AE features during loading. Although four damage modes were identified in a previous study and were shown to be influenced by specific indenter geometries [30], the novelty of this work is its analysis of the effects of seawater on the structural health of FRPSS.

2. Materials, Experiment, and Methodology

2.1. Materials and Specimens

Face sheets of the studied FRPSS were fabricated from E-glass plain-weave fabric with a weight of 160 g/m^2, sourced from Samson Composites Ltd. (Shenzhen, China). The matrix was epoxy resin obtained from EPOCHEM Ltd. (Lagos, Nigeria), with a volumetric ratio of 2:1. As for the core material, EASYCell 75 closed-cell PVC foam composites from Easycomposites Ltd. (Stoke-on-Trent, UK), were employed. The samples were fabricated at room temperature (27 °C) and 48% humidity in Nigeria with a curing time of 18 h. The process involved hand lay-up and vacuum-bagging techniques, resulting in sample plates with dimensions of 300 mm × 300 mm. The samples had three different core configurations, distinguished by the presence or otherwise and positions of additional adhesive layers connecting parts of the foam core (Figure 1). This fabrication method was chosen for its cost-effectiveness, simplicity, and versatility. Detailed mechanical parameters of the constituents and sample designations can be found in Tables 1 and 2, respectively.

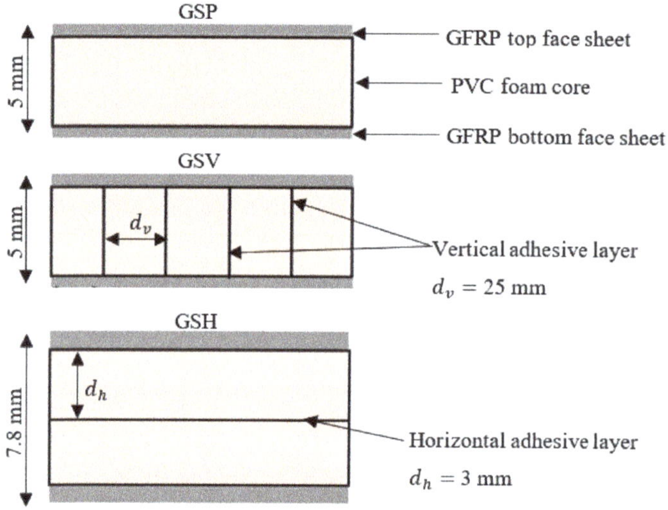

Figure 1. FRPSS fabrication configurations.

Table 1. Mechanical properties of materials [27,29].

Material	Young's Modulus (GPa)	Shear Modulus (GPa)	Tensile Strength (MPa)	Poisson Ratio	Density (g/cm^3)
	E	G_{12}		ν_{12}	ρ
E-glass fabric	72.39	8.27	3100–3800	0.26	2.25
PVC foam	0.075	0.028	1.89	-	0.075
Epoxy matrix	3.2–3.5	-	70–80	0.29	1.16

Table 2. Sample nomenclature.

Designation	Description
GSPC	Glass sandwich with pure core/conical indenter
GSVC	Glass sandwich with vertical adhesive layered core/conical indenter
GSHC	Glass sandwich with horizontal adhesive layered core/conical indenter
GSPH	Glass sandwich with pure core/hemispherical indenter
GSVH	Glass sandwich with vertical adhesive layered core/hemispherical indenter
GSPS	Glass sandwich with pure core/flat indenter
GSVS	Glass sandwich with vertical adhesive layered core/flat indenter
GSHS	Glass sandwich with horizontal adhesive layered core/flat indenter

Note: Below, notation _c and _s after sample designations is used for unexposed and exposed samples, respectively.

2.2. Sea Water Exposure and Moisture Absorption

To analyse the environmental effect, some manufactured samples were placed inside an Ascott S450 salt spray chamber (Figure 2), Ascott Analytical Equipment Ltd., Tamworth, UK, with salt spray at a salinity of 3.5%, following ASTM B117-19 [31] standards. Both the saturation and cabinet temperatures of the chamber were adjusted to 40 °C. This temperature setting was specifically selected as it below the glass transition temperature of the materials, thereby preventing any temperature-induced damage during the exposure.

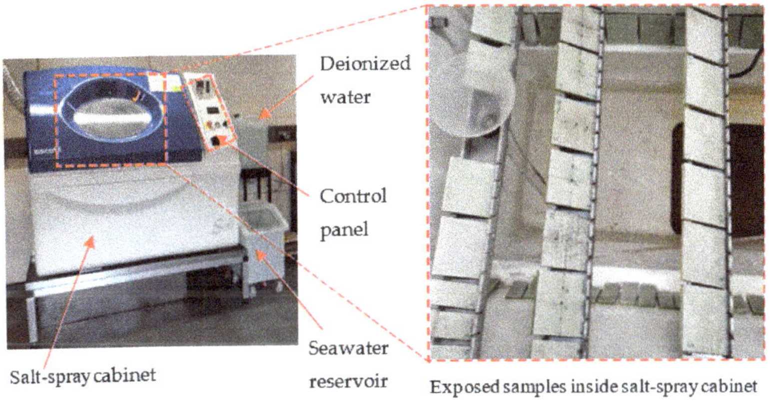

Figure 2. Experimental setup for salt fog spray.

Similarly, maintaining a constant pH level was crucial to prevent chemical reactions with the samples. Moisture absorption in composite structures occurs through three mechanisms: (i) moisture ingress into manufacturing-induced defects [25]; (ii) capillary/wicking along the fibre/matrix interface; (iii) combination of water molecules and hydrophilic resin groups [30]. Gravimetric measurements were conducted to determine the moisture gain in the exposed samples [15]. The average moisture uptake (M_t) increases with immersion time as defined below:

$$M_t = \frac{M_1 - M_0}{M_0} \times 100\% \quad (1)$$

where M_t is the percentage of moisture gained, while M_0 is the initial (dry) mass of the sample and M_1 is the mass of the wet sample at a specific time. The water uptake adheres to Fickian law, i.e., it is a function of the square root of immersion time [15]. It is important to highlight that, for this study, the impact of edge corrections on moisture absorption was presumed to be minimal. This is because during fabrication, the aspect ratio of the samples was within acceptable limits (>4.5) and emphasis was placed on the meso- and macro-scale damage morphologies. A schematic of the research framework is shown in Figure 3.

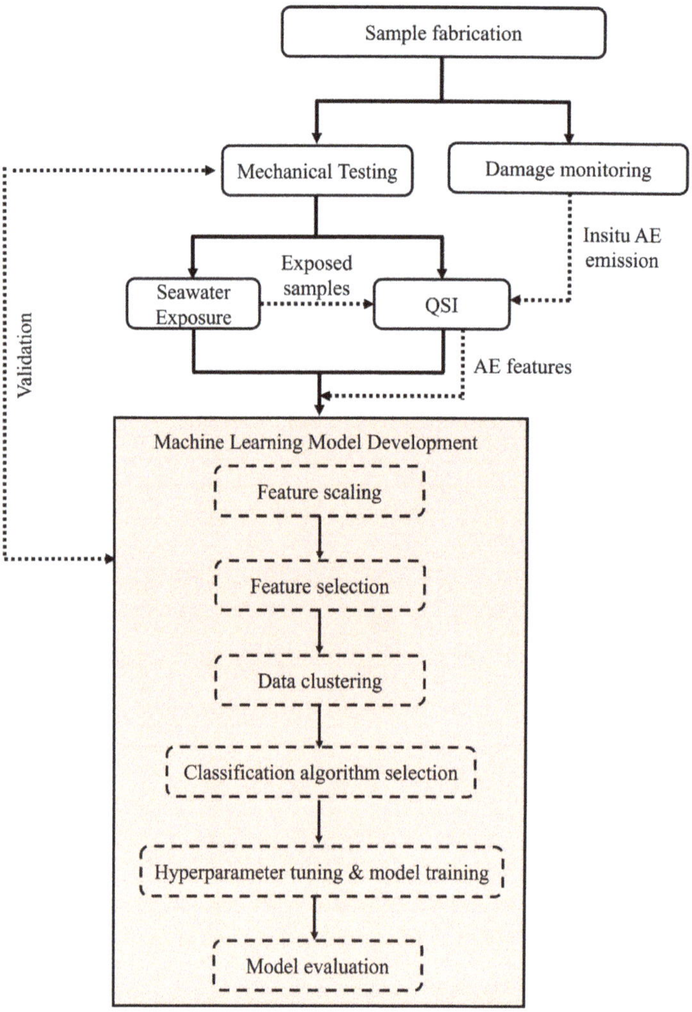

Figure 3. Schematic of research methodology.

2.3. Quasi-Static Indentation Tests

Quasi-static indentation (QSI) tests were conducted using various indenter shapes with a minimum of 5 samples per configuration to assess the materials' damage tolerance and provide insights into the sequence of damage. The experiments followed the ASTM D6264/D6264M-17 standard, employing a displacement control of 1 mm/min, while the vertical displacement was measured with a linear variable differential transformer (LVDT) [25]. An Instron 3369 universal testing machine with a 50 kN load cell (Instron Corporation, Norwood, MA, USA) was used for the tests, with a fixture constructed from steel plates (Figure 4). Three different indenter types made from stainless steel—flat (diameter of 9.5 mm), hemispherical (diameter of 9.5 mm), and conical (shaft diameter of 8.5 mm)—were used (Figure 5). These indenters were chosen to investigate the indentation effects of blunt, semi-blunt, and sharp foreign objects, respectively, on the FRPSS specimens. During the test, the indenter was aligned with the centre of the specimen with an offset of no more than 0.01 mm and was then applied until complete perforation of the sample.

Further details regarding the QSI methodology employed can be found in our previous study [30].

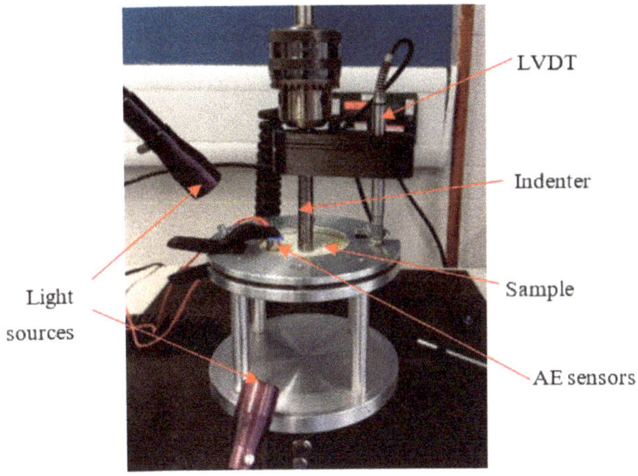

Figure 4. QSI experimental setup.

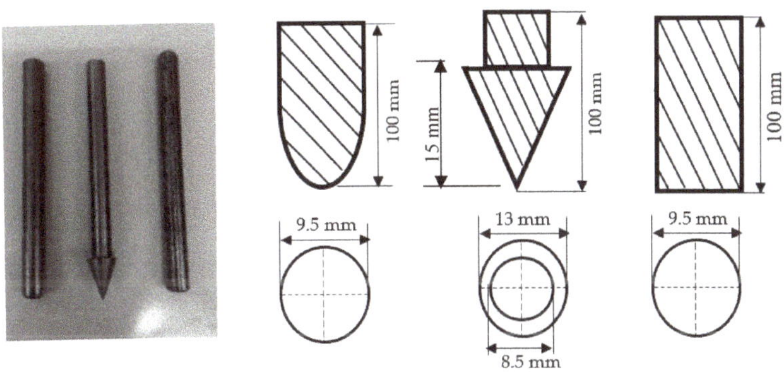

Figure 5. Indenter types with their dimensions.

2.4. Acoustic Emission

Several AE signal parameters, encompassing both time- and frequency-domain features, were analysed in this study. These parameters include Time (s), Class ID, Channel, Parametric, Risetime, Counts to Peak, Counts, Energy (J), Duration (s), Amplitude (dBae), ASL, Threshold, Average Frequency (Hz), RMS, Signal Strength, Absolute Energy (J), Frequency Centroid, and Peak Frequency (Hz). To capture these parameters, a structural health monitoring system, Micro-SHM, Physical Acoustics Corporation, West Windsor Township, NJ, USA), with a frequency band of 1 kHz–1 MHz equipped with 4 AE channels and 2 parametric channels was employed. However, only two channels connected to the Nano-30 AE sensors (125 kHz–750 MHz), mounted on the top and bottom face sheets (see Figure 4) were utilized.

2.5. ML Approach

The ML approach includes critical steps that contribute to the development of predictive models. These steps include feature scaling and selection, data clustering, classification algorithm selection, hyperparameter tuning, model training, and performance evaluation. Each of these components is essential in enhancing the accuracy, reliability, and robustness of the predictive models developed with the ML process; they are briefly discussed below.

2.5.1. Feature Scaling

Feature scaling is performed on a dataset to prevent dominance issues and reduce calculation complexities. The process also helps to mitigate the impact of outliers, while enhancing the convergence and compatibility of ML algorithms. Accordingly, the time- and frequency-domain features (i.e., $X = [x_1, \ldots, x_N]^T \in \mathbb{R}^{n \times m}$) were transformed into a range between 0 and 1, and scaled as follows:

$$X_{scaled} = \frac{X - X_{min}}{X_{max} - X_{min}} \tag{2}$$

where X_{min} and X_{max} are the minimum and maximum values of the features in the dataset, respectively, while X_{scaled} denotes the scaled feature value after normalization.

2.5.2. Feature Selection

Before model training is undertaken, it is important to identify the relevant information features. Accordingly, a feature selection process was carried out by choosing a subset of features or variables from a larger set within a dataset. The process is governed by criteria such as the relevance of the features to the response variable, the predictive strength, or the capacity of such features to enhance model performance. The primary goal of feature selection is to reduce the dimensionality of the input features and tackle the influence of highly correlated ones, to reduce the computational cost, and in some cases to enhance the efficiency and efficacy of prediction algorithms. In this work, the permutation feature importance (PFI) method was adopted for the feature selection process, which is provided with the ML classifiers, thereby making them computationally efficient and robust to outliers and noisy features [32,33]. PFI measures the contribution of each feature to the statistical performance of a model for a given dataset by randomly shuffling the values of a single feature and examining the consequent decrease in the performance score. By disrupting the connection between the feature and the value to be predicted, PFI assesses the extent to which the model depends on that specific feature. The main steps of the PFI approach are summarized below.

For instance, if M is a fitted predictive model and X a feature matrix with n samples and m features, then $X \in \mathbb{R}^{n \times m}$, and y is the target vector corresponding to n samples, such that $y \in \mathbb{R}^n$.

For each feature j in X, y;

For each repetition k in $1, \ldots, K$;

Randomly shuffle j of X, y to generate a corrupted version of $X, y = \tilde{X}_{k,j}, y$;

Compute reference score, $s_{k,j}$ of M on $\tilde{X}_{k,j}, y$;

Compute importance i_j for j defined as:

$$i_j = s - \frac{1}{K}\sum_{k=1}^{K} s_{k,j} \tag{3}$$

PFI overcomes limitations of the impurity-based feature importance since it does not have a bias towards high-cardinality features and can be computed on a held-out validation set [34].

2.5.3. Data Clustering

The k-means++ algorithm is a widely recognized unsupervised ML method for tackling clustering problems. In this work, the algorithm was employed to identify the dominant damage mechanisms. Given $X = [x_1, \ldots, x_N]$ with d-dimensional Euclidean space $\mathbb{R}^d$, $A = [a_1, \ldots, a_c]$ is the c cluster centres, and $z = [\mathcal{Z}_{ik}]_{nxc}$, where $\mathcal{Z}_{ik} \in \{0, 1\}$ indicates if datapoint x_i belongs to the kth cluster, $k = 1, \ldots, c$. Therefore, the objective function is:

$$J(z, A) = \sum_{i=1}^{n} \sum_{k=1}^{c} \mathcal{Z}_{ik} \|x_i - a_k\|^2 \tag{4}$$

such that the k-means++ algorithm iteratively optimizes $J(z, A)$ by updating cluster centres and memberships according to specific conditions, such as

$$a_k = \frac{\sum_{i=1}^{n} z_{ik} x_{ij}}{\sum_{i=1}^{n} z_{ik}} \text{ and}$$

$$z_{ik} = \begin{cases} 1 & \text{if} \|x_i - a_k\|^2 = \min\limits_{1 \le k \le c} \|x_i - a_k\|^2 \\ 0 & \text{otherwise} \end{cases}$$

The challenge of k-means++ is usually the need to specify the number of clusters a priori [23,35,36]. It is noteworthy that the optimum clusters for the k-means++ were identified in the previous work [30] using two cluster-validity indices, namely the normalized Calinski–Harabasz index and the Davies–Bouldin index.

2.5.4. Classification Algorithm Selection

Various ML classification algorithms are available to develop prediction models. These algorithms, however, outclass each other based on the peculiarity of the dataset and prediction intentions. The Lazy Predict method was selected in this study to facilitate the selection of appropriate ML classifiers for the development of the prediction model. The method is designed for the efficient training and evaluation of multiple ML models by employing default configurations and hyperparameters for the models. This method enabled the identification and prediction of damage mechanisms by deploying a total of thirty ML classification algorithms, with the top-performing classifiers across the 9 configurations (3 × unexposed and 3 × exposed samples subjected to 3 different indenter types) selected for subsequent model development. Ensemble learning algorithms featured among the top-performing classifiers using the Lazy Predict method. These algorithms leverage multiple weak learners to make a resilient predictor. They also address overfitting and handle complex data interactions more effectively, while exhibiting resilience to noise and outliers. Additionally, they excel in scalability and efficiency when dealing with large datasets, as obtained in this study, making them suitable for the development of prediction models in diverse ML tasks. Considering the imbalanced distribution of damage-mode class data, the robustness of ensemble learners helps to prevent the majority class from dominating the learning process, allowing the model to give more attention to the minority class. This could offer ensemble learners, such as light gradient boosting machine (LightGBM), random forest (RF), and extreme gradient boosting (XGBoost), significant advantages over other algorithms in class imbalance problems [37–39]. RF is an ensemble learning classifier, which produces results by aggregating predictions from numerous decision trees, thereby augmenting prediction accuracy through averaging. An illustration of RF is depicted in Figure 6.

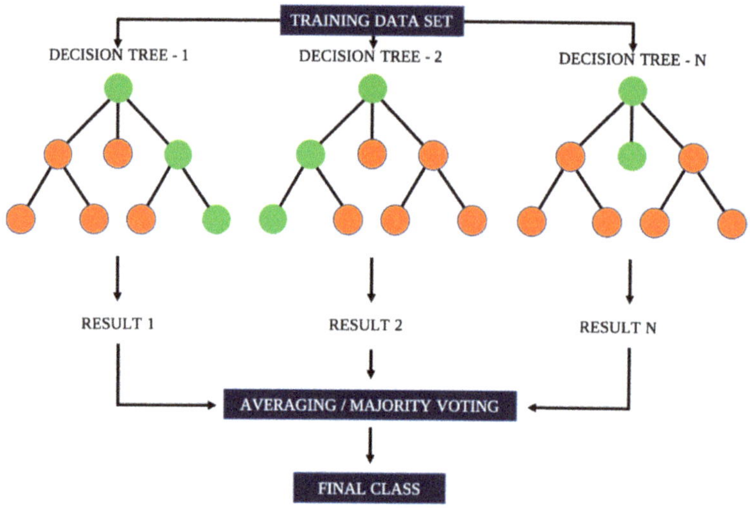

Figure 6. Illustration of classification using RF.

2.5.5. Hyperparameter Tuning and Model Training

Following the selection of the appropriate ML classifiers, a tuning process for the respective hyperparameters of the classifier is necessary to obtain the optimal set of hyperparameters to enhance the predictive performance of the model on unseen data. They can also be employed to deal with overfitting and control the computational cost of model training. Typical hyperparameters in the featured classifiers include the number of estimators, the maximum depth of the trees and the number of features to be considered in the quest for an optimal fit [40]. In this work, the GridSearch cross-validation (CV) [34] was utilized for hyperparameter tuning. It involves creating a grid of all possible hyperparameter combinations and partitioning the dataset into multiple k-folds/subsets (for instance, $k = 5$), where $k - 1$ folds are used to train the data and the rest are used to evaluate the model. Subsequently, the folds are rotated so that all folds are featured in the model training and testing processes. Hence, the model performance is the average mean accuracy score for each of the hyperparameter combinations, with the combination that delivers the optimal performance score adopted for the final model training on a held-out set. The model development process and the hyperparameter tuning phase for one hyperparameter combination are presented in Figure 7. To reduce the risk of bias in the developed model, a random shuffle of the data is undertaken with the true class structure preserved. Shuffling also prevents certain patterns in the original ordered data (such as timestamps) from being learned, thereby forcing the model to learn more generalized patterns rather than specific patterns related to the order of data points. Only 80% of the samples for each condition were used for the hyperparameter tuning and model training process, while the remaining 20% were used to evaluate/validate the performance of the prediction model [41].

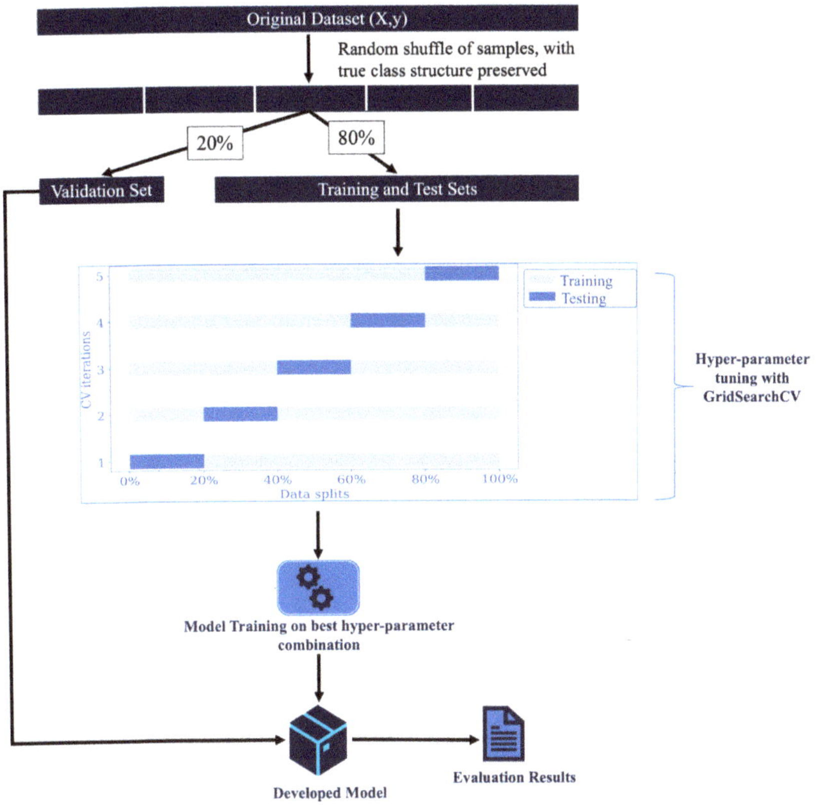

Figure 7. Model development process based on one hyperparameter combination.

2.5.6. Model Evaluation

In assessing the performance of predictive models in detecting the damage conditions of the dataset, four machine learning metrics, namely accuracy, precision, recall, and F1-score [42], were employed. These four metrics were used because they provide more comprehensive evaluation of the classifier's performance [43,44]. Accuracy measures the proportion of correct predictions out of the total cases, while precision and recall help us to understand how well the model performs for individual classes. Precision signifies the accuracy of positive predictions and recall indicates the model's ability to correctly identify positive instances. F1-score is the harmonic mean of precision and recall, with both metrics contributing equally to the score. Our multidimensional approach ensures a more nuanced understanding of the classifier's effectiveness across various scenarios and classes. The employed metrics are expressed as follows:

$$\text{Accuracy} = \frac{\text{TP} + \text{TN}}{\text{TP} + \text{TN} + \text{FP} + \text{FN}} \quad (5)$$

$$\text{Precision} = \frac{\text{TP}}{\text{TP} + \text{FP}} \quad (6)$$

$$\text{Recall} = \frac{\text{TP}}{\text{TP} + \text{FN}} \quad (7)$$

$$\text{F1-score} = \frac{2 \cdot \text{TP}}{2 \cdot \text{TP} + \text{FP} + \text{FN}} \quad (8)$$

where True Positives (TP) and True Negatives (TN) indicate the correct predictions by the model for positive and negative classes, respectively. Also, False Positives (FP) and False Negatives (FN) represent the instances where the model's predictions were incorrect for positive and negative classes, respectively. The values for these metrics range between 0 and 1, with values close to unity indicating better predictive performance of the model.

3. Results and Discussion

3.1. Moisture Uptake

The moisture absorption curve obtained from the tests adhered to Fickian law and exhibited a significant increase in the initial period of exposure (Figure 8). The FRPSS without additional adhesive layers (GSP) demonstrated a higher amount of moisture uptake than their GSV and GSH counterparts. This difference might have arisen from the lack of obstruction to the diffusion process once plasticisation of the matrix commenced, contrasting with the other samples that required more epoxy plasticisation over the exposure period. Additionally, the reduced moisture absorption observed in the GSH samples indicates that the water ingress primarily occurred in the in-plane direction, with a limited through-thickness effect (edge effect). As can be seen for all the samples, the moisture uptake continued until the saturation stage was reached, which is represented by the plateau in the curves. It should be noted that this stage was reached for GSP after a longer period of exposure while GSV and GSH attained saturation almost simultaneously. This shows that the samples with adhesive cores had similar moisture absorption properties due to the effects of the epoxy layers.

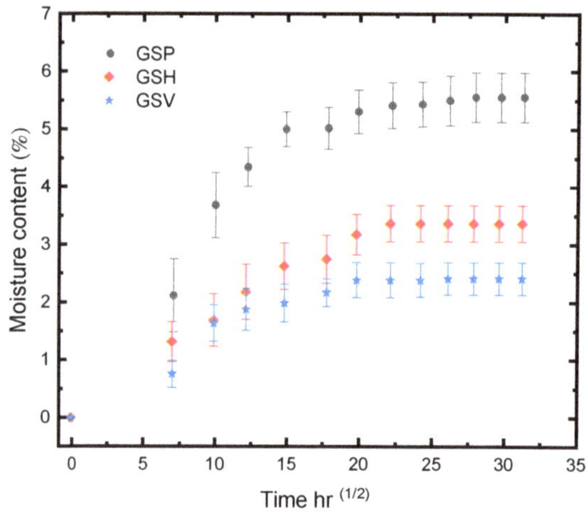

Figure 8. Moisture uptake of FRPSS.

3.2. Quasi-Static Out-of-Plane Behaviour

3.2.1. Failure Load

The mechanical performance of the FRPSS depended on both environmental exposure and indenter geometry (Figure 9). Overall, it was evident that samples subjected to the conical indenter demonstrated the lowest failure loads when compared to those subjected to flat and hemispherical indenters across all of the sample types. This phenomenon could be attributed to an early onset of localized damage at the point of contact caused by the sharp indenter, which resulted in enhanced matrix shear cracking and fibre fracture. Conversely, in cases with larger contact areas (as for hemispherical and flat indenters), more of the reinforcements came into contact with the indenter, thus enhancing the damage resistance of the loaded structure and indicating a higher load-bearing capacity until the point of sudden failure when the shear strength was exceeded. It is noteworthy that this behaviour varied with the core configuration, as samples with adhesive layers (both vertical and horizontal) exhibited greater levels of damage resistance, with the horizontal adhesive-layered samples displaying a more brittle failure character.

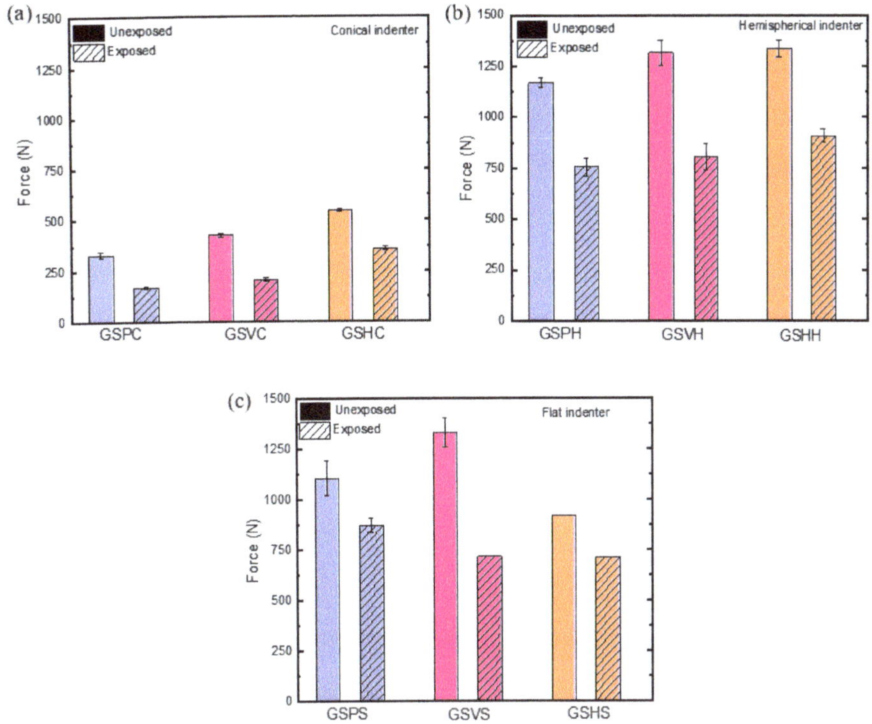

Figure 9. Failure load of FRPSS caused by different indenters: (**a**) conical; (**b**) hemispherical; (**c**) flat.

After the seawater exposure, the test results indicated a general decline in the load-bearing capacity for all the tested samples. Specifically, samples subjected to conical indenters exhibited the highest decrease in the failure load, with reductions of 48.9%, 51.5%, and 34.1% for the GSP, GSV, and GSH specimens, respectively. This decline can be attributed to a combination of factors, including the sharp nose of the indenter and the ageing of the structure due to seawater exposure. The conical indenter geometry led to an early onset of damage, while the seawater induced plasticisation and swelling of the epoxy matrix, affecting the adhesive bonding between the constituents of the FRPSS. At the microscale, this deterioration was characterized by a weaker fibre/matrix interface, as well as reduced intra- and interlaminar bonding, and debonding between the face sheet and the core at the macroscale [30]. In the case of indentation with larger contact areas, such as with hemispherical and flat indenters, the GSV samples were the most affected by the seawater exposure, experiencing decreases in the failure load of 38.8% and 46.1%, respectively. This could be attributed to the overall weakening of bond strength at the intersection between the top face sheet, the core, and the vertical adhesive layer, as well as the low compressive strength of the epoxy layer. Interestingly, the GSH samples performed slightly better, with a decrease of 32.1% (compared to 35.4%) for GSP under the hemispherical indenters, while there was a negligible difference for the flat indenters. This highlights that, while the core configuration could lead to enhanced damage resistance for FRPSS, it could also have significantly adverse effects for maritime applications.

3.2.2. Energy Absorption Capability

Energy absorption serves as a crucial metric in comprehending the damage behaviour of composites. Hence, a comparison of the energy absorption properties of the samples under different indenter configurations can be based on the total absorbed energy E_a obtained by integration of the area under the force–displacement curve:

$$E_a = \int_{x_0}^{x_1} F(x)dx \tag{9}$$

Like the trend for the failure load, it is noticeable that E_a decreased after seawater exposure for all of the samples (Figure 10). Furthermore, specimens loaded with the conical indenter (GSPC, GSVC and GSHC) exhibited the lowest energy absorption capability and experienced the largest decline after the seawater exposure, with reductions of 66.9%, 65.2%, and 41.4%, respectively. As previously explained, this could be attributed to the easier penetration of the sharp nose and accelerated degradation due to seawater ageing. The energy performance of specimens subjected to hemispherical and flat indenters varied in a way similar to the failure load trend.

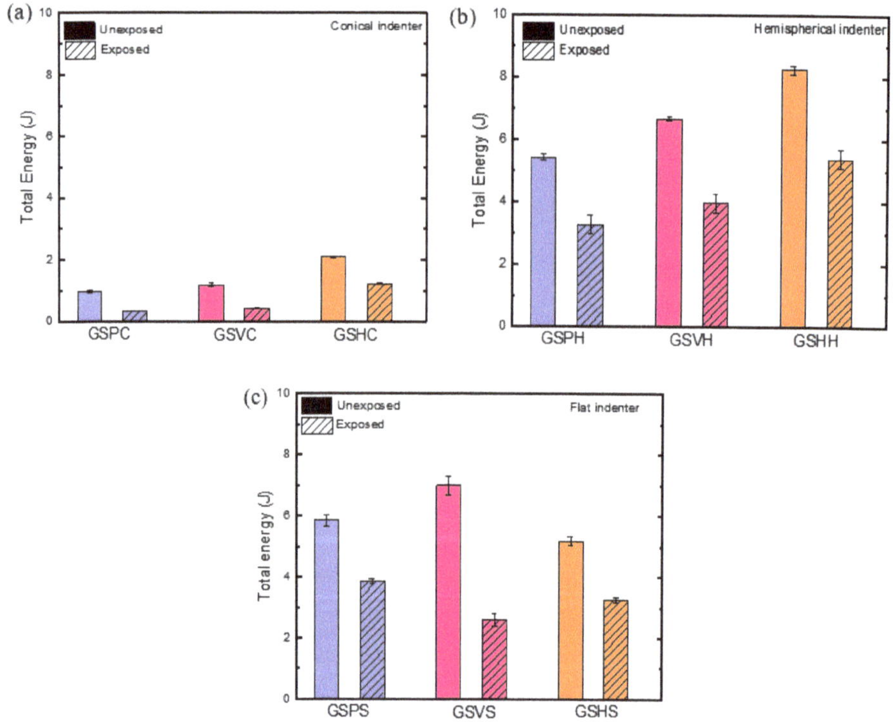

Figure 10. Total absorbed energy for FRPSS under different indenters: (**a**) conical; (**b**) hemispherical; (**c**) flat.

3.3. AE Results

It has been shown that the behaviour of AE features, such as cumulative counts, can offer crucial insights into assessing the damage mechanisms and failure characteristics of composite materials. This is due to their ability to facilitate a comprehensive classification of distinct damage zones under quasi-static loading conditions [21–23]. These distinct zones can be identified when there is a change in the gradient of the cumulative count's slope (Figure 11—zoomed-in section), indicating a transition in the load-bearing capacity of the sample characterized by the presence of a damage sequence [22]. Considering that the displacement speed for all samples was kept constant, the time change would provide an insight into the effects of moisture uptake on the rate of damage at critical loads. A comparison of the force–displacement plots and the AE cumulative counts for the FRPSS samples, both unexposed and exposed to seawater, is depicted in Figures 11–13.

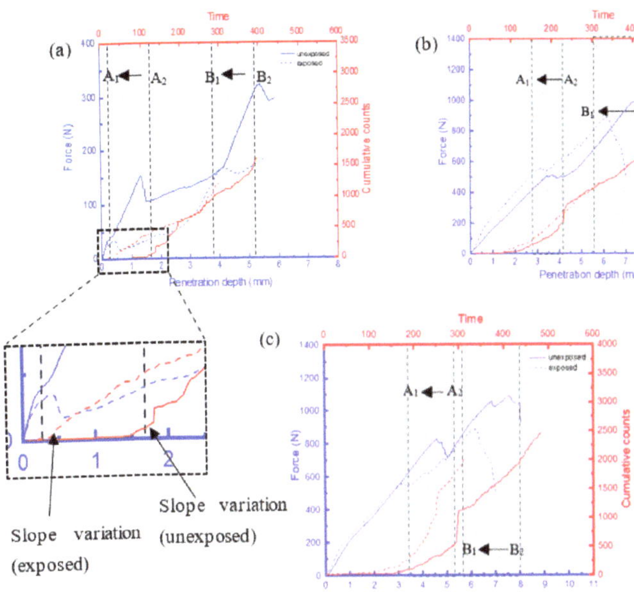

Figure 11. Relationship between load and AE cumulative counts of GSP specimen with different indenters: (**a**) conical; (**b**) hemispherical; (**c**) flat.

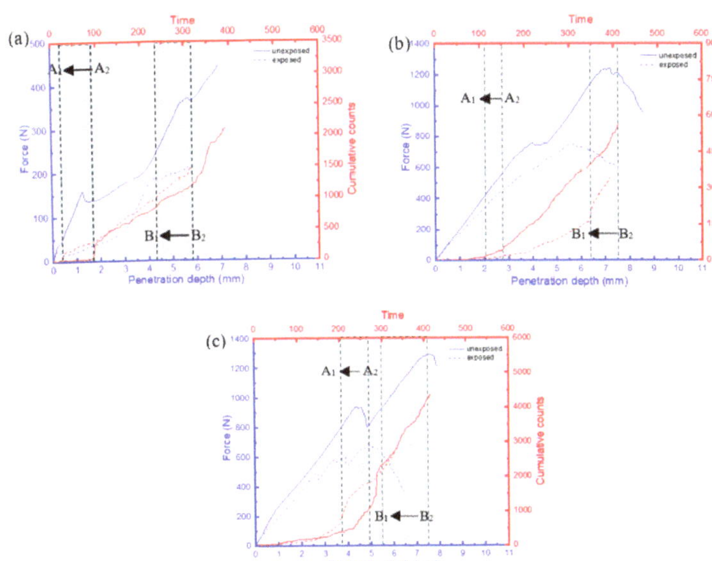

Figure 12. Relationship between load and AE cumulative counts for GSV specimen with different indenters: (**a**) conical; (**b**) hemispherical; (**c**) flat.

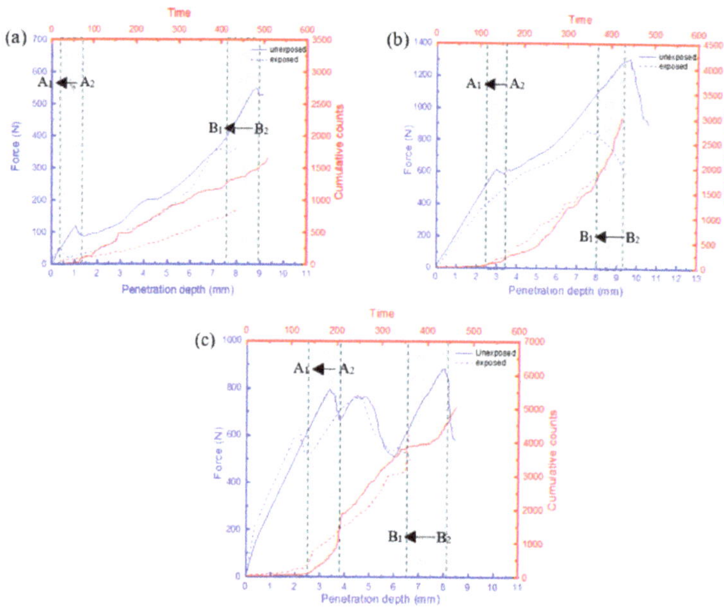

Figure 13. Relationship between load and AE cumulative counts for GSH specimen with different indenters: (**a**) conical; (**b**) hemispherical; (**c**) flat.

There was a reduction in the time taken for the change in the cumulative count slopes of the samples after seawater exposure to occur for penetrations of the top and bottom face sheets. This time difference can be calculated by $A_2 - A_1$ for the top face sheet penetration and $B_2 - B_1$ for the bottom face sheet penetration. Here A_2 and B_2 represent the penetration time for the top face sheet for the unexposed specimen while A_1 and B_1 denote the penetration time for the exposed samples, respectively. GSP samples subjected to conical, hemispherical, and flat indenters (Figure 11) experienced drops of 80%, 75%, and 16% in the top face sheet and 28%, 28%, and 27% in the bottom one, respectively. For GSV, the values were 78%, 27%, and 26% for the top face sheet and 25%, 17%, and 27% for the bottom face sheet, respectively, (Figure 12) while the respective values for GSH were 73%, 31%, and 29% and 15%, 15%, and 20% (Figure 13). It can be seen that the conical indenters caused the largest drop in the time in the top face sheet for all samples which reinforces the heightened damage onset of sharp indenters on exposed samples. Interestingly, the time interval for complete failure ranged from 15% to 28% as a result of the varying combination of the damage resistances of the core, indenter shape, and friction in the through-thickness direction. Consequently, it can be concluded that the time taken for the onset of damage for FRPSS after seawater exposure to the top face sheet was primarily driven by the indenter geometry, while the other constituents played a contributory role in the subsequent damage sequence.

3.4. ML Setup

The primary tools utilized in the ML part of this study were the Python programming language along with the *pandas*, *NumPy*, *joblib*, *scikit-learn (sklearn)*, and *Matplotlib* libraries. These libraries facilitated essential tasks that were necessary for model development, such as data manipulation, numerical computation, the implementation of ML algorithms, and visualisation. Owing to the large dataset size, *joblib* aided in parallelising the execution capabilities, thereby reducing the cost of computationally expensive tasks such as cross-validation and hyperparameter tuning. These computations were carried out using the ALICE high-performance computing facility at the University of Leicester, UK. For each specimen type (e.g., GSPH), the experimental data for both control and exposed samples were uploaded and combined into one data frame, and the normalization process was applied using the Min–Max scaler, transforming each value of each feature to be within the range of 0 to 1. The resulting dataset was randomly shuffled using a predefined random seed and split in ratio of 4:1. Thus, 80% of the data were used for model training, with while the remaining 20% were used for evaluating the model performance. During the splitting process, the data were stratified based on the response class labels, so that the samples of all class labels were equally distributed to alleviate issues arising from class imbalance during the model training process. Thereafter, a clustering analysis was initiated based on the *k*-means++ algorithm using the amplitude and peak frequency features. The number of clusters to identify as well as the number of centroids to generate was set at $n = 4$ in our previous study [30].

For computations, a predefined random state was used to allow for reproducibility. The resulting predicted clusters served as the response variable. Thus, representing the damage modes for the respective specimens. Consequently, the Lazy Predict algorithm was initialized to assess the model performance of thirty ML classification algorithms on their default configurations and hyperparameters for the models. For each specimen, the dataset was reshuffled based on the NumPy random state generator, with Lazy Predict providing the preliminary assessment for model performance. The average accuracy results from five iterations led to the identification of the three top-performing classifiers across the eight damage modes that were predicted (four each for exposed and unexposed sample). These included LightGBM, RF, and XGBoost. Hyperparameter tuning was subsequently undertaken with GridSearch CV to determine the best hyperparameter combination that offered the best generalisation performance on the respective classifiers. In the tuning process, $k = 5$ cross-validation folds were selected with an accuracy set as the baseline scoring parameter. PFI was then carried out by shuffling the features 10 times, and the model refitted to estimate the importance of the feature based on the mean decrease in accuracy. The features were then sorted based on their importance and stacked into a two-dimensional array, in which only the highest ranked feature was in the first row and all the features in the fifteenth row. Using a joblib parallelization, each of the rows of features was fitted on their respective classifiers, and the row which delivered the best model performance was evaluated on the initially held-out validation set based on the four performance metrics. It is noteworthy that the weighted averages were specified for the parameters of the precision, recall, and F1-score matrices, which accounted for the class imbalance of the samples. Finally, the confusion matrices of the best-performing model of the three classifiers were computed and discussed in the subsequent sections.

3.5. Feature Analysis

Generally, the obtained results indicated that the amplitude, frequency centroid, and peak frequency features were the major signals contributing to the accuracy of the predictive ML models (Figure 14) for all samples. Furthermore, the frequency centroid was more dominant among them, as shown in Table 3.

Using Table 3 (in which the highest values of AE features for each sample are highlighted), an analysis of the mean accuracy decrease (MAD) value for GSPC AE features shows that the removal of the AE amplitude feature from the dataset decreased the mean accuracy by 42%. Furthermore, it can also be observed that, in the LightGBM and XGBoost models, the FC was the most important feature, contributing to about 45–70% of the predictive power.

For specimens loaded with the hemispherical indenter, the PF contributed significantly to about 30–50% of the MAD. On the contrary, it was observed that, for the flat and conical indenters, peak frequency had a mostly insignificant influence on the predictive power of the models. This can be linked to the material's behaviour at the transition points (elastic to plastic) underneath these indenters. The hemispherical indenter offers a more consistent transition gradient at the critical load points than the conical and flat indenters, which caused an abrupt failure at these critical points, thereby creating a larger scatter of data points. A similar occurrence for the ML regression algorithm was reported in [45,46]. Therefore, for composite structures loaded with conical and flat indenters, the peak frequency cannot be relied upon alone to assess the damage evolution using AE signals. In practical terms, structural health monitoring of composite materials for marine applications with AE could be achieved for similar sharp and blunt incident objects using the frequency centroid and amplitude. For semi-blunt objects, peak frequency inputs could also provide valuable data for failure monitoring. For the models built on the RF classifier, the amplitude, frequency centroid, and peak frequency were the main contributors to the MAD. This could indicate a higher level of robustness to outliers of RF in damage prediction for data sets with varying levels of fluctuation. Generally, the threshold feature did not contribute to the MAD, implying its irrelevance in the development of a predictive model for damage assessment based on ensemble learning algorithms. Since energy had a negligible value in the MAD, it was not included in the analysis.

Table 3. Results of AE feature analysis.

Sample	Model	AE Features								
		Amp	FC	PF	ASL	RT	D	C	AE	IF
GSPC	LightGBM	**0.42**	0.07	0.00	0.06	0.00	0.01	0.02	0.03	0.01
	RF	0.35	**0.41**	0.50	0.06	0.01	0.01	0.01	0.01	0.02
	XGBoost	0.37	**0.67**	0.00	0.03	0.04	0.03	0.02	0.01	0.01
GSPH	LightGBM	0.43	**0.46**	0.45	0.10	0.01	0.05	0.06	0.07	0.01
	RF	0.46	0.41	**0.50**	0.06	0.01	0.01	0.02	0.01	0.01
	XGBoost	0.49	0.43	**0.51**	0.08	0.01	0.02	0.02	0.03	0.01
GSPS	LightGBM	0.42	**0.70**	0.00	0.06	0.00	0.02	0.02	0.03	0.01
	RF	**0.42**	0.35	0.38	0.03	0.00	0.00	0.01	0.00	0.01
	XGBoost	0.43	**0.70**	0.00	0.05	0.00	0.02	0.02	0.04	0.01
GSVC	LightGBM	0.43	**0.59**	0.00	0.16	0.04	0.03	0.04	0.02	0.04
	RF	**0.40**	0.30	0.26	0.13	0.02	0.02	0.03	0.00	0.03
	XGBoost	0.43	**0.59**	0.00	0.16	0.04	0.04	0.04	0.03	0.04
GSVH	LightGBM	**0.48**	0.39	0.33	0.04	0.01	0.01	0.01	0.01	0.01
	RF	**0.48**	0.34	0.31	0.03	0.01	0.01	0.01	0.00	0.00
	XGBoost	0.49	0.43	**0.51**	0.08	0.01	0.02	0.02	0.03	0.01
GSVS	LightGBM	**0.44**	0.70	0.00	0.03	0.00	0.02	0.05	0.03	0.01
	RF	**0.43**	0.31	0.37	0.05	0.00	0.02	0.03	0.00	0.00
	XGBoost	0.43	**0.70**	0.00	0.07	0.00	0.03	0.05	0.03	0.00
GSHC	LightGBM	0.46	**0.69**	0.00	0.01	0.01	0.03	0.02	0.03	0.00
	RF	**0.41**	0.36	0.28	0.01	0.01	0.02	0.01	0.02	0.01
	XGBoost	0.46	**0.69**	0.00	0.01	0.01	0.03	0.02	0.03	0.01
GSHH	LightGBM	0.36	**0.47**	0.31	0.05	0.01	0.01	0.02	0.01	0.01
	RF	0.38	**0.46**	0.22	0.03	0.01	0.01	0.01	0.00	0.01
	XGBoost	0.38	**0.45**	0.32	0.05	0.01	0.02	0.02	0.02	0.01
GSHS	LightGBM	0.35	**0.63**	0.00	0.09	0.03	0.03	0.04	0.02	0.02
	RF	**0.33**	**0.33**	0.28	0.07	0.02	0.01	0.02	0.01	0.02
	XGBoost	0.34	**0.63**	0.00	0.09	0.03	0.03	0.04	0.02	0.02

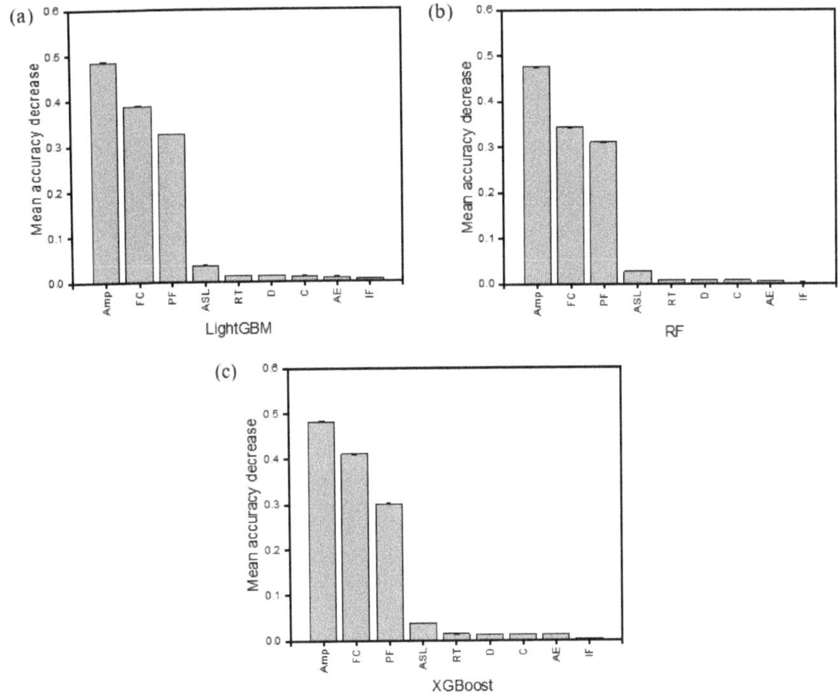

Figure 14. Sample feature analysis of GSVH with best performing algorithm: (**a**) LightGBM; (**b**) RF; (**c**) XGBoost (Amp—amplitude; FC—frequency centroid; PF—peak frequency; ASL—average signal level; RT—rise time; D—duration; C—counts; AE—absolute energy; IF—initiation frequency).

3.5.1. Hyperparameter Tunning Results

The hyperparameters, range, selected parameters, and mean cross-validation scores of GSVC that were used to identify the damage modes are presented in Table 4. Other hyperparameters not shown in the table were used in their default states.

Table 4. Hyperparameters of classifiers and their studied ranges for GSVC.

Model	Attribute	Range	Selected Value	Mean Score
LightGBM	colsample_bytree	[0.7, 0.8, 0.9, 1]	0.9	0.8836
	learning rate	[0.01, 0.1, 0.2]	0.01	
	max_depth	[−1, 5, 10, 15]	10	
	min_child_weight	[0.001, 0.01, 0.1]	0.001	
	n_estimators	[100, 200, 250, 300]	300	
RF	criterion	['gini', 'entropy', 'log_loss']	'entropy'	0.8944
	max_depth	[None, 2, 3, 4, 5, 7, 8, 10, 20]	None	
	max_features	[None, 'sqrt', 'log2']	'sqrt'	
	min_samples_leaf	[1, 2, 3, 4, 5]	1	
	min_samples_split	[2, 4, 5, 7, 8, 10]	7	
	n_estimators	[100, 150, 200, 250, 300]	300	
XGBoost	colsample_bytree	[0.7, 0.8, 0.9, 1]	0.9	0.8955
	gamma	[0, 0.1, 0.2]	0	
	learning rate	[0.01, 0.1, 0.2]	0.01	
	max_depth	[3, 6, 9]	6	
	n_estimators	[100, 200, 300]	200	

Following this, the models were trained using the selected hyperparameter values (which provided an optimal performance while dealing with overfitting) for the respective classifier attributes to identify the damage sequence.

3.5.2. Identification of Damage Sequence

The confusion matrices of the prediction models for the GSP, GSV, and GSH samples allow us to visualize their performance by comparing the true and the predicted labels (Figures 15–17, respectively). For the sake of analytical comparisons and interpretability, the individual quantitative values were normalized, demonstrating the results of clustering (C1, C2, C3, C4, S1, S2, S3 and S4), with C1–C4 and S1–S4 representing the four damage modes (matrix cracking, delamination, fibre breakage, and core damage) for the unexposed and exposed samples, respectively. The identification of the damage modes for FRPSS was discussed in previous studies [27]. Generally, the developed models could clearly distinguish between unexposed and exposed specimens, with the control samples exhibiting a higher extent of correct predictions between 66 and 100%. This is because, as demonstrated previously, seawater exposure led to the degradation of the mechanical performance of the constituents of the FRPSS and the entire structures. This degradation caused a faster onset of damage in the matrix due to plasticisation, as well as a more pronounced weakening at the constituent's interface (matrix–fibre interface, intra-laminate/interlaminate, and face sheet/core). It can be observed that the correct classifications of the GSP samples loaded with the hemispherical and flat indenters had higher values than those loaded with the conical one, with the lowest being 74.1% and 79.9% for GSPC-S2 and GSPC-C1, respectively (Figure 15a). For the GSV samples, the lowest classification results were GSVH-S4 and GSVS-C4 at 77.6% (Figure 16b) and 65.8% (Figure 16c), respectively. Lastly, GSHC-C3 and GSHH-S1 showed the lowest values of 77.8% (Figure 17a) and 74.2% (Figure 17b) for the GSH samples, respectively. These results demonstrate the ability of the developed models to deal with damage induced by moisture uptake and the dominant influence of seawater in the prediction capabilities of the ML algorithm. A description of the class clustering data for exposed and unexposed samples is given in Table 5.

In terms of the average performances for the eight damage modes, C1–C4 and S1–S4, Table 6 shows the model performance values for each sample loaded with the various indenters with the highest values highlighted. The models with the highest performance values for each sample had a range of 86.4%–95.9% across the selected performance indicators (namely, accuracy, precision, recall, and F1-score). Although the four metrics yielded similar scores in terms of their predictive capabilities on each sample, Friedman tests taken at the level of $\alpha = 0.05$ resulted in F-statistic and p-values of 8.181 and 0.043 for LightGBM; 9.243 and 0.026 for RF; and 7.950 and 0.047 for the XGBoost models, respectively. The tests, therefore, indicated that there was a statistically significant difference in performance between the metrics and the models. Accordingly, at least one of the models (LightGBM, RF or XGBoost) performed differently from the others on the four evaluated metrics. Thus, the XGBoost model demonstrated the best overall performance across all the studied samples. This was noted on the GSVS sample, achieving accuracy, precision, recall, and F1-score values of 0.9587, 0.9592, 0.9587, and 0.9587, respectively. Specifically, the LightGBM model exhibited the highest performance for the GSV sample, while the XGBoost model demonstrated the highest performance for the GSH sample, across all three of the indenter types employed.

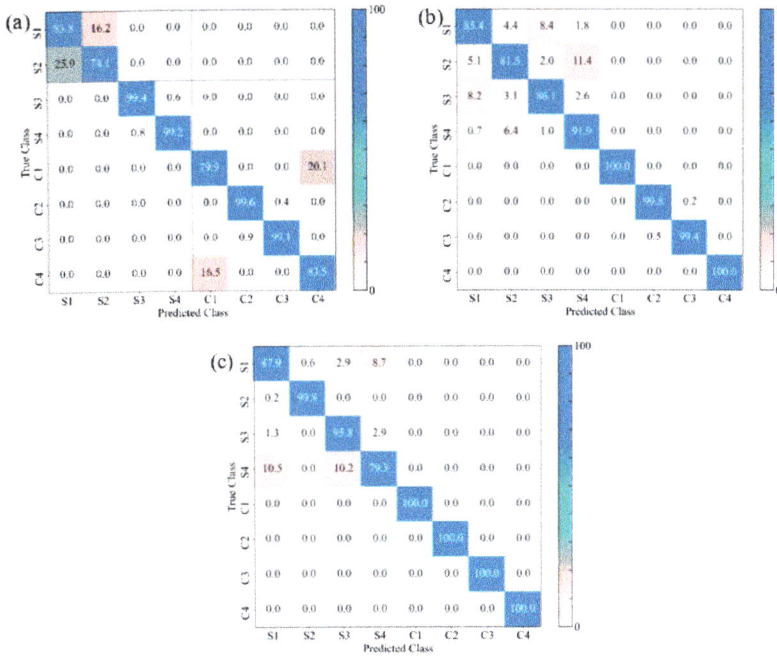

Figure 15. Confusion matrices for different FRPSS: (**a**) GSPC; (**b**) GSPH; (**c**) GSPS.

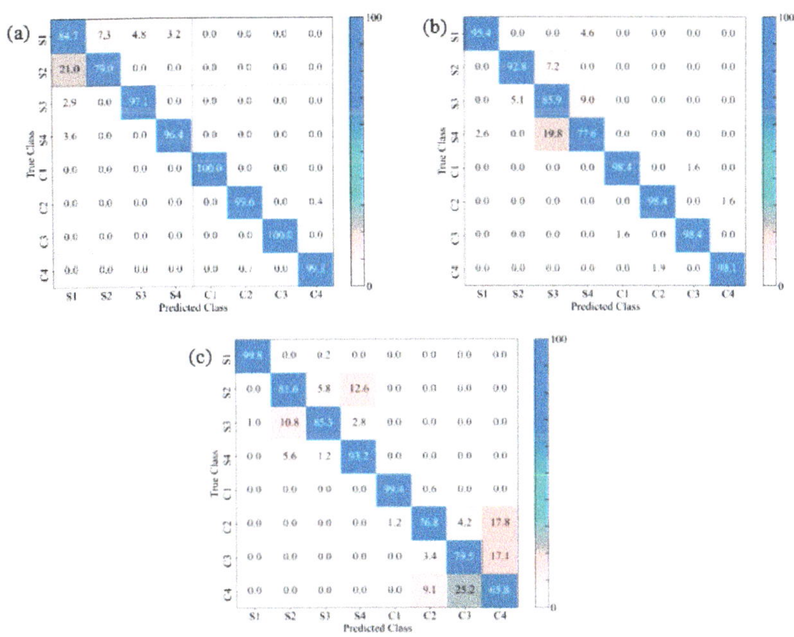

Figure 16. Confusion matrices for different FRPSS: (**a**) GSVC; (**b**) GSVH; (**c**) GSVS.

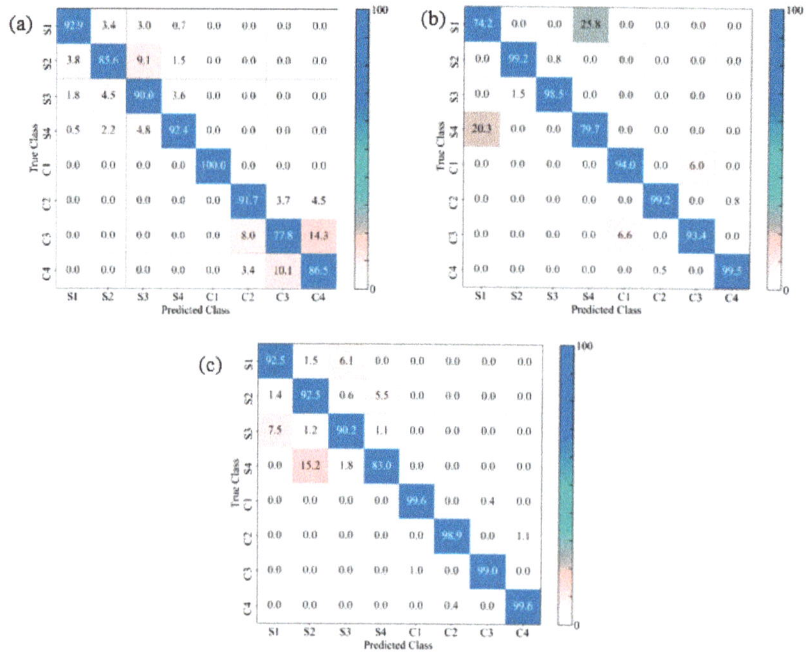

Figure 17. Confusion matrices for different FRPSS: (**a**) GSHC; (**b**) GSHH; (**c**) GSHS.

Table 5. Class clustering data description.

Class Clustering Data Description								
Sample	**Cluster**	**Value**	**Sample**	**Cluster**	**Value**	**Sample**	**Cluster**	**Value**
				GSP				
GSPC_c1	C1	16,381	GSPH_c1	C1	47,633	GSPS_c1	C1	45,924
GSPC_c2	C2	15,177	GSPH_c2	C2	52,810	GSPS_c2	C2	69,598
GSPC_c3	C3	32,798	GSPH_c3	C3	38,213	GSPS_c3	C3	72,779
GSPC_c4	C4	16,988	GSPH_c4	C4	54,834	GSPS_c4	C4	45,146
GSPC_s1	S1	17,142	GSPH_s1	S1	57,804	GSPS_s1	S1	26,283
GSPC_s2	S2	17,898	GSPH_s2	S2	70,847	GSPS_s2	S2	53,096
GSPC_s3	S3	18,911	GSPH_s3	S3	85,418	GSPS_s3	S3	70,230
GSPC_s4	S4	18,655	GSPH_s4	S4	88,616	GSPS_s4	S4	36,562
				GSV				
GSVC_c1	C1	22,860	GSVH_c1	C1	25,789	GSVS_c1	C1	28,432
GSVC_c2	C2	21,091	GSVH_c2	C2	34,000	GSVS_c2	C2	27,548
GSVC_c3	C3	25,386	GSVH_c3	C3	21,235	GSVS_c3	C3	27,665
GSVC_c4	C4	16,755	GSVH_c4	C4	26,295	GSVS_c4	C4	23,663
GSVC_s1	S1	19,879	GSVH_s1	S1	53,640	GSVS_s1	S1	53,889
GSVC_s2	S2	18,700	GSVH_s2	S2	61,286	GSVS_s2	S2	36,701
GSVC_s3	S3	14,859	GSVH_s3	S3	51,369	GSVS_s3	S3	35,678
GSVC_s4	S4	19,492	GSVH_s4	S4	81,788	GSVS_s4	S4	61,885
				GSH				
GSHC_c1	C1	20,454	GSHH_c1	C1	40,256	GSHS_c1	C1	47,697
GSHC_c2	C2	13,528	GSHH_c2	C2	31,579	GSHS_c2	C2	32,055
GSHC_c3	C3	35,224	GSHH_c3	C3	47,866	GSHS_c3	C3	17,419
GSHC_c4	C4	17,772	GSHH_c4	C4	27,235	GSHS_c4	C4	36,516
GSHC_s1	S1	26,507	GSHH_s1	S1	17,311	GSHS_s1	S1	45,513
GSHC_s2	S2	22,917	GSHH_s2	S2	59,395	GSHS_s2	S2	24,719
GSHC_s3	S3	23,774	GSHH_s3	S3	35,561	GSHS_s3	S3	33,314
GSHC_s4	S4	12,656	GSHH_s4	S4	70,114	GSHS_s4	S4	41,140

Table 6. Summary of best-performing ML models.

Sample	Model	Performance Indicators			
		Accuracy	Precision	Recall	F1-Score
GSPC	LightGBM	0.9103	0.9106	0.9103	0.9102
	RF	0.9106	0.9108	0.9106	0.9105
	XGBoost	**0.9111**	**0.9114**	**0.9111**	**0.9110**
GSPH	LightGBM	0.9357	0.9356	0.9357	0.9355
	RF	0.9434	0.9433	0.9434	0.9432
	XGBoost	**0.9453**	**0.9453**	**0.9453**	**0.9452**
GSPS	LightGBM	0.9562	0.9556	0.9562	0.9557
	RF	**0.9568**	**0.9563**	**0.9568**	**0.9563**
	XGBoost	0.9568	0.9562	0.9568	0.9563
GSVC	LightGBM	0.8974	0.8973	0.8974	0.8972
	RF	0.8971	0.8969	0.8971	0.8968
	XGBoost	**0.8996**	**0.8995**	**0.8996**	**0.8995**
GSVH	LightGBM	0.9429	0.9429	0.9429	0.9429
	RF	0.9421	0.9422	0.9421	0.9421
	XGBoost	**0.9436**	**0.9437**	**0.9436**	**0.9435**
GSVS	LightGBM	0.9584	0.959	0.9584	0.9583
	RF	0.9556	0.956	0.9556	0.9555
	XGBoost	**0.9587**	**0.9592**	**0.9587**	**0.9587**
GSHC	LightGBM	0.9548	0.9553	0.9548	0.9548
	RF	0.9535	0.9544	0.9535	0.9535
	XGBoost	**0.9548**	**0.9554**	**0.9548**	**0.9549**
GSHH	LightGBM	**0.9389**	**0.9394**	**0.9389**	**0.9392**
	RF	0.9384	0.9391	0.9384	0.9387
	XGBoost	0.9383	0.9388	0.9383	0.9385
GSHS	LightGBM	**0.8636**	**0.8640**	**0.8636**	**0.8630**
	RF	0.8616	0.8614	0.8616	0.8612
	XGBoost	0.8632	0.8635	0.8632	0.8627

4. Conclusions

This study investigated the effect of moisture uptake on the damage to FRPSS with GFRP face sheets and PVC foam cores caused by loading with different indenters, which is relevant for marine applications. Multiple ML predictive models were applied to the data collected with AE during QSI tests for both control and seawater-treated samples. The results obtained from the study indicated the following:

- The decline in the load-bearing capacity for all samples after the seawater exposure was attributed to several factors. Samples loaded with the conical indenter experienced the highest decrease in the maximum load, with reductions of 48.9%, 51.5%, and 34.1% for GSP, GSV, and GSH specimens, respectively. For indentation cases with a higher contact area, the GSV samples were notably impacted by seawater exposure, with reductions of 38.8% and 46.1%, while the GSH samples showed a slightly better performance, decreasing by 32.1% (compared to 35.4% for GSP) under the hemispherical indenter and exhibiting a negligible difference under the flat indenter.
- In terms of energy absorption, a similar trend was observed, with this parameter demonstrating the largest decrease for the samples loaded with the conical indenter—66.9%, 65.2%, and 41.4% (respectively, for GSP, GSV, and GSH) after seawater exposure. This could be attributed to the easier penetration of the sharp (conical) indenter and accelerated degradation due to seawater ageing. Furthermore, the energy performance of specimens subjected to hemispherical and flat indenters varied in a way similar to the failure load trend.
- The AE amplitude, frequency centroid, and peak frequency were the major signals contributing to the accuracy of the predictive ML models for all samples, with the FC being dominant. The MAD values for GSPC, GSVC, and GSHC decreased the mean accuracy after the removal of the AE amplitude by 42%, 52%, and 69%, respec-

tively, with LightGBM models. The FC was the most important feature, contributing 45–70% to predictive power in LightGBM and XGBoost models. For the hemispherical indenter, the PF contributed significantly (30–50% MAD). The models showed high performances (86.4–95.9%) in distinguishing between the unexposed (control) specimens and the exposed ones. The lowest correct classification rates were observed for samples loaded with the conical indenter.

Overall, the developed ML models could clearly distinguish between unexposed and exposed specimens, with the control samples exhibiting a higher extent of correct predictions between 66 and 100%. The reason for this, as shown by the experimental results, was the effect of moisture uptake on the degradation of mechanical properties of the constituents in the FRPSS. Also, it was observed that the XGBoost model performed best overall, achieving a 95.9% accuracy for the GSVS samples. This paper thus demonstrated the potential of ML techniques for damage prediction in marine structures and components and the viability of using these techniques with data from in situ AE inputs, which could be important for various industrial applications.

Author Contributions: Conceptualization, N.O.-u., V.V.S. and E.D.; methodology, N.O.-u., V.V.S. and E.D.; software, N.O.-u. and A.G.U.; formal analysis, N.O.-u. and A.G.U.; resources, N.O.-u., K.P.B. and E.D.; data curation, N.O.-u. and A.G.U.; writing—original draft preparation, N.O.-u. and A.G.U.; writing—review and editing, V.V.S., K.P.B. and E.D.; visualization, N.O.-u. and A.G.U.; supervision, V.V.S. and E.D. All authors have read and agreed to the published version of the manuscript.

Funding: This research was funded by the Nigerian Air Force grant number OPS/1282DTG27145AJUL21 and K.P.B. acknowledges the support from the Royal Society, grant number RGS\R1\221368.

Institutional Review Board Statement: Not applicable.

Informed Consent Statement: Not applicable.

Data Availability Statement: Dataset available on request from the authors.

Conflicts of Interest: The authors declare no conflicts of interest.

References

1. Altin Karataş, M.; Gökkaya, H. A Review on Machinability of Carbon Fiber Reinforced Polymer (CFRP) and Glass Fiber Reinforced Polymer (GFRP) Composite Materials. *Def. Technol.* **2018**, *14*, 318–326. [CrossRef]
2. Mouritz, A.P.; Gellert, E.; Burchill, P.; Challis, K. Review of Advanced Composite Structures for Naval Ships and Submarines. *Compos. Struct.* **2001**, *53*, 21–41. [CrossRef]
3. Manalo, A.C.; Aravinthan, T.; Karunasena, W.; Islam, M.M. Flexural Behaviour of Structural Fibre Composite Sandwich Beams in Flatwise and Edgewise Positions. *Compos. Struct.* **2010**, *92*, 984–995. [CrossRef]
4. Langdon, G.S.; von Klemperer, C.J.; Rowland, B.K.; Nurick, G.N. The Response of Sandwich Structures with Composite Face Sheets and Polymer Foam Cores to Air-Blast Loading: Preliminary Experiments. *Eng. Struct.* **2012**, *36*, 104–112. [CrossRef]
5. Siriruk, A.; Jack Weitsman, Y.; Penumadu, D. Polymeric Foams and Sandwich Composites: Material Properties, Environmental Effects, and Shear-Lag Modeling. *Compos. Sci. Technol.* **2009**, *69*, 814–820. [CrossRef]
6. Osa-uwagboe, N.; Silberschimdt, V.V.; Aremi, A.; Demirci, E. Mechanical Behaviour of Fabric-Reinforced Plastic Sandwich Structures: A State-of-the-Art Review. *J. Sandw. Struct. Mater.* **2023**, *23*, 109963622311704. [CrossRef]
7. Bakalarz, M.M.; Kossakowski, P.G. Numerical, Theoretical, and Experimental Analysis of LVL-CFRP Sandwich Structure. *Materials* **2023**, *17*, 61. [CrossRef] [PubMed]
8. Studzinski, R.; Pozorski, Z.; Garstecki, A. Sensitivity Analysis of Sandwich Beams and Plates Accounting for Variable Support Conditions. *Bull. Pol. Acad. Sci. Tech. Sci.* **2013**, *61*, 201–210. [CrossRef]
9. Prasad, S.; Carlsson, L.A. Debonding and Crack Kinking in Foam Core Sandwich Beams-II. Experimental Investigation. *Eng. Fract. Mech.* **1994**, *47*, 825–841. [CrossRef]
10. Gargano, A.; Galos, J.; Mouritz, A.P. Importance of Fibre Sizing on the Seawater Durability of Carbon Fibre Laminates. *Compos. Commun.* **2020**, *19*, 11–15. [CrossRef]
11. Idrisi, A.H.; Mourad, A.H.I.; Abdel-Magid, B.M.; Shivamurty, B. Investigation on the Durability of E-Glass/Epoxy Composite Exposed to Seawater at Elevated Temperature. *Polymers* **2021**, *13*, 2182. [CrossRef] [PubMed]
12. Barreira-Pinto, R.; Carneiro, R.; Miranda, M.; Guedes, R.M. Polymer-Matrix Composites: Characterising the Impact of Environmental Factors on Their Lifetime. *Materials* **2023**, *16*, 3913. [CrossRef] [PubMed]
13. El-Hassan, H.; El-Maaddawy, T.; Al-Sallamin, A.; Al-Saidy, A. Performance Evaluation and Microstructural Characterization of GFRP Bars in Seawater-Contaminated Concrete. *Constr. Build. Mater.* **2017**, *147*, 66–78. [CrossRef]

14. Ghabezi, P.; Harrison, N.M. Hygrothermal Deterioration in Carbon/Epoxy and Glass/Epoxy Composite Laminates Aged in Marine-Based Environment (Degradation Mechanism, Mechanical and Physicochemical Properties). *J. Mater. Sci.* **2022**, *57*, 4239–4254. [CrossRef]
15. Pandiyan, V.; Wróbel, R.; Leinenbach, C.; Shevchik, S. Optimizing In-Situ Monitoring for Laser Powder Bed Fusion Process: Deciphering Acoustic Emission and Sensor Sensitivity with Explainable Machine Learning. *J. Mater. Process Technol.* **2023**, *321*, 118144. [CrossRef]
16. Li, S.; Chen, B.; Tan, C.; Song, X. In Situ Identification of Laser Directed Energy Deposition Condition Based on Acoustic Emission. *Opt. Laser Technol.* **2024**, *169*, 110152. [CrossRef]
17. Pal, A.; Kundu, T.; Datta, A.K. Assessing the Influence of Welded Joint on Health Monitoring of Rail Sections: An Experimental Study Employing SVM and ANN Models. *J. Nondestr. Eval.* **2023**, *42*, 102. [CrossRef]
18. Garbowski, T.; Cornaggia, A.; Zaborowicz, M.; Sowa, S. Computer-Aided Structural Diagnosis of Bridges Using Combinations of Static and Dynamic Tests: A Preliminary Investigation. *Materials* **2023**, *16*, 7512. [CrossRef] [PubMed]
19. Monaco, E.; Rautela, M.; Gopalakrishnan, S.; Ricci, F. Machine Learning Algorithms for Delaminations Detection on Composites Panels by Wave Propagation Signals Analysis: Review, Experiences and Results. *Progress Aerosp. Sci.* **2024**, *146*, 100994. [CrossRef]
20. Perfetto, D.; Rezazadeh, N.; Aversano, A.; De Luca, A.; Lamanna, G. Composite Panel Damage Classification Based on Guided Waves and Machine Learning: An Experimental Approach. *Appl. Sci.* **2023**, *13*, 10017. [CrossRef]
21. Almeida, R.S.M.; Magalhães, M.D.; Karim, M.N.; Tushtev, K.; Rezwan, K. Identifying Damage Mechanisms of Composites by Acoustic Emission and Supervised Machine Learning. *Mater. Des.* **2023**, *227*, 111745. [CrossRef]
22. Lee, I.Y.; Roh, H.D.; Park, H.W.; Park, Y. Bin Advanced Non-Destructive Evaluation of Impact Damage Growth in Carbon-Fiber-Reinforced Plastic by Electromechanical Analysis and Machine Learning Clustering. *Compos. Sci. Technol.* **2022**, *218*, 109094. [CrossRef]
23. Barile, C.; Pappalettera, G.; Paramsamy Kannan, V.; Casavola, C. A Neural Network Framework for Validating Information–Theoretics Parameters in the Applications of Acoustic Emission Technique for Mechanical Characterization of Materials. *Materials* **2023**, *16*, 300. [CrossRef]
24. Guo, F.; Li, W.; Jiang, P.; Chen, F.; Yang, C. Deep Learning for Time Series-Based Acoustic Emission Damage Classification in Composite Materials. *Russ. J. Nondestruct. Test.* **2023**, *59*, 665–676. [CrossRef]
25. *ASTM D6264/D6264M-17*; Standard Test Method for Measuring the Damage Resistance of a Fiber-Reinforced Polymer-Matrix Composite to a Concentrated Quasi-Static Indentation Force. ASTM International: West Conshohocken, PA, USA, 2018; Volume 98, pp. 1–12. [CrossRef]
26. Dai, L.; Wu, X.; Zhou, M.; Ahmad, W.; Ali, M.; Sabri, M.M.S.; Salmi, A.; Ewais, D.Y.Z. Using Machine Learning Algorithms to Estimate the Compressive Property of High Strength Fiber Reinforced Concrete. *Materials* **2022**, *15*, 4450. [CrossRef]
27. Jefferson Andrew, J.; Arumugam, V.; Ramesh, C.; Poorani, S.; Santulli, C. Quasi-Static Indentation Properties of Damaged Glass/Epoxy Composite Laminates Repaired by the Application of Intra-Ply Hybrid Patches. *Polym. Test.* **2017**, *61*, 132–145. [CrossRef]
28. Oh, H.T.; Won, J.I.; Woo, S.C.; Kim, T.W. Determination of Impact Damage in Cfrp via Pvdf Signal Analysis with Support Vector Machine. *Materials* **2020**, *13*, 5207. [CrossRef] [PubMed]
29. Geren, N.; Acer, D.C.; Uzay, C.; Bayramoglu, M. The Effect of Boron Carbide Additive on the Low-Velocity Impact Properties of Low-Density Foam Core Composite Sandwich Structures. *Polym. Compos.* **2021**, *42*, 2037–2049. [CrossRef]
30. Osa-uwagboe, N.; Udu, A.G.; Silberschmidt, V.V.; Baxevanakis, K.P.; Demirci, E. Damage Assessment of Glass-Fibre-Reinforced Plastic Structures under Quasi-Static Indentation with Acoustic Emission. *Materials* **2023**, *16*, 5036. [CrossRef]
31. *ASTM B117-19*; ASTM International Standard Practice for Modified Salt Spray (Fog) Testing 1. ASTM International: West Conshohocken, PA, USA, 2019. [CrossRef]
32. Udu, A.G.; Lecchini-Visintini, A.; Dong, H. *Feature Selection for Aero-Engine Fault Detection*; Springer Nature: Cham, Switzerland, 2023; Volume 3, ISBN 9783031398476. [CrossRef]
33. Li, J.; Cheng, K.; Wang, S.; Morstatter, F.; Trevino, R.P.; Tang, J.; Liu, H. Feature Selection: A Data Perspective. *ACM Comput. Surv.* **2017**, *50*, 1–45. [CrossRef]
34. Pedregosa, F.; Varoquaux, G.; Gramfort, A.; Michel, V.; Thirion, B.; Grisel, O.; Blondel, M.; Prettenhofer, P.; Weiss, R.; Dubourg, V.; et al. Scikit-Learn: Machine Learning in Python. *J. Mach. Learn. Res.* **2011**, *127*, 2825–2830. [CrossRef]
35. Sinaga, K.P.; Yang, M.S. Unsupervised K-Means Clustering Algorithm. *IEEE Access* **2020**, *8*, 80716–80727. [CrossRef]
36. Yoder, J.; Priebe, C.E. Semi-Supervised k-Means++. *J. Stat. Comput. Simul.* **2017**, *87*, 2597–2608. [CrossRef]
37. Sagi, O.; Rokach, L. Ensemble Learning: A Survey. *Wiley Interdiscip. Rev. Data Min. Knowl. Discov.* **2018**, *8*, e1249. [CrossRef]
38. Khan, A.A.; Chaudhari, O.; Chandra, R. A Review of Ensemble Learning and Data Augmentation Models for Class Imbalanced Problems: Combination, Implementation and Evaluation. *Expert. Syst. Appl.* **2024**, *244*, 122778. [CrossRef]
39. Udu, A.G.; Lecchini-Visintini, M.; Ghalati, M.; Dong, H. Addressing Class Imbalance in Aero Engine Fault Detection. In Proceedings of the 22nd IEEE International Conference Machine Learning and Applications ICMLA, Jacksonville, FL, USA, 15–17 December 2023. [CrossRef]
40. Bischl, B.; Binder, M.; Lang, M.; Pielok, T.; Richter, J.; Coors, S.; Thomas, J.; Ullmann, T.; Becker, M.; Boulesteix, A.L.; et al. Hyperparameter Optimization: Foundations, Algorithms, Best Practices, and Open Challenges. *Wiley Interdiscip. Rev. Data Min. Knowl. Discov.* **2023**, *13*, e1484. [CrossRef]

41. Xie, Q.; Suvarna, M.; Li, J.; Zhu, X.; Cai, J.; Wang, X. Online Prediction of Mechanical Properties of Hot Rolled Steel Plate Using Machine Learning. *Mater. Des.* **2021**, *197*, 109201. [CrossRef]
42. Rezazadeh, N.; de Oliveira, M.; Perfetto, D.; De Luca, A.; Caputo, F. Classification of Unbalanced and Bowed Rotors under Uncertainty Using Wavelet Time Scattering, LSTM, and SVM. *Appl. Sci.* **2023**, *13*, 6861. [CrossRef]
43. De Diego, I.M.; Redondo, A.R.; Fernández, R.R.; Navarro, J.; Moguerza, J.M. General Performance Score for Classification Problems. *Appl. Intell.* **2022**, *52*, 12049–12063. [CrossRef]
44. Guo, F.; Li, W.; Jiang, P.; Chen, F.; Liu, Y. Deep Learning Approach for Damage Classification Based on Acoustic Emission Data in Composite Materials. *Materials* **2022**, *15*, 4270. [CrossRef]
45. Udu, A.G.; Osa-uwagboe, N.; Olusanmi, A.; Aremu, A.; Khaksar, M.; Dong, H. A Machine Learning Approach to Characterise Fabrication Porosity Effects on the Mechanical Properties of Additively Manufactured Thermoplastic Composites. *J. Reinf. Plast. Compos.* **2024**, *online first*. [CrossRef]
46. Osa-uwagboe, N.; Udu, A.G.; Ghalati, M.K.; Silberschmidt, V.V.; Aremu, A.; Dong, H.; Demirci, E. A Machine Learning-Enabled Prediction of Damage Properties for Fiber-Reinforced Polymer Composites under out-of-Plane Loading. *Eng. Struct.* **2024**, *308*, 117970. [CrossRef]

materials

MDPI

Article

Monitoring Fatigue Damage of Orthotropic Steel Decks Using Nonlinear Ultrasonic Waves

Jiahe Liu [1], Fangtong Zheng [2], Wei Shen [3] and Dongsheng Li [1,4,*]

1 School of Civil Engineering, Dalian University of Technology, Dalian 116024, China; 1104060917@mail.dlut.edu.cn
2 China Northeast Architectural Design & Research Institute Co., Ltd., Shenyang 110000, China; zhengfangtong@126.com
3 State Key Laboratory of Featured Metal Materials and Life-Cycle Safety for Composite Structures, School of Civil Engineering and Architecture, Guangxi University, Nanning 530004, China; shenwei431@gxu.edu.cn
4 State Key Laboratory of Coastal and Offshore Engineering, Dalian University of Technology, Dalian 116024, China
* Correspondence: lidongsheng@dlut.edu.cn

Abstract: Orthotropic steel decks (OSDs) are commonly used in the construction of bridges due to their load-bearing capabilities. However, they are prone to fatigue damage over time due to the cyclic loads from vehicles. Therefore, the early structural health monitoring of fatigue damage in OSDs is crucial for ensuring bridge safety. Moreover, Lamb waves, as elastic waves propagating in OSD plate-like structures, are characterized by their long propagation distances and minimal attenuation. This paper introduces a method of emitting high-energy ultrasonic waves onto the OSD surface to capture the nonlinear Lamb waves formed, thereby calculating the nonlinear parameters. These parameters are then correlated with the fatigue damage endured, forming a damage index (DI) for monitoring the fatigue life of OSDs. Experimental results indicate that as fatigue damage increases, the nonlinear parameters exhibit a significant initial increase followed by a decrease. The behavior is distinct from the characteristic parameters of linear ultrasound (velocity and energy), which also exhibit changes but to a relatively smaller extent. The proposed DI and fatigue life based on nonlinear parameters can be fitted with a Gaussian curve, with the R-squared value of the fitting curve being close to 1. Additionally, this paper discusses the influence of rib welds within the OSDs on the DI, whereby as fatigue damage increases, it enlarges the value of the nonlinear parameters without altering their trend. The proposed method provides a more effective approach for monitoring early fatigue damage in OSDs.

Keywords: orthotropic steel decks (OSDs); structural health monitoring; ultrasonic; fatigue; nonlinear parameters

1. Introduction

Orthotropic steel decks (OSDs) form a composite structural system that is reinforced by welding top plates, longitudinal ribs, and transverse stiffeners. Originating in Germany during the 1930s [1], OSD technology has since been adopted worldwide, exemplified by its use in the Bronx–Whitestone Bridge in the United States, the Daishi Bridge in Japan, and the Humen Bridge in China [2]. The widespread popularity of OSDs can be attributed to several key advantages, including their lightweight nature, exceptional load-bearing capacity, minimal joint requirements, and ease of construction [3,4].

However, with the increasing service life and number of OSD bridges, it has become evident that OSDs are not without their issues [1,5]. For instance, they require complex welding details and are prone to intricate stress patterns during the initial design phase [6,7]. Over time, they are susceptible to the frequent passage of overweight vehicles, which significantly heightens the risk of fatigue damage, posing a direct threat to the overall safety of the bridge. The previously mentioned Bronx–Whitestone Bridge, Daishi Bridge, and Humen Bridge have all experienced fatigue-induced cracks [2], necessitating regular maintenance and inspections. Therefore, real-time and effective monitoring of the OSD's fatigue life is crucial, especially for the early detection of fatigue damage.

Currently, various methods for identifying fatigue damage have been developed, including ultrasonic testing [2,8–10], strain measurement [11–13], acoustic emission [14], electro-mechanical impedance [6], infrared thermography [15], magnetic flux leakage [16], and prediction methods based on temperature and traffic data [17]. Strain measurement employs strain gauges to detect deformation at specific locations [13], while infrared thermography captures images based on temperature variations at crack sites [15]. Although infrared thermography and magnetic flux leakage methods provide a certain level of visual detection, their sensitivity in the early stages of crack and damage detection is insufficient, making it difficult to capture the appearance and development of small cracks in a timely manner. Prediction methods based on temperature and traffic data establish a nonlinear relationship between traffic information, temperature variations of the bridge deck, and the fatigue life of the OSD [17]; however, they rely on extensive historical data and complex algorithms. This makes them highly dependent on data completeness and model accuracy. In practical applications, the variability of traffic flow and environmental temperature often compromises the applicability and precision of these prediction models.

In contrast, ultrasonic testing requires fewer sensors and offers longer propagation distances [18], making it more promising for early fatigue prediction in OSDs [2,8–10]. Shi et al. [2,10] experimentally and numerically validated the fact that reflected Lamb waves can be used to monitor the geometric shape of cracks that are caused by fatigue in OSDs, highlighting the importance of sensor placement for effective monitoring. Gao et al. [8] utilized a pitch-catch sensor configuration to analyze the energy of ultrasonic backscattering, enabling the early warning of fatigue damage in welded steel plates. The aforementioned ultrasonic monitoring methods for fatigue damage rely on linear ultrasonic evaluation metrics, such as changes in the energy and velocity of scattered waves.

The early cracks in OSDs are better explained by contact acoustic nonlinearity theory [10], as linear velocity and energy tend to plateau during the initial stages of micro-damage [19], limiting their sensitivity to early fatigue damage. Nonlinear ultrasound, in contrast, is more sensitive in characterizing fatigue micro-damage. Within a certain amplitude range, nonlinear ultrasound generates higher harmonics through interaction with the material's nonlinearity [20,21]. These harmonics arise from the coupling of the fundamental wave with microstructures, defects, and lattice dislocations within the material. Lee et al. [20] quantified the direction and length of cracks using nonlinear parameters through numerical simulation. These nonlinear parameters are not only applicable to steel plates but are also effective in assessing early micro-damage in aluminum plates [21] that has been caused by fatigue. Compared to ordinary steel or aluminum plates, OSDs feature U-ribs and weld seams. The presence of weld seams inevitably induces nonlinear effects [22], posing greater challenges for monitoring fatigue damage in OSDs. Therefore, investigating the applicability of nonlinear ultrasound technology for monitoring the fatigue life of OSDs is of significant interest.

This paper makes the following key contributions: it is the first to explore the interaction between nonlinear ultrasonics and fatigue cracks in OSDs, with particular attention to the impact of weld seams at the OSD ribs on nonlinear ultrasonic signals. A novel detection method is introduced, capable of quantitatively analyzing the remaining fatigue life of OSDs by fitting the DI based on nonlinear parameters to a Gaussian curve, which will be validated on seven OSD specimens. The method's efficacy is thoroughly compared with traditional linear ultrasonic evaluation methods.

The structure of this paper is as follows: Section 2 introduces the evaluation methods and the preparation of the required materials, including the derivation of the DI based on nonlinear ultrasound, the fabrication of OSDs, fatigue testing, and ultrasound testing configurations; Section 3 presents the changes in linear and nonlinear ultrasound signals before and after OSD fatigue, along with the fitting analysis of DI and fatigue life; Section 4 compares the proposed method with linear ultrasound evaluation methods and examines the impact of weld seams on nonlinear parameters; and Section 5 concludes the paper.

2. Materials and Methods

2.1. Nonlinear Ultrasonics

The British physicist Sir Horace Lamb first described Lamb waves in 1917 [23]. These waves can propagate in plate-like structures with parallel free boundaries, similar to steel plates. Lamb waves exhibit two fundamental modes—the symmetric mode (S-mode) and the antisymmetric mode (A-mode) [24]. In the S-mode, the plate vibrates simultaneously outward or inward, while in the A-mode, the vibrations on either side of the plate are opposite. Moreover, their properties vary with frequency. The relationship between the velocity and frequency of Lamb waves in OSDs can be determined through the Rayleigh–Lamb equation [25], as illustrated in Figure 1. It is evident that at any given frequency in OSD, both modes will be present. However, since the subsequent ultrasonic excitation is in an antisymmetric manner, the A-mode is expected to be the predominant mode of vibration [24].

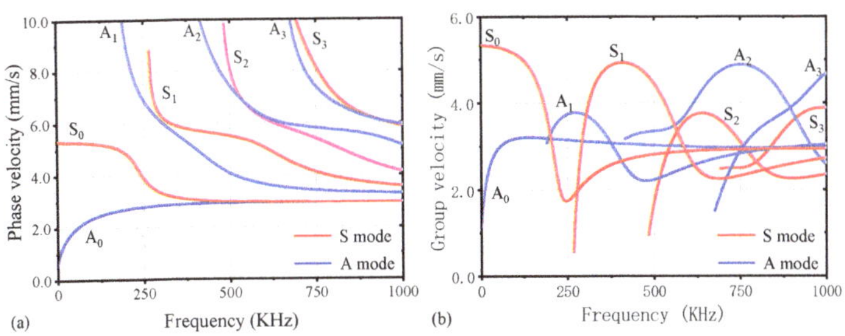

Figure 1. Lamb waves dispersion curves of OSDs: (**a**) phase velocity and (**b**) group velocity.

Fatigue cracks within OSDs may exhibit an alternating open-and-close state under cyclic loads of a certain amplitude, a phenomenon vividly termed the "breathing effect" [20]. This effect results in local nonlinear interactions between the Lamb waves propagating in the OSD and the vicinity of the fatigue cracks, leading to waveform distortions and the emergence of second-order or higher-order harmonics. The internal stress endured by the contact surfaces of the breathing cracks can be represented using a simplified one-dimensional model, as follows:

$$\sigma = E\varepsilon(1 + \beta\varepsilon + \cdots) \quad (1)$$

where $E(pa)$ is the modulus of elasticity; β represents the second-order elastic coefficient; and $\sigma(pa)$ and ε denote stress and strain, respectively. By neglecting the higher-order terms beyond the second order in Equation (1) and utilizing the relationship between particle displacement and strain, substitution into the one-dimensional wave equation yields the following result [26]:

$$\frac{\partial^2 u}{\partial t^2} = c^2\frac{\partial^2 u}{\partial x^2} + 2c^2\rho\frac{\partial u}{\partial x}\frac{\partial^2 u}{\partial x^2} \quad (2)$$

where $u(\mathrm{m})$ represents displacement; $\rho(\mathrm{kg/m^3})$ is the density of the medium; $x(\mathrm{m})$ is the distance over which the wave propagates; $t(\mathrm{s})$ stands for time; and $c(\mathrm{m/s})$ is the wave speed.

Based on perturbation theory [26], it is assumed that the displacement u is composed of a linear response, u_l, and a nonlinear response, u_{nl}. By synthesizing both the multiscale method and the trial solution method, as well as neglecting higher-order small quantities during the solution process, an approximate analytical solution to Equation (2) can be obtained, as follows:

$$u = u_l + u_{nl} = U_1 cos(kx - \omega t) - U_2 sin[2(kx - \omega t)] \quad (3)$$

where $k(m^{-1})$ represents the wavenumber, and $\omega(\text{rad/s})$ denotes the angular frequency. The variables U_1 and U_2, respectively, stand for the amplitudes of the fundamental wave and the second harmonic, with the following relationship established between them:

$$U_2 = \frac{\beta}{8} U_1^2 k^2 x \quad (4)$$

where $\beta/8$, k^2, x can be uniformly regarded as describing the nonlinearity coefficient of the medium [20,27] β_0, facilitating subsequent DI construction.

$$\beta_0 = \frac{U_2}{U_1^2} \quad (5)$$

In actuality, the forces exerted on the surfaces of fatigue cracks are exceedingly complex, and the one-dimensional simplified model mentioned above struggles to precisely represent the nonlinear interaction process of fatigue cracks [26]. Nonetheless, the nonlinear parameters deduced from various theories can all be formulated as expressed in Equation (5). Consequently, this paper opts for the nonlinear parameter as the principal characterizing parameter for fatigue cracks and proceeds to construct the DI based on this.

Given the presence of OSDs with varying degrees of fatigue, to facilitate the comparison of nonlinear parameters between specimens with different identifiers, the DI is further introduced as follows:

$$DI = \frac{\beta_0'}{\beta_{0(o\%)}} \quad (6)$$

where β_0' represents the nonlinear parameter for different OSDs under varying fatigue lives, and $\beta_{0(o\%)}$ denotes the nonlinear parameter for OSDs when not subjected to any load. A larger DI indicates a stronger nonlinear phenomenon of Lamb waves within the OSD. Furthermore, attempting to fit the DI and damage degree through linear or nonlinear curves is a key step in the proposed methods and is an essential part of non-destructive testing methods [8,20].

2.2. Experimental Specimens

Scaled dimensions were utilized for the fabrication of the OSD. Both the top plate and U-ribs were constructed from Q345b steel, which has a density of 7850 kg/m^3 and a Poisson's ratio of 0.3. Material tests on the steel top plates used yielded an average elastic modulus of 218.3 GPa and an average yield strength of 400.6 MPa. The dimensions of the scaled OSD are illustrated in Figure 2a. A bevel groove weld was employed between the U-ribs and the top plate. To ensure a uniform welding quality, an automatic submerged arc welding method was used, followed by cutting. The weld penetration was no less than 75% of the plate thickness, and burn-through was not permitted. The assembly gap, b, was less than or equal to 0.5 mm, with a bevel angle of the top plate α being 50 ± 2^0. After welding, a pre-fabricated crack with a depth of 1 mm and a width of less than or equal to 0.3 mm was created at the weld joint using wire cutting. The details of the top plate and U-rib welding, as well as the actual welding effect, are shown in Figure 2b. Ultimately, seven sets of specimen entities were obtained, as shown in Figure 2c, and they were respectively labeled as OSD specimens S1, S2, S3, S4, S5, S6, and S7.

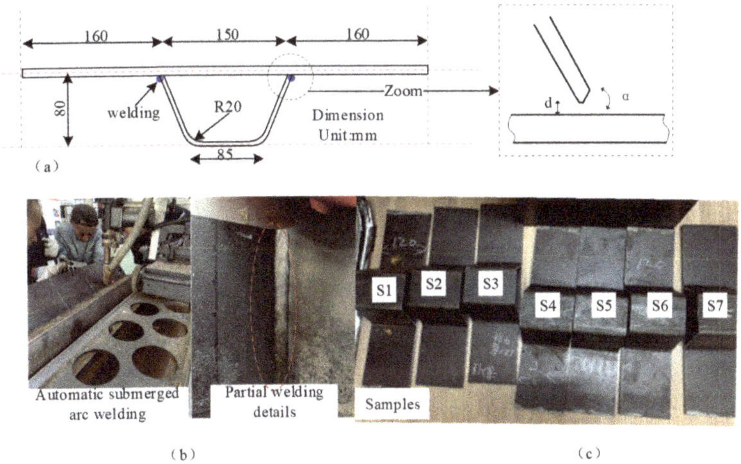

Figure 2. (**a**) OSD processing dimension diagram; (**b**) welding process schematic; (**c**) OSD specimen labeling diagram.

2.3. Fatigue Loading Configuration

Fatigue loading tests were conducted at the Vibration and Strength Testing Center of Dalian University of Technology, using the PLG-200C high-frequency tension-compression fatigue testing machine, as shown in Figure 3. The tests employed a four-point bending loading method. Before setting the mean stress and stress amplitude, a simple bending normal stress test was conducted on an OSD specimen, determining the mean stress for fatigue loading to be 30 kN and the stress amplitude to be 6 kN. Under these experimental conditions, the processed and scaled OSD was subjected to fatigue loading, with the testing machine's operating frequency being essentially stable at approximately 89.3 Hz, as depicted in Figure 3. The OSD S7 specimen was loaded until significant fracture deformation was observed, with the fatigue cycle count reaching 820,000. When noticeable deformation occurred, the testing machine's operating frequency decreased to 88.9 Hz. Additionally, to compare the macroscopic metallographic states of the fatigue life plates, the OSD specimens S1–S6 corresponded to total cycle counts of 200,000; 300,000; 400,000; 500,000; 600,000; and 700,000, respectively. It should be noted that during the test process, all OSD specimens were paused and unloaded every 100,000 cycles for ultrasonic testing.

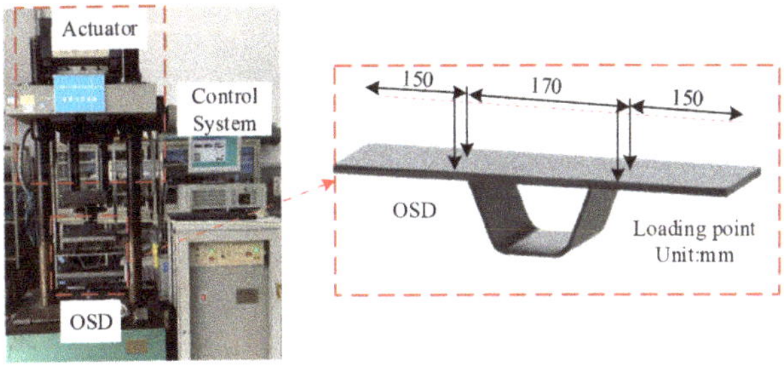

Figure 3. Schematic diagram of the OSD fatigue test loading process.

2.4. Ultrasonic Testing Configuration

Figure 4 presents a schematic diagram of the ultrasonic testing process for OSDs. The excitation probe of the ultrasonic transducer is placed at the center of the plate, 120 mm from the left boundary of the top plate, while the receiving probe is positioned at the midline. Ultrasonic waves are emitted using the RITEC SNAP (model RAM-5000); then, they are passed through an attenuator and a low-pass filter to reach the transmitting probe resonating at 250 kHz. This setup is intended to better detect the second harmonic response. The receiving end employs an ultrasonic probe with a resonant frequency of 500 kHz, and the probe sends the fundamental wave signal directly back to the main system. Since the received signal's fundamental frequency amplitude is between 10 and 100 times that of the harmonic frequency, a 500 kHz band-pass filter is set in the receiving circuit to isolate the second harmonic signal. The filters used in this study were provided by RITEC (model FDK, serial No. 9074). The low-pass filter is a 10th-order Butterworth design with a 3 dB attenuation at 250 kHz and an input impedance of 50 ohms. The band-pass filter is a 5th-order design with a 40 dB out-of-band attenuation. The excitation voltage is set at 640 V, with an attenuation of 4 dB and a gain of 40 dB. The gate amplifier is set to an output level of 32. During the nonlinear ultrasonic detection process, an oscilloscope is used to obtain the time-domain graphs of the fundamental wave signal and the second harmonic signal. The RAM-5000 SNAP system extracts the first wave signal from the time-domain graphs of both the fundamental and second harmonic signals to perform a sweep and to obtain their spectrum graphs.

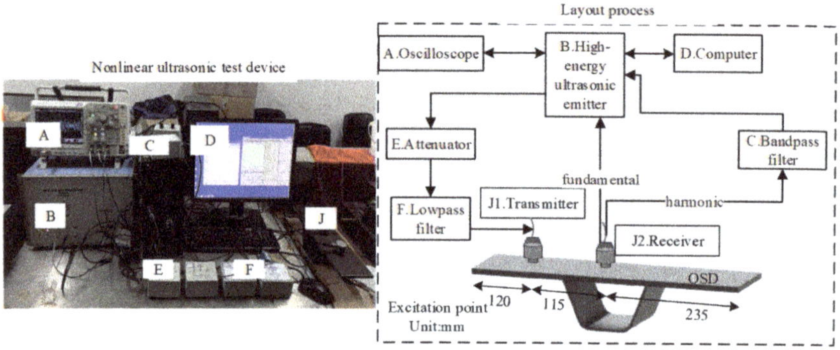

Figure 4. Schematic diagram of the ultrasonic testing connection for OSDs.

As shown in Figure 1, an excitation value near 250 kHz can relatively weaken the ultrasonic frequency dispersion linearity and reduce the number of modes, matching the phase velocity with the second harmonic [28]. Using a total of 10 cycles ensures the concentration of the ultrasonic spectrum signal towards the central frequency, further weakening the dispersion effect [29]. Therefore, the excitation signal uses a ten-cycle sinusoidal modulation signal with a central frequency of 250 kHz, as shown in the following expression:

$$w(t) = \begin{cases} 0.5\,[1 - cos(2\pi f_c t/n)]\,sin(2\pi f_c t) & t \in (0, n/f_c) \\ 0 & otherwise \end{cases} \quad (7)$$

where $w(t)$ is the excitation signal; f_c (kHz) is the central frequency; and n is the number of modulation periods of the Hanning window, with its time-domain and frequency-domain expressions shown in Figure 5. The Hanning window serves to reduce the spectral leakage of the excitation signal [30,31].

To provide a more intuitive overview of the construction process of the proposed evaluation method, a flowchart is illustrated in Figure 6. This includes the materials, experimental procedures, data acquisition, quantitative analysis, and two extended discussions on the evaluation method. As shown in Figure 6, the prepared OSD specimens first undergo fatigue loading. After reaching the specified number of fatigue cycles, ultrasonic testing is performed. If the specimens do not meet the fatigue cycle requirement, fatigue testing continues after the ultrasonic test. The ultrasonic test results are presented in the form of linear and nonlinear parameters. In addition to performing nonlinear curve fitting on the seven specimens for quantitative analysis, two discussions on the evaluation method are also included. These discussions focus on the sensitivity of nonlinear ultrasonic indicators in comparison to their linear counterparts, as well as the impact of weld seams on nonlinear ultrasonic parameters in OSDs.

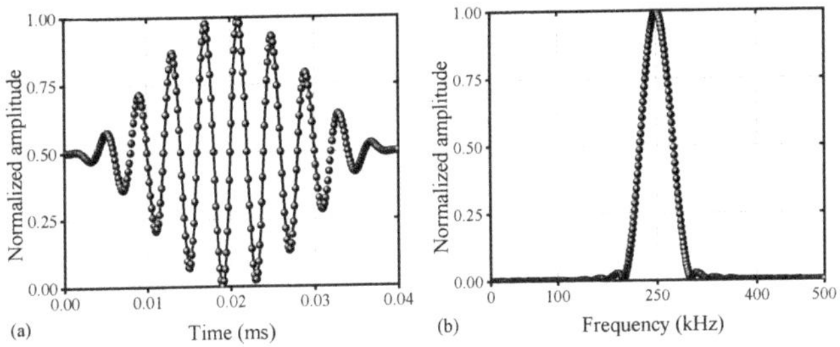

Figure 5. Schematic diagram of the excitation signal (**a**) time-domain and (**b**) frequency-domain signal.

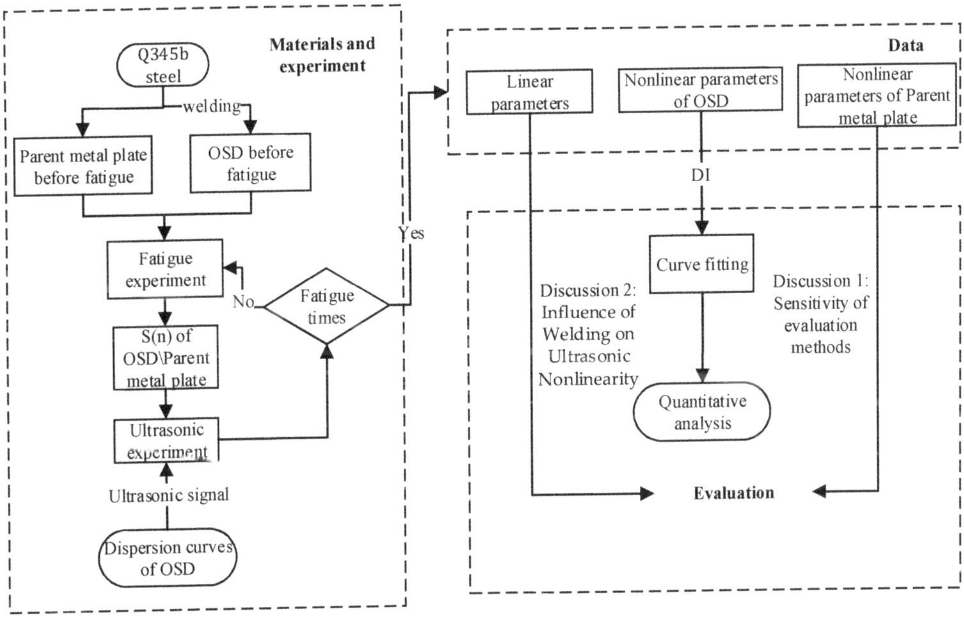

Figure 6. Flowchart of the evaluation method construction process for fatigue damage detection in OSDs.

3. Results

To enhance the reliability of the test data for the fundamental wave amplitude and the second harmonic amplitude, the coupling agent was reapplied for each ultrasonic monitoring session, and the test was conducted three times. The sweep amplitudes of the fundamental and second harmonics were obtained by measuring and taking the average value, which was then used to calculate the nonlinear parameter. Figure 7 summarizes the DI values for different specimens.

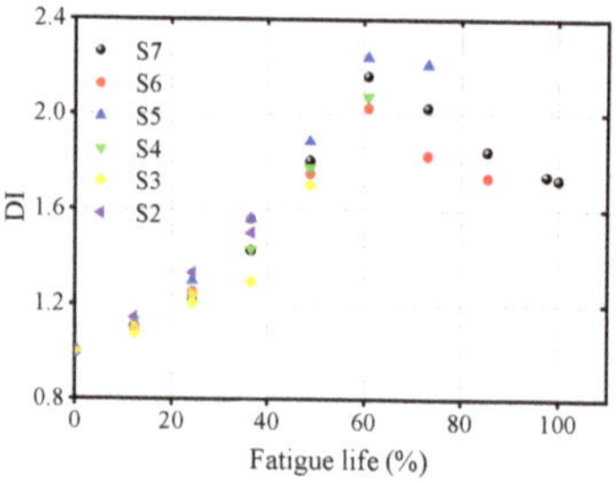

Figure 7. Variation of DI values for different specimens with fatigue life.

3.1. Changes in Ultrasonic Signals before and after Fatigue Loading

Due to the largest variation in the DI values of S5 (as shown in Figure 7), and since the trend of ultrasonic signal changes in other specimens is similar, this section takes S5 as an example to demonstrate the changes in the ultrasonic nonlinear effects before and after OSD fatigue loading. The fundamental wave amplitude values of S5 before and after fatigue loading are shown in Figures 8a and 9a, respectively. The results indicate that the energy of the initial fundamental wave is not significantly different. The transmission times of the first wave are 0.0364 ms and 0.0403 ms, respectively, with sound speeds of 3159 m/s and 2854 m/s, respectively. The difference in sound speed is also minimal and close to the theoretical sound speed of the A0 mode in the group velocity dispersion curve shown in Figure 1 (A0: 3159 m/s), thereby confirming that the first signal to arrive is primarily in the antisymmetric mode, consistent with the excitation method of the ultrasonic signal.

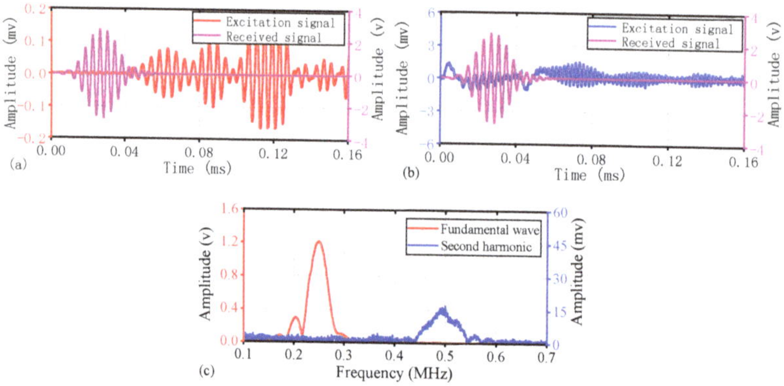

Figure 8. Experimental results for undamaged condition of S5: (**a**) Time-domain plot of the fundamental wave; (**b**) Time-domain plot of the harmonic wave; (**c**) Spectrogram of the fundamental and harmonic frequencies.

Figures 8b and 9b illustrate the capture of harmonic signals through a 500 kHz band-pass filter. The measurement of the harmonic wave speed is not sufficiently precise, and a rough estimate from the time-domain graph suggests a decrease in the harmonic wave speed after fatigue loading. This may be related to the apparent diffraction of the defect signal post-fatigue loading. However, the amplitude of the first wave of the second harmonic after loading was significantly increased compared to before fatigue loading, indicating that fatigue induces a strong nonlinear response in OSD.

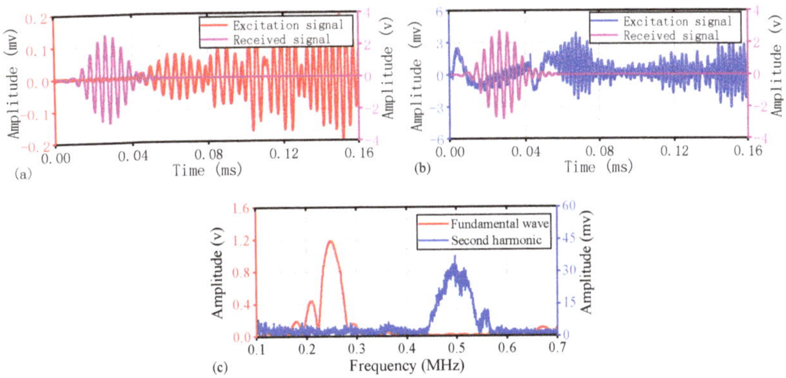

Figure 9. Experimental results for fatigue condition of S5: (**a**) Time-domain plot of the fundamental wave; (**b**) Time-domain plot of the harmonic wave; (**c**) Spectrogram of the fundamental and harmonic frequencies.

Figure 8c shows that before the fatigue loading of S5, the peak of the fundamental wave spectrum is around 249 kHz, while the peak of the harmonic spectrum is around 494 kHz. The harmonic frequency being a multiple of the fundamental frequency indicates that this second harmonic is a generated nonlinear signal. From the spectrum, the peak amplitude of the fundamental wave is 1.206 V, while the peak amplitude of the harmonic is 0.0154 V. The nonlinear parameter is calculated to be 0.0106; at this time, the harmonic is primarily caused by the nonlinearity of the welding structure. Post-fatigue loading, the spectrum of S5, as depicted in Figure 9c, shows the fundamental wave spectrum peak at approximately 247 kHz and the harmonic spectrum peak at around 491 kHz, with almost no change in the peak frequency values. The peak amplitude of the fundamental wave through the path of the crack is 1.171 V, while the peak amplitude of the harmonic is 0.0321 V. The nonlinear parameter is calculated to be 0.0234, indicating a significant increase due to the local nonlinearity caused by the fatigue crack.

3.2. Evaluation of OSD Fatigue Life

Figure 10 reveals that as the fatigue cycle count increases for the seven specimens of OSDs, the DI value continuously rises, reaching a peak at approximately 61.0% of the fatigue life before subsequently decreasing. The early increase in DI may be attributed to the accumulation of dislocations at the wire-cutting sites during the initial stages of fatigue, which gradually evolve into micro-closed fatigue cracks, resulting in a 'breathing' phenomenon and a significant increase in nonlinear response. This process implies that the normalized nonlinear parameter can characterize the accumulation of dislocations; however, once the fatigue cracks propagate to a certain extent, the 'breathing' phenomenon diminishes with the widening of the crack opening, leading to a weakened nonlinear response, as indicated by a reduction in the nonlinear parameter. Additionally, it can be seen that the DI derived from nonlinear parameters does not exhibit a linear or even monotonic relationship with fatigue life. This observation is consistent with the findings of Lee et al. [20], who noted similar phenomena when analyzing fatigue cracks in steel plates using nonlinear ultrasonics. This indicates that linear or monotonic nonlinear curves are no longer applicable in this context. In contrast, a nonlinear curve based on a Gaussian model [32] can effectively represent the trend where the DI initially increases and then decreases. Therefore, to quantitatively analyze the relationship between fatigue life and DI value, a Gaussian curve is chosen for fitting, with the expression for the fitting curve as follows:

$$y = y_0 + Ae^{-\frac{(x-x_c)^2}{2w^2}} \tag{8}$$

where y_0, A, ω, x_c represents the shape parameters of the Gaussian function, which determine the specific form of the Gaussian curve. The numerical values of the shape parameters for the fitting curves of the specimens, other than S1, can be found in Table 1, while the correspondence between the fitting curves and the actual data is displayed in Figure 10. Due to the short fatigue loading cycle of S1, the number of data points is insufficient to satisfy the unknown values required for curve fitting, thus precluding an effective fatigue life fit for S1. However, the fitting curves for the other six groups of specimens will be validated using the data from S1.

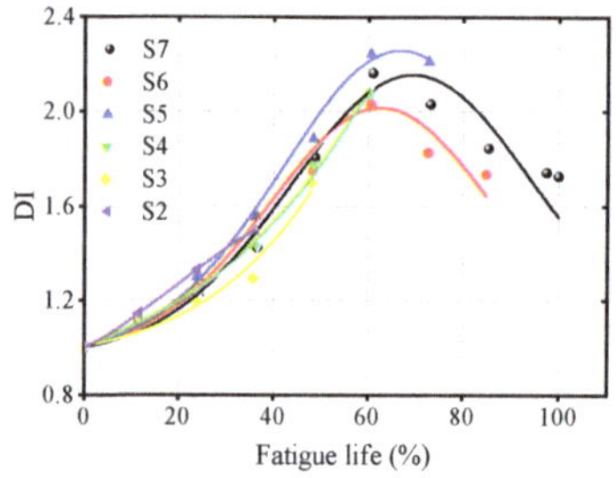

Figure 10. Gaussian fitting curve of OSDs' fatigue life and DI.

Table 1. Parameter values of different OSD fitting curves.

Specimens	y_0	A	ω	x_c
S2	0.97042	0.69347	0.25813	1.18431
S3	0.97996	0.63097	0.23621	1.03675
S4	0.97254	0.67169	0.25296	1.28564
S5	0.83359	2.12618	0.74694	9.94756
S6	0.96443	1.24339	0.42064	3.52888
S7	0.86297	0.48607	0.26865	0.70411

Figure 11 illustrates the applicability of a single fitting curve derived from specimens S2 to S7 across all other specimens. The results indicate that for specimens experiencing fewer fatigue cycles, such as S2 to S5, the fitting curve's applicability to other specimens is lower, with variance even reaching negative values, suggesting a suboptimal fit. In contrast, for S7, the fitting curve obtained after enduring 100% fatigue loading, when applied to other specimen groups, yielded variance values exceeding 0.81. Notably, the fitting variance for each individual specimen falls within the range of 0.97 to 1, further validating the efficacy of the Gaussian curve in predicting the relationship between fatigue life and DI in OSDs. The fitting curve extends the quantitative analysis of OSD fatigue damage assessment into the field of nonlinear ultrasonics.

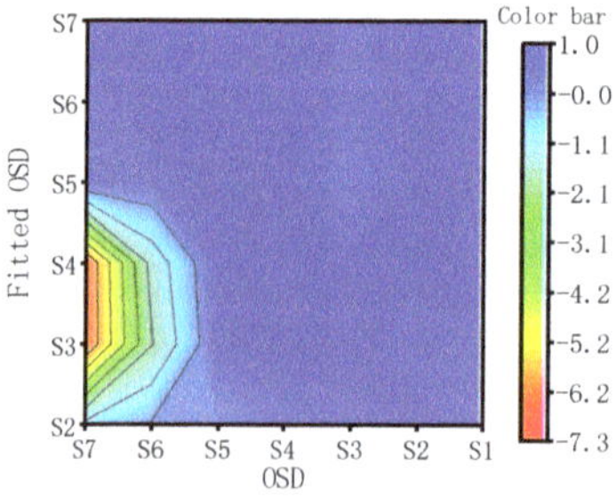

Figure 11. The R-squared values of the fitted curve for other OSD fittings.

4. Discussion

4.1. Comparison with Linear Ultrasonics

In linear ultrasonic methods, velocity and energy are commonly used metrics to assess material conditions [2,8,9]. Signal energy is calculated by squaring the amplitude of the sample signals and summing them, given that the energy density of ultrasonic waves is proportional to the square of the amplitude. This section examines changes in energy and velocity for specimen S7, as depicted in Figure 12. S7, having undergone a complete fatigue life cycle, serves as a representative data set for comparative analysis.

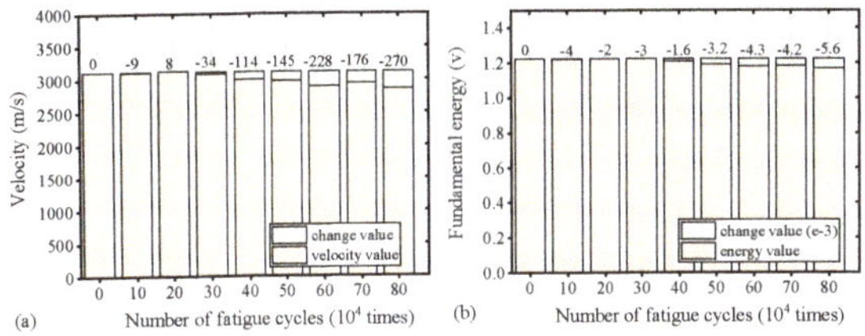

Figure 12. Linear ultrasonic change of S7 with different fatigue life: (**a**) velocity and (**b**) energy.

Figure 12 demonstrates that as fatigue cycles increase, the sound velocity and fundamental wave energy of S7 both show a slight and insignificant downward trend. The relative change rate for both parameters did not exceed 1%, suggesting a substantial potential for random error. Therefore, sound velocity and energy are not highly sensitive indicators of fatigue life and are prone to significant errors. Conversely, as shown in Figure 10, the DI based on nonlinear parameters effectively characterizes fatigue life with a broad range of variation.

4.2. Influence of Welding on Ultrasonic Nonlinearity

The OSD features a U-rib structure, with the top plate being connected to the U-ribs through welding. Even advanced welding techniques cannot entirely prevent the formation of microscopic defects in the welded areas, which contribute to a nonlinear ultrasonic response. To ensure the credibility of the results, the influence of welding on ultrasonic nonlinearity must be considered.

A parent metal plate was fabricated using 10 mm thick steel from the same batch as the OSD. Except for the welding process, the fabrication flow and ultrasonic monitoring methods were consistent with those used for the OSD preparation. Figure 13 compares the nonlinear parameters of the OSD before and after fatigue loading with those of the parent metal plate specimens. Before fatigue loading, the nonlinear parameters of the parent metal plate showed a similar trend to S7 due to fatigue damage. However, the nonlinear parameters of S7 were noticeably higher than those of the parent metal plate specimens, indicating that the welded structure within the OSD induces a significant nonlinear ultrasonic response.

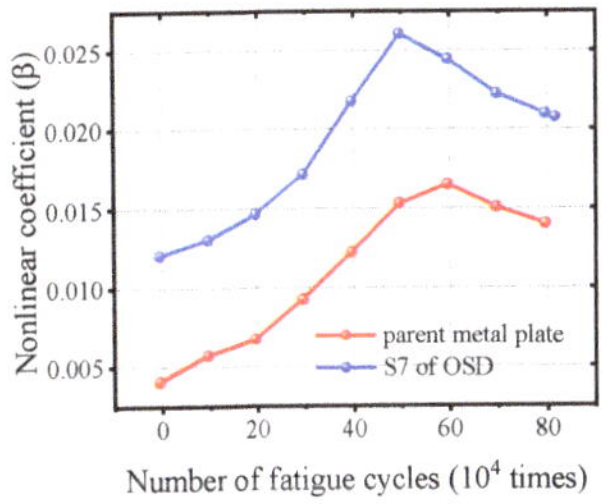

Figure 13. Comparison of nonlinear parameters of parent metal plate base metal and S7 before and after fatigue loading.

Furthermore, prior to fatigue loading, the excitation probe was placed at varying distances from the left end of the weld seam on seven sets of OSD specimens. The receiving probe was correspondingly moved to maintain a total transmission distance of 115 mm for testing to investigate the weld seam's impact on the nonlinear parameters. Figure 14 illustrates the average nonlinear parameters at different distances from the weld seam. It shows that the closer the excitation probe is to the center of the weld seam, the greater the nonlinear parameters and the more pronounced the nonlinear response is. This suggests that the welded structure produces a certain degree of nonlinear response.

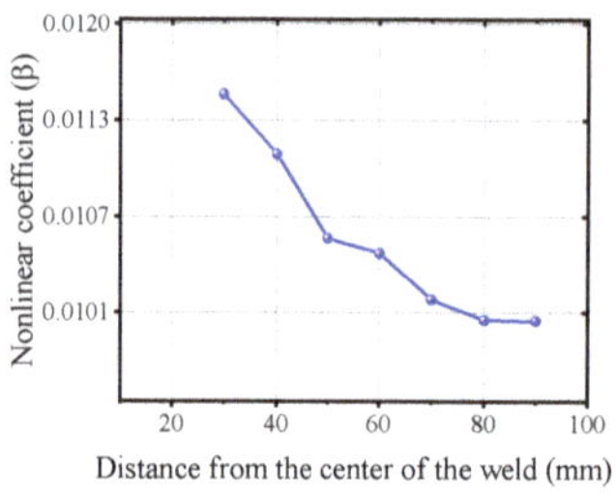

Figure 14. Nonlinear parameter change curve with different distance from weld.

As the excitation probe moves further from the center of the weld seam, the nonlinear parameters gradually decrease. However, when the distance becomes sufficiently large, the change in nonlinear parameters is minimal. This indicates that part of the nonlinear response in the OSD fatigue damage test is due to the weld seam. Therefore, ultrasonic detection results should not be attributed solely to fatigue loading, as this may introduce a degree of error. This influence is also why different Gaussian curve shape parameters were observed among the seven sets of OSD specimens. Unfortunately, due to the presence of weld seams, the proposed method cannot achieve a baseline-free characteristic, which poses challenges for assessing existing bridges. Nonetheless, it remains suitable for evaluating newly constructed bridges.

5. Conclusions

This study proposes a method for detecting the fatigue life of OSDs based on nonlinear ultrasonic parameters. By employing a nonlinear ultrasonic system, tests were conducted on seven OSD specimens and a parent plate. It was observed that in the early stages of fatigue, the nonlinear effects in the OSDs significantly increased, peaking at 61% of the fatigue life, after which they began to decline. Concurrently, the study compared the performance of linear ultrasonic indicators in assessing OSDs' fatigue life and found no significant changes under fatigue loading. In contrast, the DI based on nonlinear parameters could be well-fitted to the remaining fatigue life of OSDs using a Gaussian curve, enabling the quantitative analysis of OSDs' fatigue life. Additionally, the change trend of the nonlinear parameters of OSDs was consistent with the increase in fatigue life when compared to steel plates of the same thickness, confirming that weld seams do not affect the effectiveness of the nonlinear ultrasonic method in evaluating fatigue life. This research lays the groundwork for the practical application of nonlinear ultrasonic parameters in predicting the fatigue life of OSDs.

However, the nonlinear response induced by the welded structures within the OSDs leads to variations in the fitting curve parameters among different specimens. This nonlinear response from the weld seams must be considered in ultrasonic testing to avoid misinterpreting the results as solely being caused by fatigue loading. This limitation prevents the evaluation method from being baseline-free, thereby constraining the assessment of existing bridges. Addressing this issue remains a focal point for future research.

Author Contributions: Conceptualization, J.L., F.Z. and W.S.; methodology, D.L.; software, F.Z.; validation, J.L., W.S. and F.Z.; formal analysis, J.L.; investigation, F.Z.; resources, J.L.; data curation, D.L.; writing—original draft preparation, F.Z.; writing—review and editing, J.L.; visualization, J.L.; supervision, D.L.; project administration, F.Z.; funding acquisition, D.L. and W.S. All authors have read and agreed to the published version of the manuscript.

Funding: The authors are grateful for the financial support from the National Natural Science Foundation of China (NSFC) under Grant No. 52278294, Guangxi Science & Technology Bases and Talents Special Program (Grant Nos. Guike AD21220050 and AA23026011).

Institutional Review Board Statement: Not applicable.

Informed Consent Statement: Not applicable.

Data Availability Statement: The data presented in this study are available on request from the corresponding author.

Conflicts of Interest: Author Fangtong Zheng was employed by the company China Northeast Architectural Design & Research Institute Co., Ltd. The remaining authors declare that the research was conducted in the absence of any commercial or financial relationships that could be construed as a potential conflict of interest.

References

1. Kozy, B.M.; Connor, R.J.; Paterson, D.; Mertz, D.R. Proposed Revisions to Aashto-Lrfd Bridge Design Specifications for Orthotropic Steel Deck Bridges. *J. Bridge Eng.* **2011**, *16*, 759–767. [CrossRef]
2. Shi, L.; Cheng, B.; Xiang, S.; Li, D.; Liu, T. Monitoring for Fatigue Crack Geometry in Orthotropic Steel Bridge Decks by Application of Reflected Lamb Waves. *Thin-Walled Struct.* **2023**, *192*, 111170. [CrossRef]
3. Zeng, Y.; He, H.; Qu, Y.; Sun, X.; Tan, H.; Zhou, J. Numerical Simulation of Fatigue Cracking of Diaphragm Notch in Orthotropic Steel Deck Model. *Materials* **2023**, *16*, 467. [CrossRef] [PubMed]
4. Liu, Y.; Huang, W.; Yu, B.; Chen, Z.; Wang, P. Fatigue Performance Analysis of Welded T-Joints in Orthotropic Steel Bridge Decks with Ultrasonic Impact Treatment. *Materials* **2023**, *16*, 6196. [CrossRef]
5. Abdelbaset, H.; Zhu, Z. Behavior and Fatigue Life Assessment of Orthotropic Steel Decks: A State-of-the-Art-Review. *Structures* **2024**, *60*, 105957. [CrossRef]
6. Fisher, J.W.; Barsom, J.M. Evaluation of Cracking in the Rib-to-Deck Welds of the Bronx–Whitestone Bridge. *J. Bridge Eng.* **2016**, *21*, 04015065. [CrossRef]
7. Tecchio, G.; Lorenzoni, F.; Caldon, M.; Donà, M.; da Porto, F.; Modena, C. Monitoring of Orthotropic Steel Decks for Experimental Evaluation of Residual Fatigue Life. *J. Civ. Struct. Health Monit.* **2017**, *7*, 517–539. [CrossRef]
8. Gao, S.; Zhang, R.; Fan, Z.; Li, N.; Yue, Y.; Xie, L. Online Monitoring of Fatigue Damage in Welded Joints Using Diffuse Ultrasound. *Ultrasonics* **2024**, *138*, 107191. [CrossRef]
9. Pahlavan, L.; Blacquière, G. Fatigue Crack Sizing in Steel Bridge Decks Using Ultrasonic Guided Waves. *NDT E Int.* **2016**, *77*, 49–62. [CrossRef]
10. Shi, L.; Cheng, B.; Li, D.; Xiang, S.; Liu, T.; Zhao, Q. A Cnn-Based Lamb Wave Processing Model for Field Monitoring of Fatigue Cracks in Orthotropic Steel Bridge Decks. *Structures* **2023**, *57*, 105146. [CrossRef]
11. Deng, Y.; Li, A.; Feng, D. Fatigue Reliability Assessment for Orthotropic Steel Decks Based on Long-Term Strain Monitoring. *Sensors* **2018**, *18*, 181. [CrossRef] [PubMed]
12. Di, J.; Ruan, X.; Zhou, X.; Wang, J.; Peng, X. Fatigue Assessment of Orthotropic Steel Bridge Decks Based on Strain Monitoring Data. *Eng. Struct.* **2021**, *228*, 111437. [CrossRef]
13. Wei, S.; Zhang, Z.; Li, S.; Li, H. Strain Features and Condition Assessment of Orthotropic Steel Deck Cable-Supported Bridges Subjected to Vehicle Loads by Using Dense Fbg Strain Sensors. *Smart Mater. Struct.* **2017**, *26*, 104007. [CrossRef]
14. Li, D.; Nie, J.-H.; Ren, W.-X.; Ng, W.-H.; Wang, G.-H.; Wang, Y. A Novel Acoustic Emission Source Location Method for Crack Monitoring of Orthotropic Steel Plates. *Eng. Struct.* **2022**, *253*, 113717. [CrossRef]
15. Sakagami, T. Remote Nondestructive Evaluation Technique Using Infrared Thermography for Fatigue Cracks in Steel Bridges. *Fatigue Fract. Eng. Mater. Struct.* **2015**, *38*, 755–779. [CrossRef]
16. Yamada, T.; Shiraishi, A.; Okuno, M.; Sugiyama, H.; Kanjo, N.; Tsukamoto, S.; Yamagami, T. Application of Electromagnetic Testing to Orthotropic Steel Deck. In Proceedings of the 6th International Conference on Bridge Maintenance, Safety and Management, Lake Maggiore, Italy, 8–12 July 2012.
17. Farreras-Alcover, I.; Chryssanthopoulos, M.K.; Andersen, J.E. Data-Based Models for Fatigue Reliability of Orthotropic Steel Bridge Decks Based on Temperature, Traffic and Strain Monitoring. *Int. J. Fatigue* **2017**, *95*, 104–119. [CrossRef]
18. Liu, J.; Li, T.; Li, D.; Shen, W. Experimental and Numerical Validation of Guided Wave Based on Time-Reversal for Evaluating Grouting Defects of Multi-Interface Sleeve. *Smart Struct. Syst.* **2024**, *33*, 41.

19. Bjørheim, F.; Siriwardane, S.C.; Pavlou, D. A Review of Fatigue Damage Detection and Measurement Techniques. *Int. J. Fatigue* **2022**, *154*, 106556. [CrossRef]
20. Lee, Y.F.; Lu, Y. Advanced Numerical Simulations Considering Crack Orientation for Fatigue Damage Quantification Using Nonlinear Guided Waves. *Ultrasonics* **2022**, *124*, 106738. [CrossRef]
21. Hu, B.; Amjad, U.; Kundu, T. Monitoring Fatigue Cracks in Riveted Plates Using a Sideband Intensity Based Nonlinear Ultrasonic Technique. *Ultrasonics* **2024**, *141*, 107335. [CrossRef]
22. Wang, X.; Wang, X.; Niu, X.-G.; Xiao, D.-M.; Hu, X.-L. Application of Nonlinear Ultrasonic Technique to Characterize the Creep Damage in Asme T92 Steel Welded Joints. *NDT E Int.* **2018**, *98*, 8–16. [CrossRef]
23. Worlton, D.C. Experimental Confirmation of Lamb Waves at Megacycle Frequencies. *J. Appl. Phys.* **1961**, *32*, 967–971. [CrossRef]
24. Zhang, H.; Wang, F.; Lin, J.; Hua, J. Lamb Wave-Based Damage Assessment for Composite Laminates Using a Deep Learning Approach. *Ultrasonics* **2024**, *141*, 107333. [CrossRef] [PubMed]
25. Chen, H.; Zeng, J.; Wang, J.; Xu, B.; Mo, Y. Multiscale Homogenization Numerical Study on the Mechanism of Interface Debonding Detection for Steel–Concrete Composite Structures with Multichannel Surface Wave Measurements. *Constr. Build. Mater.* **2023**, *368*, 130386. [CrossRef]
26. Wang, K.; Liu, M.; Su, Z.; Yuan, S.; Fan, Z. Analytical Insight into "Breathing" Crack-Induced Acoustic Nonlinearity with an Application to Quantitative Evaluation of Contact Cracks. *Ultrasonics* **2018**, *88*, 157–167. [CrossRef] [PubMed]
27. Nilsson, M.; Huttunen-Saarivirta, E.; Bohner, E.; Ferreira, M. Non-Destructive Evaluation of Corrosion in Steel Liner Plates Embedded in Concrete Using Nonlinear Ultrasonics. *Constr. Build. Mater.* **2023**, *408*, 133691. [CrossRef]
28. Kanakambaran, K.V.; Balasubramaniam, K. Frequency Sweep Study on the Generation of Dual-Mode Second Harmonics (Dmsh) on an Isotropic Nonlinear Elastic Cylindrical Rod by T(0,1) Mode. *J. Sound Vib.* **2024**, *579*, 118374. [CrossRef]
29. Wang, K.; Liu, M.; Su, Z.; Guo, S.; Cui, F. Mode-Mismatching Enhanced Disbond Detection Using Material Nonlinearity in Guided Waves at Low Frequency. *J. Sound Vib.* **2021**, *490*, 115733. [CrossRef]
30. Asokkumar, A.; Jasiūnienė, E.; Raišutis, R.; Kažys, R.J. Comparison of Ultrasonic Non-Contact Air-Coupled Techniques for Characterization of Impact-Type Defects in Pultruded Gfrp Composites. *Materials* **2021**, *14*, 1058. [CrossRef]
31. Liu, J.; Jun, Y.; Li, D.; Cui, X.; Zhou, J. Combined Two-Level Guided Wave Structural Health Monitoring Strategy Using Multifeature Integration and Machine Learning: Application to Early-Age Grouted Sleeves. *Smart Mater. Struct.* **2023**, *32*, 095026. [CrossRef]
32. Motulsky, H.J.; Lennart, A.R. Fitting Curves to Data Using Nonlinear Regression: A Practical and Nonmathematical Review. *FASEB J.* **1987**, *1*, 365–374. [CrossRef] [PubMed]

materials

MDPI

Article

Low-Power Field-Deployable Interdigital Transducer-Based Scanning Laser Doppler Vibrometer for Wall-Thinning Detection in Plates

To Kang [1], Soonwoo Han [1], Yun-Taek Yeom [2,*] and Ho-Yong Lee [3]

1 Korea Atomic Energy Research Institute, Daejeon 34057, Republic of Korea; tkang@kaeri.re.kr (T.K.); swhan@kaeri.re.kr (S.H.)
2 Department of Smart Mechanical Engineering, Dongyang University, Yeongju 36040, Republic of Korea
3 Ceracomp Co., Ltd., Cheonan 31094, Republic of Korea; hlee.ceracomp@gmail.com
* Correspondence: ytyeom@dyu.ac.kr; Tel.: +82-054-630-1821

Abstract: Lamb waves have become a focal point in ultrasonic testing owing to their potential for long-range and inaccessible detection. However, accurately estimating the flaws in plates using Lamb waves remains challenging because of scattering, mode conversion, and dispersion effects. Recent advances in laser ultrasonic wave techniques have introduced innovative visualization methods that exploit the dispersion effect of Lamb waves to visualize defects via, for example, acoustic wavenumber spectroscopy. In this study, we developed an interdigital transducer (IDT)-based scanning laser Doppler vibrometer (SLDV) system without a power amplifier using a low-power IDT fabricated from lead magnesium niobate–lead zirconate titanate single crystals. To validate the proposed low-power IDT-based SLDV, four different defective plates were measured for defects. A comparison between a conventional IDT-based SLDV, a dry-coupled IDT-based SLDV, and the proposed method demonstrated that the latter is highly reliable for measuring thin plate defects.

Keywords: Lamb wave; low-power interdigital transducer; scanning laser Doppler vibrometer; wavenumber sensitivity; thin-wall plate detection

1. Introduction

Lamb waves have garnered significant interest from researchers owing to their capacity to be used in inaccessible or long-range detection applications. These waves offer a unique advantage in ultrasonic testing by providing information about specimens from scattered waves, which is challenging to achieve via traditional methods. However, the data obtained from Lamb waves regarding specimen defects, particularly from scattered waves, are often insufficient for accurately estimating the defects in plates. Consequently, many studies have aimed to quantitatively analyze Lamb wave properties, including reflection, mode conversion, and the interactions between symmetric, anti-symmetric, and shear horizontal waves [1–4]. However, the complex behavior of these waves, which is characterized by scattering, diffraction, and mode conversion, poses significant challenges in detecting minor pitting defects or gradual wall thinning. Additionally, the dispersion effect, which results in the spreading of Lamb waves over time and distance as they traverse a plate, is a critical issue [5–8].

In response to these challenges, laser ultrasonic wave techniques have been adopted [9–28] to enable defects to be visualized through signal processing methods without the need for intricate Lamb wave analysis. Quantitative visualization imaging algorithms such as acoustic wavenumber spectroscopy (AWS) [18], local wavenumber domain estimation [15,29–31], local wavenumber mapping [32,33], and 2D wavenumber estimation [34] have been employed. Moreover, the dispersion effect of Lamb waves, which has been traditionally viewed as a drawback because of its impact on wave propagation, is used in these visualization methods. Specifically, the phase velocity changes in Lamb waves passing through structures provide valuable thickness information, transforming the dispersion effect from a disadvantage into a useful tool for quantitatively assessing structures.

Previous studies on the use of exciters and receivers for ultrasonic wave generation and detection in plates have explored several combinations, including pulsed lasers, piezoelectric transducers (PZTs), and air-coupled transducers used as exciters and laser Doppler vibrometers, PZTs, or air-coupled transducers used as receivers. Notable work on the optimization of exciter and receiver configurations was conducted by An et al. [10], who identified that the combination of scanning laser excitation with fixed-point PZT sensing yielded the highest signal-to-noise ratio among the laser ultrasonic scanning setups explored, which included fixed-point PZT excitation with scanning laser sensing, scanning laser excitation with fixed-point PZT sensing, fixed-point laser excitation with scanning laser sensing, and scanning laser excitation with fixed-point laser sensing. Their research primarily focused on the analysis of propagated and scattered waves from defects in a plate. Addressing a different aspect, Flynn and Jarmer introduced a method for the high-speed imaging of plate defects using steady-state ultrasonic vibration excited by a continuous sinusoidal wave [35]. This approach, particularly the fixed-point PZT excitation with a scanning laser sensing configuration, allowed for high energy acquisition owing to energy efficiently pumping into the structures, the absence of measurement delays, and reduced acquisition times. This was achieved while ensuring operational safety using class 1–2 lasers with lower power compared to those used in propagation wave analysis methods. Consequently, continuous-based excitation using scanning laser Doppler vibrometers (SLDVs) has emerged as a powerful tool for visualizing and detecting defects in plates.

Despite the advantages of SLDVs over scanning laser excitation methods, using the A_0 Lamb wave mode complicates the detection of shallow wall thinning in plates. Early damage detection is crucial for preventing catastrophic failures in industrial facilities, such as nuclear and thermal power plants. Traditional SLDV approaches, which use A_0 mode for wall-thinning detection, must contend with the mode's dispersive nature, which impacts the accuracy of evaluating the thickness of wall thinning. This limitation restricts the wall-thinning detection capability of SLDV to thicknesses higher than 30% in plates and pipes. To address this challenge, Kang et al. [36,37] and Moon et al. [38] explored the detection of shallow wall thinning using an interdigital transducer (IDT)-based SLDV [36–38], which offers a promising avenue for enhancing early damage detection capabilities.

IDTs are manufactured using a picosecond laser machining process and are engineered to excite modes sensitive to shallow wall thinning, which are characterized by a high wavenumber sensitivity. In the studies of Kang et al. and Moon et al. [36–38], IDT-based SLDVs were proven capable of detecting wall thinning higher than 5%, aligning with the requirements of KEPIC 3521 for nuclear power plants. A notable limitation of both PZT and IDT technologies is their dependence on a couplant, such as oil or water, which complicates and encumbers the inspection process. To address this, Kang et al. [39] introduced a dry-coupled IDT-based SLDV system. This innovative approach used a $\lambda/4$ impedance transformer with alumina for dry coupling and successfully detected wall thinning over 3% in plates without requiring any couplant.

Despite these advances, and even though the dry-coupled IDT-based SLDV system is more amenable to field deployment, it still faces challenges that need resolution. One significant issue is the power consumption of the exciter. Typically, a power amplifier is used to enable the IDT to generate a high-power sinusoidal wave in IDT-based SLDV operations. For instance, the HSA4052 model from the NF Corporation(Yokohama, Japan), with its maximum voltage of ± 75 V at 25 Ω, is commonly used.

In this study, we propose a novel methodology for operating the IDT-based SLDV without a power amplifier. This approach involves replacing the conventional IDT material with a CSL2 (PMN-PZT) to enhance the sensitivity of IDT sensors, thereby developing a highly reliable system that does not require a power amplifier. The rest of this paper is structured as follows. Section 2 describes the development of the newly designed low-power IDT-based SLDV. Section 3 evaluates the performance of the low-power IDT-based SLDV through a comparative analysis; it also elaborates on the performance metrics of the IDT-based SLDV, dry-coupled IDT-based SLDV, and low-power IDT-based SLDV, highlighting their differences and capabilities. Section 4 concludes the paper with a summary that encapsulates the findings and potential implications of this research.

2. Low-Power IDT-Based SLDV

2.1. Fabrication of Lead Magnesium Niobate–Lead Zirconate Titanate Single Crystals

In this study, the solid-state single crystal growth (SSCG) method was used to synthesize lead magnesium niobate–lead zirconate titanate (PMN-PZT) [$Pb(Mg_{1/3}Nb_{2/3})O_3$-$Pb(Zr,Ti)O_3$] piezoelectric single crystals (Product Code: CSL2) at Ceracomp Co., Ltd. (Cheonan, Republic of Korea) [40–44]. Figure 1 displays a schematic representation of the SSCG method.

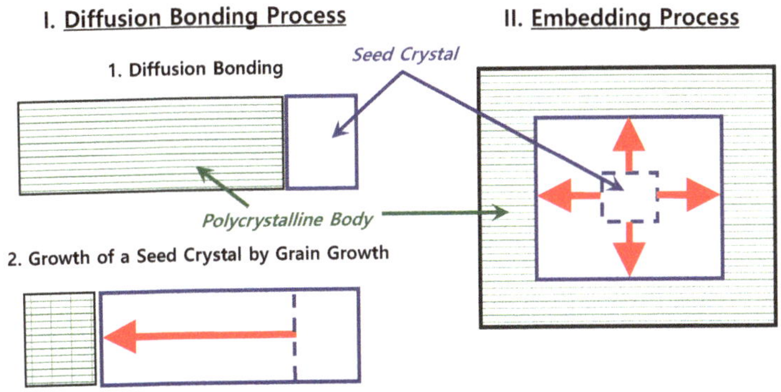

Figure 1. Solid-state single crystal growth (SSCG) method: illustration of single crystal development within a polycrystalline matrix via grain growth.

Among the various kinds of PMN-PZT single crystals, the PMN-PZT (CSL2) single crystal was selected because it has high piezoelectric constants, has a composition of "40PMN-25PZ-35PT + 1.0La_2O_3 [mol%]", and exhibits a rhombohedral phase at room temperature. Compared to conventional single-crystal methods, such as the flux and Bridgman methods, the SSCG method is more cost-effective and suitable for mass production because the difficult steps of melting and solidification in the conventional methods can be completely avoided in the SSCG process. Because of its advantages, the SSCG method is the most effective way of growing high-performance piezoelectric single crystals with very complicated chemical compositions.

High-purity raw materials, including Pb_3O_4 (99.9%, Alfa Aesar, Ward Hill, MA, USA), $MgNb_2O_6$ (99.9%, H. C. Starck GmbH, Newton, MA, USA), and TiO_2 (99.99%, Ishihara, San Francisco, CA, USA), were selected for the fabrication process. After precisely measuring each raw material, the powders were subjected to ball milling for 24 h, then dried and calcined at 800 °C. The calcinated powders were further processed via secondary ball milling, with the addition of excess PbO powder, and were subsequently dried and sieved to produce the final powder. This powder was then uniaxially hot pressed at high temperatures to form a dense, primary sintered body. For the SSCG process, a $Ba(Zr_{0.1}Ti_{0.9})O_3$ seed single crystal was positioned on top of the ceramic sintered body, after which a specialized heat treatment was performed. To mitigate the loss of volatile PbO, the double-crucible method was used during the SSCG heat treatment. This process facilitated the continuous growth of the $Ba(Zr_{0.1}Ti_{0.9})O_3$ seed single crystal within the polycrystalline ceramic matrix, culminating in the production of single crystals measuring 40 mm × 40 mm × 10 mm, as shown in Figure 2. A notable advantage of the SSCG method is the absence of composition gradients within the resulting single crystal, which ensures chemical uniformity. This uniformity is attributed to the fact that the PMN-PZT phase does not melt during the fabrication process, thereby preserving the integrity of the single crystal structure.

Figure 2. Image of PMN-PZT piezoelectric single crystal grown in polycrystalline ceramics via the SSCG method.

Figure 2 shows a photograph of a PMN-PZT piezoelectric single crystal produced using the SSCG method.

2.2. Design and Fabrication of Low-Power IDT

In linear piezoelectric materials, the interactions between electrical and mechanical variables conform to linear relationships as defined by the ANSI/IEEE Standard 176-1987 [45]. The constitutive relationships that govern the interaction between these variables are

$$\boldsymbol{S} = s^E \boldsymbol{T} + d \tag{1}$$

$$\boldsymbol{D} = d\boldsymbol{T} + \varepsilon^T \boldsymbol{E} \tag{2}$$

where s^E is the elastic compliance under a constant electric field, $\boldsymbol{T}$ is the stress, d is the piezoelectric charge constant, $\boldsymbol{E}$ is the electric field, $\boldsymbol{S}$ is the strain, ε^T is the permittivity of the ceramic material, and $\boldsymbol{D}$ is the electric flux density. An increase in the strain necessitates a corresponding increase in the elastic compliance, stress, piezoelectric charge constant, or the electric field. However, given that s^E, $\boldsymbol{T}$, and $\boldsymbol{E}$ remain constant, the variable d becomes the focal point for adjustment.

Table 1 compares the performance metrics of the developed PMN-PZT (CSL2) single crystals with those of APC 850 (PZT) polycrystalline ceramics. CSL2 was categorized into two subtypes based on the crystallographic orientation: CSL2 (001) exhibited higher k_{33} and d_{33} constants (appropriate for the longitudinal vibration mode), and CSL2 (011) exhibited higher k_{31} and d_{31} constants (appropriate for the lateral vibration mode).

Table 1. Specifications of CSL2 and APC 850.

Type / Parameter	CSL2 (001)	CSL2 (011)	APC 850
$tan\delta$	0.011	0.013	$\leq$2.00
k_{33}	0.91	0.88	0.72
k_{31}	0.45	0.76	0.36
d_{33} (pC/N)	1952	753	400
$-d_{31}$ (pC/N)	750	919	175

In Table 1, $tan\delta$ is the dielectric loss, k_{ij} is the electromechanical coupling factor, i represents the direction of the electric field, and j represents the direction of the vibration. The term d_{ij} corresponds to the piezoelectric charge constant previously mentioned, where polarization is induced in direction i (aligned with the piezoelectric element's polarization) per unit stress applied in direction j (perpendicular to the ceramic element's polarization) [46,47].

The CSL2 (001) subtype, operating in a longitudinal vibration mode, exhibited a d_{33} value of 1952 pC/N, significantly surpassing APC 850's d_{33} value of 400 pC/N. Similarly, the CSL2 (011) subtype in the lateral vibration mode had a d_{31} value of 919 pC/N, compared to APC 850's d_{31} value of 175 pC/N. Specifically, the d_{33} value for CSL2 (001) was 4.88 times higher than that for APC 850. In addition, the d_{31} value for CSL2 (011) was 5.25 times higher than that for APC 850. These comparisons confirmed CSL2's suitability as an IDT material for low-power applications owing to its superior d_{33} and d_{31} values relative to PZTs.

Figure 3 shows the low-power IDT produced using CSL2.

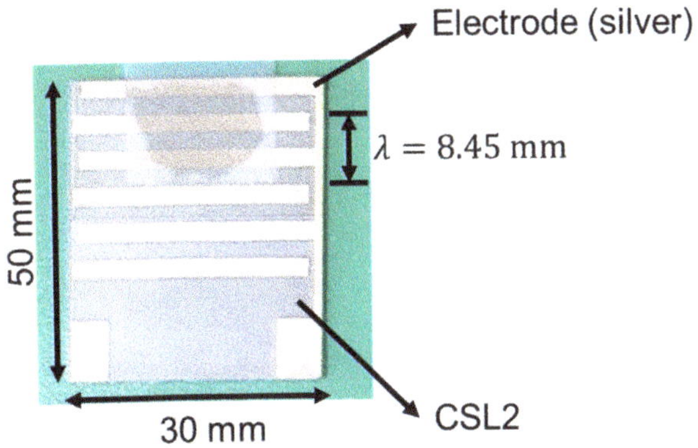

Figure 3. Prototype of low-power interdigital transducer (IDT) using CSL2.

Based on previous studies [36–39], the optimal frequency for the low-power IDT was determined, and the wavenumber sensitivity for 450 kHz was obtained by taking the derivative of the dispersion curve, as shown in Figure 4.

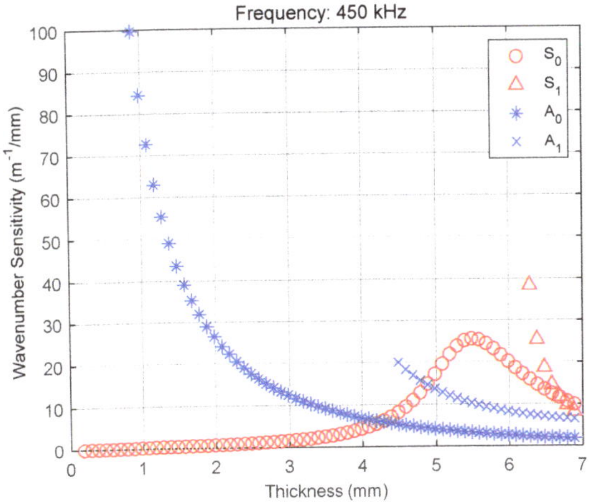

Figure 4. Wavenumber sensitivity at 450 kHz for carbon steel.

The mode of A_0 exhibited a favorable sensitivity for thicknesses of less than 2 mm but was unsuitable for plates with a thickness of 6 mm. In contrast, the S_0 mode exhibited a high sensitivity in the plates with a thickness of 6 mm, which demonstrates the precise design for efficient mode generation in low-power IDTs.

To assess the performance of the fabricated low-power IDT, the RMS value was measured under several excitation voltages. In this study, CSL2 (001) is referred to as the d_{33}-type CSL2, and CSL2 (011) as the d_{31}-type CSL2. The relationship between the RMS value and the excitation voltage was linear. Using retroreflective tape enhanced the stability of the optical measurements, resulting in more stable RMS values compared to measurements taken without the tape [44]. Figure 5 illustrates the RMS measurements for the d_{33}-type CSL2, comparing scenarios with and without the retroreflective tape.

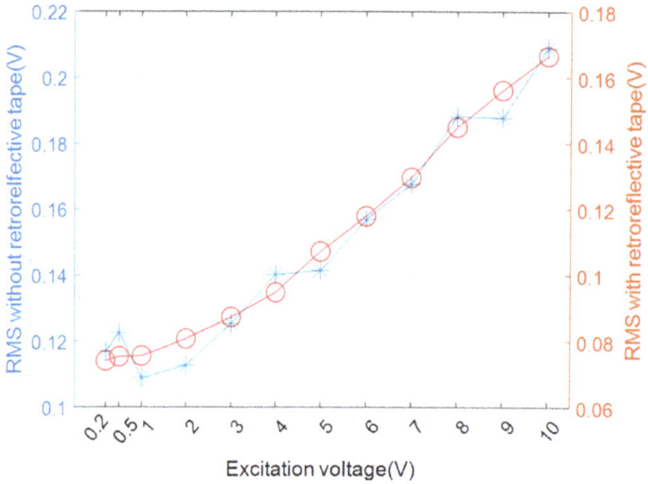

Figure 5. RMS as a function of excitation voltage for the d_{33}-type CSL2 with (orange circles) and without (blue stars) retroreflective tape.

The results indicated that the RMS values consistently exceeded 0.2 V (irrespective of the tape's presence) at an excitation voltage of 10 V. This suggests that the device can operate effectively without the need for a power amplifier. Figure 6 shows the RMS values for the d_{31}-type CSL2, for which tests were conducted exclusively without retroreflective tape.

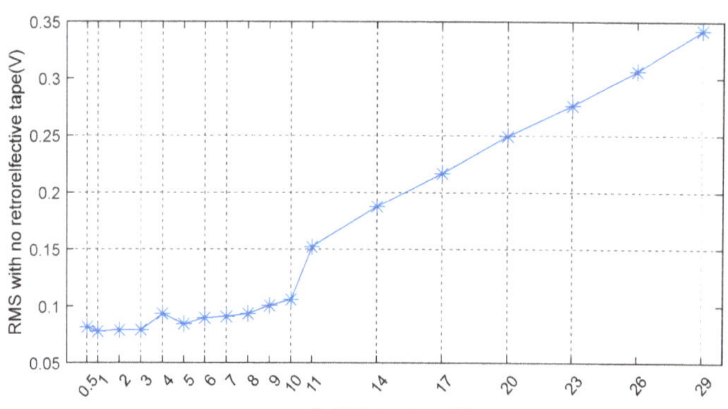

Figure 6. RMS as a function of excitation voltage for the d_{31}-type CSL2 (without retroreflective tape).

The findings demonstrated that a minimum excitation voltage of approximately 15–16 V was necessary to achieve RMS values above 0.2 V. Given that commercial function generators typically provide 10 V or less, this limitation underscores the necessity for a power amplifier for the d_{31}-type CSL2.

A comparative analysis of Figures 5 and 6 revealed that the d_{33}-type CSL2 exhibited a more efficient performance than the d_{31}-type CSL2. Considering the 10 V limit of commercial function generators, the d_{33}-type CSL2 was selected as the preferred material for low-power IDT applications.

3. Experiment

3.1. Wall-Thinning Measurement Procedure Using AWS

The modified AWS technique, as proposed by Kang et al. [37], was used to reconstruct the thickness of a plate. This modified approach builds upon the original AWS method developed by Flynn and Jarmer [35], incorporating adjustments to enhance image quality with captured signals. The procedure of the modified AWS method, which is depicted in Figure 7, emphasizes the importance of the bandpass filter in eliminating unwanted propagation modes in $v^{f_i}[x, y, f_0]$. As the cutoff frequency increases, a continuous augmentation occurs in the number of modes in the high-frequency thickness regime, indicating that the bandpass filter is essential for isolating pure propagation modes. A key innovation of this modified AWS method compared to the original method is dispersion-based thickness mapping. The fusion of images at different frequencies allows for the accurate visualization of defects at varying depths, overcoming the limitations of traditional single-mode approaches and providing intuitive thickness information that facilitates the inspection process [37].

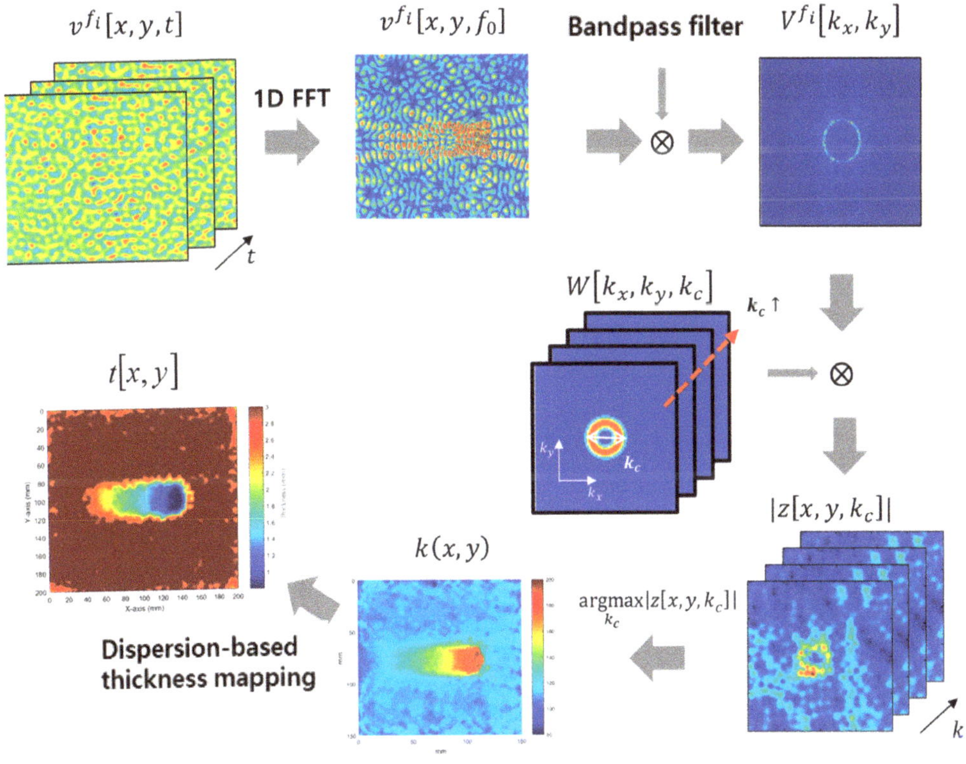

Figure 7. Procedure for measuring the thickness of plates using modified AWS.

3.2. Specimen and Experimental Setup

Figure 8 illustrates the method for measuring the thickness of a thin-wall plate with a low-powered IDT scanning laser Doppler vibrometer (SLDV).

To validate the functionality of the low-power IDT, the AWS method was used in the experimental setup. A function generator (PXI-5402) was used to produce a single-frequency sine wave, a high-speed bipolar amplifier (HSA4052) was used to deliver a high-voltage signal, and three types (standard, dry-coupled, and low-power) of IDTs were used to generate excitations. A scan head unit was used for the scanning vibrometer, which received the signal, and a data acquisition board (PCI-5124) recorded the vibration signals. The experimental arrangement is depicted in Figure 9.

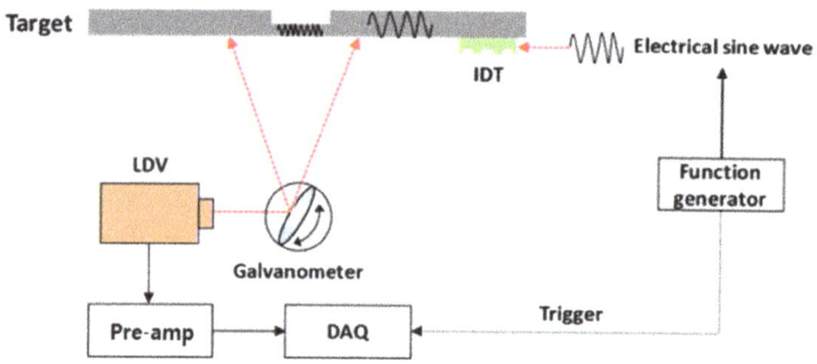

Figure 8. Schematic of low-power IDT scanning laser Doppler vibrometer (SLDV).

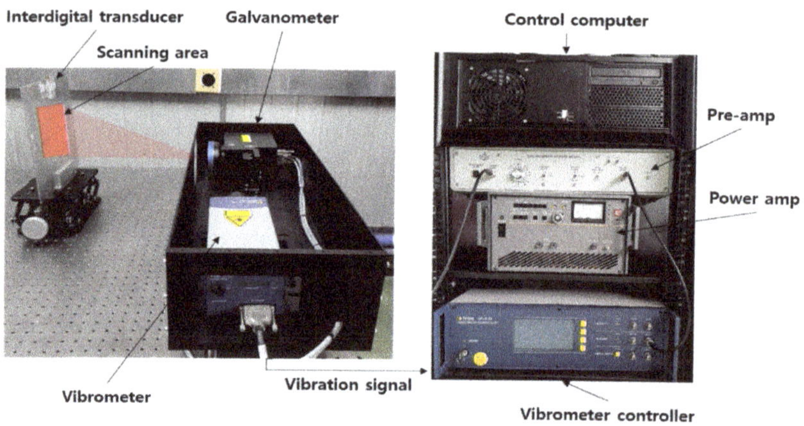

Figure 9. Setup for scanning laser Doppler vibrometer (SLDV) experiments.

The tested material was carbon steel with a thickness of 6 mm, featuring four different depths of fabricated defects. These defects were square-shaped, measuring 40 mm by 40 mm, and were located at depths of 0.3, 0.6, 0.9, and 1.2 mm, as illustrated in Figure 10. To conduct the experiment, the scan area was set to 150 mm × 150 mm in relation to the defect.

The materials of the test piece, defect size, etc., were set to the same configuration as the existing test piece to perform a comparative analysis with the results of previous research. Such defects correspond to those outlined in the KEPIC 3521 standards for nuclear power plants [36–39].

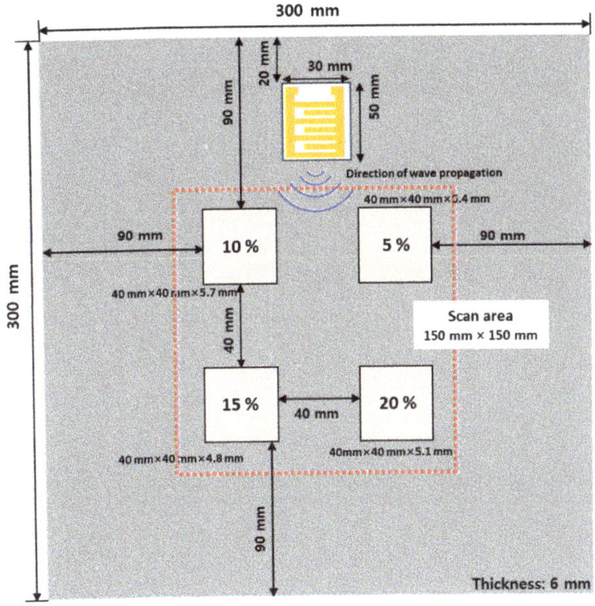

Figure 10. Carbon steel specimen showing defects at four different depths.

3.3. Experimental Results

This section presents the results of applying the d_{33}-type CSL2 IDT in both low-power IDT-based and dry-coupled IDT-based SLDV configurations. Figure 11 displays the experimental imaging results for the three SLDV techniques; panels (a) and (b) show the results of previous studies, and panel (c) displays the results of the low-power IDT-based SLDV.

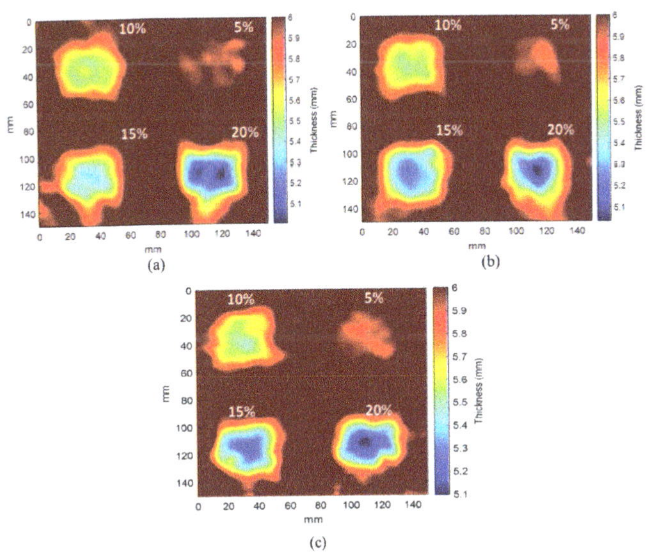

Figure 11. Comparative results for three SLDV techniques: (**a**) IDT-based SLDV, (**b**) dry-coupled IDT-based SLDV, and (**c**) low-power IDT-based SLDV.

All three techniques resulted in favorable outcomes; even the 20% depth defect was imaged with greater clarity than the other defects. This is attributable to the difference in wavenumber between the defect and sound regions. As illustrated in Figure 4, the wavenumber sensitivity was high at 6 mm, and the error rate was minimal because of the proportional relationship between the accuracy and wavenumber sensitivity.

Figure 12 displays the section thickness measurements along the y-axis for each defect depicted in Figure 11. The figure compares the performance of each SLDV technique in terms of the defects.

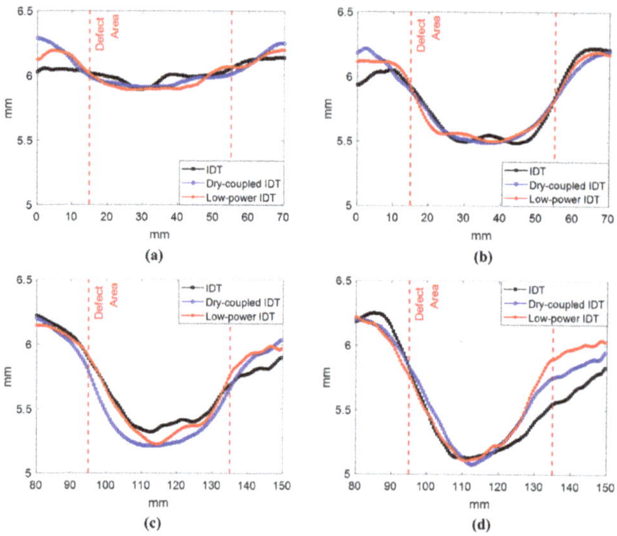

Figure 12. Thickness measurement results of three SLDV techniques for (**a**) 5%, (**b**) 10%, (**c**) 15%, and (**d**) 20% depth defects.

As illustrated in Figure 11, the results of the low-powered IDT-based SLDV method were comparable to those of the previously investigated IDT and dry-coupled IDT. This substantiates the efficacy of the low-powered IDT-based SLDV technique. The results of Figures 11 and 12 are summarized in Table 2.

Table 2. Summary of experimental SLDV results.

Depth	IDT-Based SLDV	Dry-Coupled IDT-Based SLDV	Low-Power IDT-Based SLDV
5% (Error rate)	1.77% (64.6%)	1.47% (70.6%)	1.7% (66%)
(0.3 mm)	(0.11 mm)	(0.09 mm)	(0.10 mm)
10% (Error rate)	7.52% (24.8%)	8.5% (15%)	8.37% (16.3%)
(0.6 mm)	(0.45 mm)	(0.51 mm)	(0.50 mm)
15% (Error rate)	11.17% (25.5%)	13.17% (12.2%)	13% (13.3%)
(0.9 mm)	(0.67 mm)	(0.79 mm)	(0. 78 mm)
20% (Error rate)	14.5% (27.5%)	15.5% (22.5%)	14.83% (25.8%)
(1.2 mm)	(0.87 mm)	(0.93 mm)	(0.89 mm)

The results presented in Figures 11 and 12, as well as in Table 2, demonstrate the enhanced performance of the low-power IDT SLDV technique in comparison to that of the other methods. It is evident that a 5% depth defect is associated with an elevated error rate, exceeding 65%. Furthermore, determining the thickness of thin plates with precision is a challenging task. The most effective outcomes were observed in instances where the depth defects were present at a rate of 10–15%. The low-power IDT-based SLDV demonstrated its effectiveness in accurately representing all the tested defect depths, underscoring its potential as a viable tool for detecting defects in carbon steel specimens.

4. Conclusions

In this study, a PMN-PZT [$Pb(Mg_{1/3}Nb_{2/3})O_3$-$Pb(Zr,Ti)O_3$] single-crystal (product code: CSL2) low-power IDT sensor was successfully developed. A notable advantage of the SSCG method is the absence of composition gradients within the resulting single crystal, which ensures chemical uniformity. This uniformity is attributed to the fact that the PMN-PZT phase does not melt during the fabrication process, thereby preserving the integrity of the single crystal structure. Based on a PMN-PZT, the sensor was designed for SLDV applications that do not need a power amplifier. CSL2 exhibited a higher piezoelectric charge constant compared to that of APC 850 (PZT) and functioned effectively without requiring high excitation voltages. This feature facilitated the efficient low-power detection of defects, marking a significant advance in the field of IDT sensor technology. The sensor, using the unique properties of PMN-PZT, exhibited an enhanced performance relative to both the traditional IDT-based SLDV system and those using the dry-coupled IDT configuration. The improvement was attributed to the inherent advantages offered by the CSL2 sensor's design and material composition. Experiments were conducted with four different depth specimens. In the experiments, high wavenumber sensitivity led to a low error rate, and detecting a 5% defect was more difficult than detecting a 20% defect due to the low wavenumber sensitivity of the 5% defect. These results are consistent with the results of previous papers [36–39]. The low-power IDT-based SLDV technology is anticipated to have a significant impact on non-destructive testing and evaluation, as it offers a promising solution for industrial applications in which low-power operation is essential. This study opens new avenues for efficiently and effectively monitoring structural integrity in a variety of settings, contributing to safer and more sustainable industrial practices.

Author Contributions: Conceptualization, T.K. and H.-Y.L.; methodology, T.K. and S.H.; software, T.K.; validation, T.K. and H.-Y.L.; formal analysis, T.K. and S.H.; investigation, T.K.; resources, T.K.; data curation, T.K.; writing—original draft preparation, T.K. and Y.-T.Y.; writing—review and editing, T.K. and Y.-T.Y.; visualization, T.K. and Y.-T.Y.; supervision, Y.-T.Y.; project administration, T.K.; funding acquisition, T.K. All authors have read and agreed to the published version of the manuscript.

Funding: This work was supported by a National Research Foundation of Korea (NRF) grant funded by the Korean government (No. RS-2022-00144411).

Institutional Review Board Statement: Not applicable.

Informed Consent Statement: Not applicable.

Data Availability Statement: The data that support the findings of this study are available from the corresponding author upon reasonable request.

Conflicts of Interest: Author Ho-Yong Lee was employed by the company Ceracomp Co., Ltd. The remaining authors declare that the research was conducted in the absence of any commercial or financial relationships that could be construed as a potential conflict of interest.

References

1. Lowe, M.; Diligent, O. Low-frequency reflection characteristics of the s 0 Lamb wave from a rectangular notch in a plate. *J. Acoust. Soc. Am.* **2002**, *111*, 64–74. [CrossRef] [PubMed]
2. Lowe, M.; Cawley, P.; Kao, J.; Diligent, O. The low frequency reflection characteristics of the fundamental antisymmetric Lamb wave a_0 from a rectangular notch in a plate. *J. Acoust. Soc. Am.* **2002**, *112*, 2612–2622. [CrossRef] [PubMed]
3. Hirao, M.; Ogi, H. An SH-wave EMAT technique for gas pipeline inspection. *NDT E Int.* **1999**, *32*, 127–132. [CrossRef]
4. Thon, A.; Bélanger, P. EMAT design for minimum remnant thickness gauging using high order shear horizontal modes. *Ultrasonics* **2019**, *95*, 70–78. [CrossRef]
5. Wilcox, P.; Lowe, M.; Cawley, P. The effect of dispersion on long-range inspection using ultrasonic guided waves. *NDT E Int.* **2001**, *34*, 1–9. [CrossRef]

6. Wilcox, P. A rapid signal processing technique to remove the effect of dispersion from guided wave signals. *IEEE Trans. Ultrason.* **2003**, *50*, 419–427. [CrossRef]
7. Sicard, R.; Goyette, J.; Zellouf, D. A numerical dispersion compensation technique for time recompression of Lamb wave signals. *Ultrasonics* **2002**, *40*, 727–732. [CrossRef]
8. Xu, K.; Ta, D.; Moilanen, P.; Wang, W. Mode separation of Lamb waves based on dispersion compensation method. *J. Acoust. Soc. Am.* **2012**, *131*, 2714–2722. [CrossRef]
9. Lee, J.; Takatsubo, J.; Toyama, N. Disbond monitoring at wing stringer tip based on built-in ultrasonic transducers and a pulsed laser. *Smart Mater. Struct.* **2007**, *16*, 1025–1035. [CrossRef]
10. An, Y.; Park, B.; Sohn, H. Complete noncontact laser ultrasonic imaging for automated crack visualization in a plate. *Smart Mater. Struct.* **2013**, *22*, 025022. [CrossRef]
11. Lee, C.; Park, S. Damage visualization of pipeline structures using laser-induced ultrasonic waves. *Struct. Health Monit.* **2015**, *14*, 475–488. [CrossRef]
12. He, J.; Yuan, F. Damage identification for composite structures using a cross-correlation reverse-time migration technique. *Struct. Health Monit.* **2015**, *14*, 558–570. [CrossRef]
13. Kudela, P.; Radzieński, M.; Ostachowicz, W. Identification of cracks in thin-walled structures by means of wavenumber filtering. *Mech. Syst. Signal Process.* **2015**, *50–51*, 456–466. [CrossRef]
14. Segers, J.; Hedayatrasa, S.; Poelman, G.; Van Paepegem, W.; Kersemans, M. Robust and baseline-free full-field defect detection in complex composite parts through weighted broadband energy mapping of mode-removed guided waves. *Mech. Syst. Signal Process.* **2021**, *151*, 107360. [CrossRef]
15. Rogge, M.; Leckey, C. Characterization of impact damage in composite laminates using guided wavefield imaging and local wavenumber domain analysis. *Ultrasonics* **2013**, *53*, 1217–1226. [CrossRef]
16. Tian, Z.; Howden, S.; Ma, Z.; Xiao, W.; Yu, L. Pulsed laser-scanning laser Doppler vibrometer (PL-SLDV) phased arrays for damage detection in aluminum plates. *Mech. Syst. Signal Process.* **2019**, *121*, 158–170. [CrossRef]
17. Yu, L.; Tian, Z.; Leckey, C. Crack imaging and quantification in aluminum plates with guided wave wavenumber analysis methods. *Ultrasonics* **2015**, *62*, 203–212. [CrossRef]
18. Flynn, E.; Chong, S.; Jarmer, G.; Lee, J. Structural imaging through local wavenumber estimation of guided waves. *NDT E Int.* **2013**, *59*, 1–10. [CrossRef]
19. Juarez, P.; Leckey, C. Multi-frequency local wavenumber analysis and ply correlation of delamination damage. *Ultrasonics* **2015**, *62*, 56–65. [CrossRef]
20. Kang, K.; Kim, Y.; Choi, W.; Park, K. Measurement of shallow defects using noncontact broadband leaky Lamb wave produced by pulsed laser with ultrasound microphone. *NDT E Int.* **2020**, *111*, 102224. [CrossRef]
21. Kudela, P.; Radzienski, M.; Ostachowicz, W. Impact induced damage assessment by means of Lamb wave image processing. *Mech. Syst. Signal Process.* **2018**, *102*, 23–36. [CrossRef]
22. Zhang, K.; Zhenggan, Z. Quantitative characterization of disbonds in multilayered bonded composites using laser ultrasonic guided waves. *NDT E Int.* **2018**, *97*, 42–50. [CrossRef]
23. Yu, F.; Saito, O.; Okabe, Y. Laser ultrasonic visualization technique using a fiber-optic Bragg grating ultrasonic sensor with an improved adhesion configuration. *Struct. Health Monit.* **2021**, *20*, 303–320. [CrossRef]
24. Gao, T.; Liu, X.; Zhu, J.; Zhao, B.; Qing, X. Multi-frequency localized wave energy for delamination identification using laser ultrasonic guided wave. *Ultrasonics* **2021**, *116*, 106486. [CrossRef] [PubMed]
25. Majhi, S.; Mukherjee, A.; George, N.V.; Karaganov, V.; Uy, B. Corrosion monitoring in steel bars using Laser ultrasonic guided waves and advanced signal processing. *Mech. Syst. Signal Process.* **2021**, *149*, 107176. [CrossRef]
26. Gao, T.; Wang, Y.; Qing, X. A new laser ultrasonic inspection method for the detection of multiple delamination defects. *Materials* **2021**, *14*, 2424. [CrossRef]
27. Chia, C.C.; Lee, S.Y.; Harmin, M.Y.; Choi, Y.; Lee, J.R. Guided ultrasonic waves propagation imaging: A review. *Meas. Sci. Technol.* **2023**, *34*, 052001. [CrossRef]
28. Zarei, A.; Pilla, S. Laser ultrasonics for nondestructive testing of composite materials and structures: A review. *Ultrasonics* **2024**, *136*, 107163. [CrossRef]
29. Mesnil, O.; Leckey, C.; Ruzzene, M. Instantaneous and local wavenumber estimations for damage quantification in composites. *Struct. Health Monit.* **2015**, *14*, 193–204. [CrossRef]
30. Gao, T.; Sun, H.; Hong, Y.; Qing, X. Hidden corrosion detection using laser ultrasonic guided waves with multi-frequency local wavenumber estimation. *Ultrasonics* **2020**, *108*, 106182. [CrossRef]
31. Zhao, G.; Wang, B.; Hao, W.; Luo, Y.; Chen, H. Localization and characterization of delamination in laminates using the local wavenumber method. *Compos. Struct.* **2020**, *238*, 111972. [CrossRef]
32. Jeon, J.; Gang, S.; Park, G.; Flynn, E.; Kang, T.; Han, S. Damage detection on composite structures with standing wave excitation and wavenumber analysis. *Adv. Compos. Mat.* **2017**, *26*, 53–65. [CrossRef]
33. Kang, T.; Lee, J.; Han, S.; Park, J.; Park, H.; Jeon, J. Measuring plate thickness using spatial local wavenumber filtering. *J. Korean Soc. Nondestruc. Test* **2016**, *36*, 370–376. [CrossRef]
34. Jeon, J.; Kim, D.; Park, G.; Flynn, E.; Kang, T.; Han, S. 2D-wavelet wavenumber filtering for structural damage detection using full steady-state wavefield laser scanning. *NDT E Int.* **2020**, *116*, 102343. [CrossRef]

35. Flynn, E.B.; Jarmer, G.S. High-speed, non-contact, baseline-free imaging of hidden defects using scanning laser measurements of steady-state ultrasonic vibration. *Struct. Health Monit.* **2013**, *1*, 1186–1193.
36. Kang, T.; Moon, S.; Han, S.; Jeon, J.; Park, G. Measurement of shallow defects in metal plates using inter-digital transducer-based laser-scanning vibrometer. *NDT E Int.* **2019**, *102*, 26–34. [CrossRef]
37. Kang, T.; Han, S.; Moon, S.; Han, S.; Jeon, J.; Park, G. Lamb-wave sparse-frequency interdigital-transducer-based scanning laser Doppler vibrometry for quantitative depth-wise visualization of defects in plates. *NDT E Int.* **2019**, *107*, 102137. [CrossRef]
38. Moon, S.; Kang, T.; Han, S.; Jeon, J.; Park, G. Optimization of excitation frequency and guided wave mode in acoustic wavenumber spectroscopy for shallow wall-thinning defect detection. *J. Mech. Sci. Technol.* **2018**, *32*, 5213–5221. [CrossRef]
39. Kang, T.; Han, S.; Moon, S.; Han, S.; Jeon, J.; Park, G. Measurement of defects in a plate using dry-coupled interdigital transducer–based scanning laser Doppler vibrometer. *Struct. Health Monit.* **2021**, *20*, 596–603. [CrossRef]
40. Available online: www.ceracomp.com (accessed on 1 October 2024).
41. Kang, S.; Park, J.; Ko, S.; Lee, H. Solid-state conversion of single crystals: The principle and the state-of-the-art. *J. Am. Ceram. Soc.* **2015**, *98*, 347–360. [CrossRef]
42. Zhang, S.; Lee, S.; Kim, D.; Lee, H.; Shrout, T. Temperature dependence of the dielectric, piezoelectric, and elastic constants for Pb $(Mg_{1/3}Nb_{2/3})O_3$–Pb Zr O_3–Pb Ti O_3 piezocrystals. *J. Appl. Phys.* **2007**, *102*, 114103. [CrossRef]
43. Zhang, S.; Lee, S.; Kim, D.; Lee, H.; Shrout, T. Elastic, piezoelectric, and dielectric properties of 0.71 Pb $(Mg_{1/3}Nb_{2/3})O_3$–0.29 $PbTiO_3$ crystals obtained by solid-state crystal growth. *J. Am. Ceram. Soc.* **2008**, *91*, 683–686. [CrossRef]
44. Zhang, S.; Lee, S.; Kim, D.; Lee, H.; Shrout, T. Characterization of Mn-modified Pb $(Mg_{1/3}Nb_{2/3})O_3$–PbZr O_3–PbTi O_3 single crystals for high power broad bandwidth transducers. *Appl. Phys. Lett.* **2008**, *93*, 122908. [CrossRef] [PubMed]
45. *176-1987*; IEEE Standard on Piezoelectricity. IEEE: New York, NY, USA, 1988. Available online: https://ieeexplore.ieee.org/document/26560 (accessed on 1 October 2024).
46. Xu, T. 7—Energy harvesting using piezoelectric materials in aerospace structures. *Struct. Health Monit.* **2016**, 175–215. [CrossRef]
47. Shen, Z.; Lu, J.; Tan, C.; Miao, J.; Wang, Z. d_{33} mode piezoelectric diaphragm based acoustic transducer with high sensitivity. *Sens. Actuators A Phys.* **2013**, *189*, 93–99. [CrossRef]

materials

MDPI

Article

High-Resolution Ultrasound to Quantify Sub-Surface Wrinkles in a Woven CFRP Laminate

Md Admay Amif and David A. Jack *

Department of Mechanical Engineering, Baylor University, Waco, TX 76706, USA; admay_amif1@baylor.edu
* Correspondence: david_jack@baylor.edu

Abstract: Carbon fiber reinforced polymer (CFRP) composites are popular materials in the aerospace and automotive industries because of their low weight, high strength, and corrosion resistance. However, wrinkles or geometric distortions in the composite layers significantly reduce their mechanical performance and structural integrity. This paper presents a method for non-destructively extracting the three-dimensional geometry, lamina by lamina, of a laminated composite. A method is introduced for fabricating consistent out-of-plane wrinkled CFRP laminate panels, simulating the in-service wrinkle observed in industries that utilize thick structure composites such as the vertical lift or wind power industries. The individual lamina geometries are extracted from the fabricated coupon with an embedded wrinkle from captured ultrasonic waveforms generated from single-element conventional ultrasonic (UT) scan data. From the extracted waveforms, a method is presented to characterize the wrinkle features within each individual lamina, specifically the spatially varying wrinkle height and intensity for the wrinkle. Parts were fabricated with visibly undetectable wrinkles using a wet layup process and a hot press for curing. Scans were performed in a conventional immersion tank scanning system, and the scan data were analyzed for wrinkle detection and characterization. Extraction of the layers was performed based on tracking the voltage peaks from A-scans in the time domain. Spatial Gaussian averaging was performed to smooth the A-scans, from which the surfaces were extracted for each individual lamina. The extracted winkle surface aligned with the anticipated wrinkle geometry, and a single parameter for quantification of the wrinkle intensity for each lamina is presented.

Keywords: woven fiber carbon composite; out-of-plane wrinkle; conventional ultrasound; A-scan; peak tracking

1. Introduction

Carbon fiber reinforced polymer (CFRP) composites are lightweight, strong materials used in various industries. They offer a significantly higher strength-to-weight ratio in comparison to metals. Industries such as aerospace, automotive, construction, and machinery benefit from CFRP composites [1,2]. Numerous studies have investigated aspects of carbon fiber reinforced polymer (CFRP) composites, including layup sequences [3]; ply orientations [4,5]; bond line thickness [6]; and the exploration of defects such as foreign object debris (FOD) [7], barely visible impact damage (BVID) [8], out-of-plane wrinkles [9], porosity measurement [10], interlaminar delamination [11], etc. In this study, we present a method for characterizing out-of-plane wrinkles from UT scanning data in carbon fiber reinforced polymer (CFRP) composites.

The reasons behind a wrinkle being present within a laminated composite are induced during manufacturing from a variety of sources, such as micro-scale deformations, non-uniform pressure distributions, poor co-bonding between lamina, thermal coefficient (CTE) mismatch, gaps, overlaps, etc. [12]. Hallander et al. [13] observed that the lay-up sequence significantly influenced out-of-plane wrinkling in UD prepreg laminates, even when the other variables inducing wrinkles are mitigated. Hallander et al. showed that even for quasi-isotropic, multilayer UD prepreg structures on a double curved geometry, out-of-plane wrinkling can occur.

The structural capacity of a laminated part is compromised due to the presence of a wrinkle. For example, Xie et al. [14] performed a numerical study to relate the maximum wrinkle angle and cross-sectional area with knock-down in failure stress, which can be as large as a 50% reduction in failure stress. Similarly, Hsiao and Daniel [15] provided an analytical model to predict the elastic properties of a composite part. An example is provided in Figure 1 of the change in material stiffness as a function of the wrinkle intensity I, defined as the ratio of the wrinkle height H divided by twice the winkle width $(2W)$, specifically $I = H/2W$. This figure was plot using a custom in-house code of the published Hsiao and Daniel model. Note that the point of 0.022 is highlighted as this is the actual wrinkle intensity of the part studied in the results section of the present paper. It is important to note that a wrinkle intensity of 0.022 is almost imperceptible to the trained eye, whereas it reduces the stiffness along the fiber, E_{11} by nearly 35%. Detecting and quantifying fiber waviness is crucial for maintaining quality during manufacturing, and it is necessary to have an inspection method for CFRP materials [16].

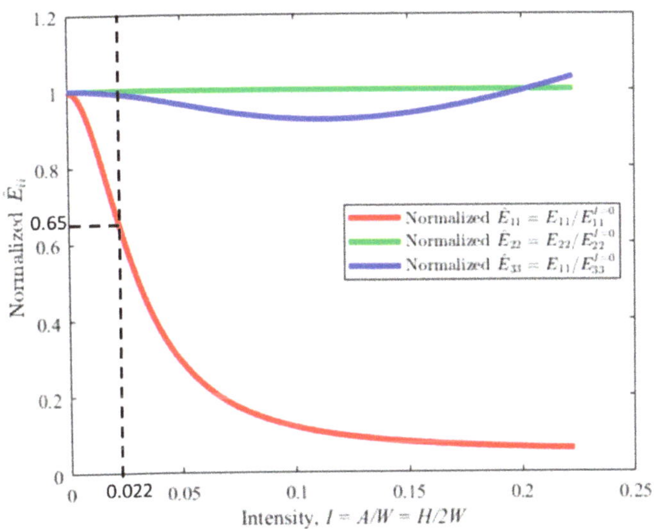

Figure 1. Effect of wrinkle intensity on normalized stiffness values along each axis using the Hsiao–Daniel model.

It is desirable to develop a non-destructive testing (NDT) approach to quantify embedded wrinkles, to prevent compromised parts from finding implementation in service (see, e.g., [9,17]). There are a variety of NDT methods commonly employed when studying structures, such as electromagnetic testing (ET), ultrasonic testing (UT), thermographic testing (TT), radiographic testing (RT), computed tomography (CT), shearography, etc., and many have found use in inspections of CFRP composites (see [18,19]). UT and X-ray CT are the two primary approaches for inspecting laminated polymer matrix composites, due to their accuracy, with the remaining aforementioned techniques finding limited use and acceptance. Due to the expense, time, limited component sizes, and complexity of CT technology, UT systems have a distinct advantage, especially with recent advances in digitizing and higher frequency waveforms [9,20]. Ultrasound has shown tremendous success when it comes to detecting and quantifying different embedded features in composite laminate parts, but most work is carried out using manual methods requiring highly trained operators to interpret the captured waveforms.

Sandhu et al. [21] presented an analytical approach to extract bulk wrinkle characteristics, i.e., wavelength and amplitude from low resolution B scans, with their algorithm converting rectangular scan coordinates into part coordinates for analysis of a curved composite part. The information generated by Sandhu et al. provides general wrinkle characteristics but does not yield quantifiable information for individual lamina. Larranaga-Valsero et al. [22] also detected wrinkle characteristics of height, severity, and maximum angle of a hybrid composite part with unidirectional carbon and woven glass fiber using the FMC/TFM method of phased array. They analyzed the instantaneous amplitude and phase analytically at the first and second resonance frequencies for optimum frequency and then simulated at the given frequency, and from the analyzed data, they were able to visually interpret the results to identify the wrinkle. Zhang et al. [23,24] observed the trend in signal intensity where inter-ply reflection kept increasing with the increase in frequency, and wrinkle defect reflection showed the opposite trend in a rich resin thick composite. They analyzed B-scan images of side drilled holes (SDH) and rich resin areas by filtering the center frequency from 2.5 MHz to 6 MHz and compared them with the simulated result. From the analyzed data, Zhang et al. were able to interpret plotted scan data to identify the wrinkle. The present authors have been unable to identify in the literature a method that automates the extraction, on a lamina-by-lamina basis in three dimensions, of the wrinkle features of each individual lamina.

In this paper, an algorithm is presented to take conventional ultrasonic data, extract the geometric position in 3D-space of each of the individual layers of a carbon fiber laminated composite, and then provide a bulk characterization of the wrinkle height and intensity for each layer. The presented method is not necessarily limited to carbon fiber filled systems, and future studies may consider alternative material systems. Figure 2 represents a wrinkle along with the coordinate system used for the results, where the x_{scan}, x_{index}, and z axes represent the scan direction, index direction, and depth from the top of the surface, respectively. The concern is that an embedded wrinkle would not be visible from the surface, thus the need for a non-destructive approach to capture the wrinkle. An additional uniqueness of this work is the application to woven lamina, whereas existing works addressed wrinkles in unidirectional laminates. In addition, the spatial variation in the wrinkle along with a layer-by-layer characterization is also unique to this work. The present study focuses on wrinkles up to the 14th layer of a laminated composite. Future studies are needed to investigate the limitations of the proposed method as composite structures become thicker. The authors have reason to believe that thicker laminates could be investigated, contingent upon there being a sufficient signal-to-noise ratio of the captured reflection wave as one penetrates deeper into the laminate.

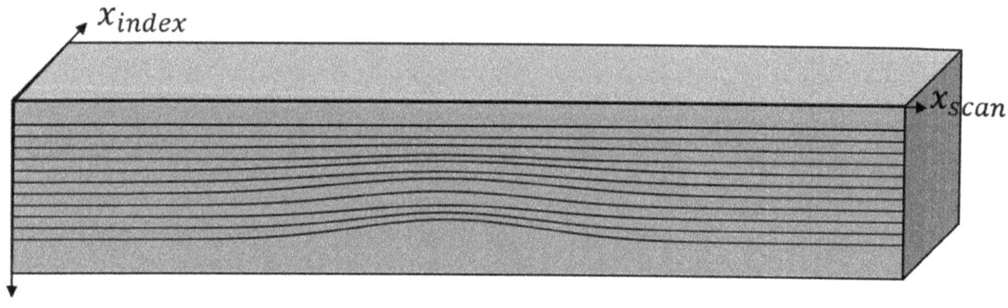

Figure 2. Coordinate system used for the scan analysis.

2. Manufacturing Method

There are several ways of fabricating a woven CFRP laminate. Laminated composites have found widespread use in the aerospace industry. In the present research, we selected a wet layup fabrication with subsequent curing utilizing a hot press. This approach allowed us to tune the wrinkle characteristics. The wet layup process continues to find its way into the manufacture of large components with complex shapes, but the methodology presented is not limited to this manufacturing process. Based on several internal studies, this approach yielded acceptable manufactured wrinkles that effectively simulated wrinkles from several proprietary parts we observed.

2.1. Wet Layup

The wet layup procedure is a hand layup technique where resin is applied to dry fiber with a brush. In this study, dry fibers were placed one after another, and resin was manually wiped onto the lamina surface before placing the next lamina. Two aluminum plates were prepared using a mold release agent, one for use as a tool on which the wet layup was performed, and another to put on top of the laid-up laminae. Then, 3K plain weave carbon fibers from ACP Composites were used for fabricating the laminate. The resin system was a Pro-Set Infusion Epoxy Resin 114 (INF-114) and a Pro-Set Infusion Epoxy 211 Hardener (INF-211) (Pro-Set, Inc., Bay City, Michigan) with a prescribed mix ratio of resin and hardener of 3.65 to 1. A FlakTek SpeedMixer (FlackTek Manufacturing Inc., Louisville, CO, USA) was used to remove bubbles and adequately mix the resin and hardener. The mixer was programmed to spin in a vacuum at 800 rpm for 30 s and then at 1500 rpm for 270 s. After laying up about half of the rectangular carbon fiber laminas, in the present study, 14 lamina multiple 3K tows were placed across the layup, as shown in Figure 3a. Then, the remaining laminae were placed on the layup, repeating the fabric and resin application process, as shown in Figure 3b.

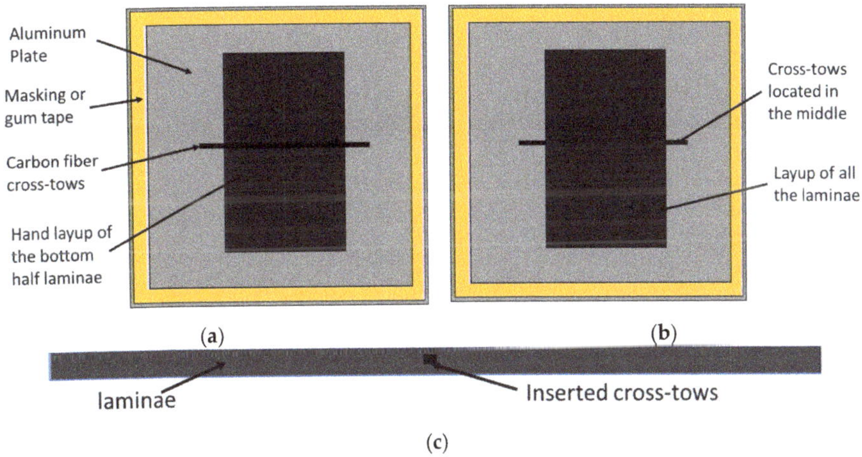

Figure 3. Schematic diagram of the fabrication process, (**a**) indicating the cross tows placed between the center lamina, (**b**) the overall layup before curing with all lamina, and (**c**) a side view of the carbon fiber laminate layup showing the embedded tows.

The cure cycle for the selected resin system calls for the part to be held under pressure for 8 h, and then placed in an elevated temperature environment to complete the curing process. In the present study, the part was kept at room temperature for 6 of the 8 h gelation time before moving the layup to the pressure step, similar to the approach suggested by a previous researcher in Minnie [9].

2.2. Curing

The recommended curing from the manufacturer requires 8 h at room temperature followed by 8 h at 82 °C. After 6 h of gelation, the laminate was kept in the hot press for 10 h. During the elevated temperature portions of the study, along with the final 2 h of the room temperature cure, a holding pressure of 276 kPa (40 psi) was applied. A programmable hot press, a Carver Auto Four/1512-PL (Carver Inc., Wabash, IN, USA), was used for curing the laminate.

3. Analysis Methods

3.1. Ultrasonic Data Collection

The ultrasonic scan system used for this study was a custom immersion tank scanning system with a single spherically focused transducer. The advantages of this type of scan include a uniformity of acoustic coupling that reduces sensitivity variations, a reduction in the scan time due to automation, and the focused immersion transducer increasing the sensitivity to small reflectors [25]. The UT scan was performed in a water tank, as shown in Figure 4, using pulse-echo scanning with access to only a single side of the laminate (see e.g., [20,25]). Based on several internal studies, a single-element 37.5 mm spherically focused probe with a 7.5 MHz frequency was found appropriate over a wide range of laminate thicknesses. The transducer was excited to 190 V, the largest value allowed by the Olympus Focus PX (Evident, Center Valley, Lehigh County, PA, USA) digitizer utilized, with a square wave pulse width of 65 ns. Two Velmex Bi-slides with a spatial resolution of 0.0025 mm, as shown in Figure 4, were used to move the transducer along the scan axis and then in the index direction following the raster pattern, as shown in Figure 5, to scan the region of interest (ROI) of the laminated part. A step size of $\Delta x = \Delta y = 0.2$ mm was used in the present study. It is noted that scans typically took 10 min in the present investigation. The scan time was inversely proportional to the scan resolution; thus, a doubling of the index size, resulted in a nearly 50% reduction in the scan time. Moreover, a subject which the authors are currently pursuing is to implement a synthetic raster using phased array, often reducing the scan time by an order of magnitude.

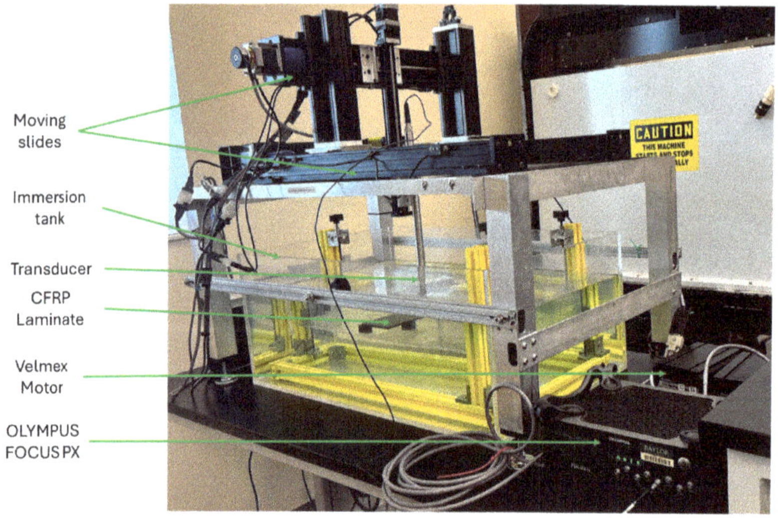

Figure 4. Immersion tank system utilized to perform scans in the present research.

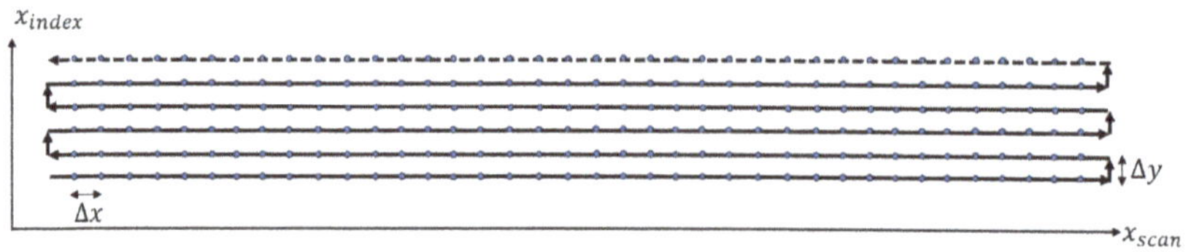

Figure 5. Typical raster pattern used to scan the region of interest.

For this study, the scan direction generally traversed transverse to the embedded cross tows, and the index direction generally aligned with the embedded cross tows. A-scans and B-scans were produced and analyzed to study the sub-surface wrinkles. An A-scan is a one-dimensional scan where the scan-echo amplitude is plotted as a function of time, and a B-scan displays a series of A-scan readings that originate in a single run along a single axis [25].

3.2. Extraction of Laminated Layers

In this study, an algorithm was constructed to extract the three-dimensional lamina position from within the laminated composite part. The algorithm allowed the quantification of the bulk wrinkle parameters, specifically the wrinkle height and wrinkle intensity, the results of which are presented in Section 4.

Figure 6 represents the algorithm of the overall layer extraction method. Data were initially captured using an Olympus Focus PX (Evident, Center Valley, Lehigh County, PA, USA) digitizer and saved in an Olympus proprietary file format, *.fpd. The raw A-scan data were then read into a MATLAB (version 2022b) script and then shifted in time such that the initial reflection wave from the front of the part was aligned across all A-scans. Next, a spatial Gaussian averaging technique (see [26] for a presentation of this algorithm) was performed to smooth the data in the plane of the laminate for a given depth. Next, the averaged A-scan over a subregion was compiled and the individual peaks in time were extracted from the waveform. The results of a typical peak detection can be seen in Figure 7. Observe that the individual peaks correlated to the interface between lamina, and these peaks were tracked between each of the A-scans within the region of interest, forming a layer-by-layer surface. Each peak indicates an individual depth of the laminated part corresponding to the depth of each individual ply. The individual peaks were then tracked in both the scan an index directions. This tracking resulted in a surface, with the vertical dimension corresponding to the lamina position, given as the time of flight, within the laminate as a function of (x_{index}, x_{scan}). This extracted surface was then smoothed using a Gaussian averaging technique, and the time of flight data were converted to depth data using the effective through thickness speed of sound of the laminate to extract the layers as a function of the individual layer depth, as shown in Figure 8 and plotted in the MATLAB environment. The effective through thickness speed of sound was obtained using the known thickness of the part times two, divided by the time of flight of the signal between the front wall and the back wall. The addition of the two was made as we utilized the pulse-echo mode of inspection.

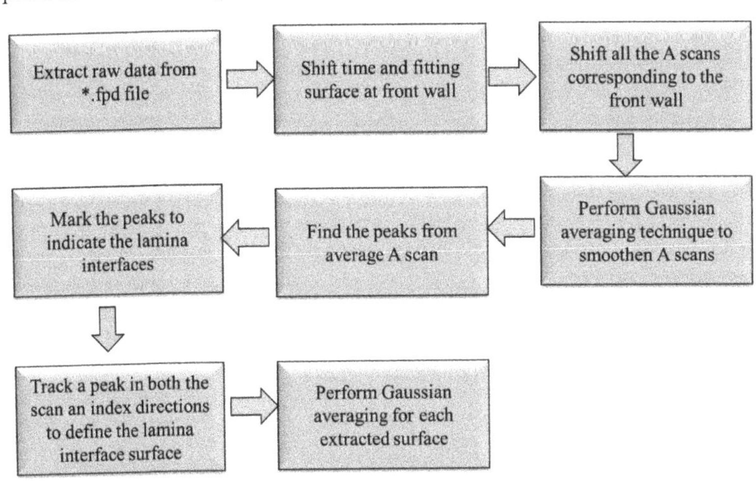

Figure 6. Flowchart of the layer extraction method, implemented in the MATLAB environment.

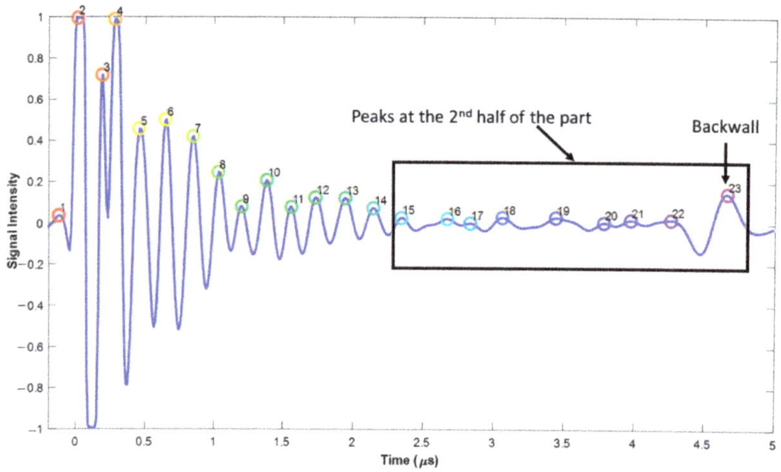

Figure 7. Average A-scan used for lamina interface detection.

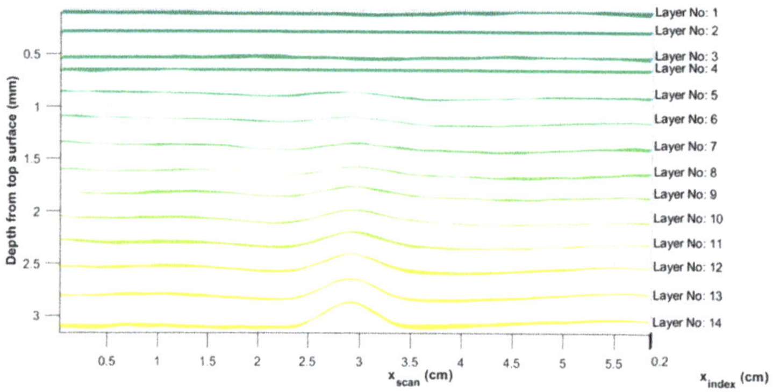

Figure 8. Extracted layer surfaces from the analysis of the first 14 lamina.

3.3. Scan Data Analysis

Figure 7 represents a typical average A-scan over a sub-region of size $x.xx$ (mm) $\times$ $y.yy$(mm) of the overall 58.4 mm $\times$ 10 mm scan. This subregion scan was taken far from the wrinkle and was used as the seed A-scan, where each of the corresponding peaks indicate the interface between the laminae. Peaks above a prescribed threshold, in the present study we used 0.02, were identified as being an interface between lamina. This worked effectively for lamina above the inserted tow layer, specifically the interface between the 14th and 15th lamina, but did not capture the interfaces for lamina past the 15th lamina. Specifically, as the synthetic part of 28 lamina had the embedded 3K tows between the 14th and 15th lamina, it did not make sense to analyze any of the lamina deeper than the 14th lamina. A manufactured component would not typically have additional tows causing a wrinkle; thus, it was not reasonable to analyze the lamina beneath the embedded tow layer. Each peak within the locally averaged A-scan was tracked across the surface of the part. It is worth noting that the signal intensity tended to decrease while penetrating the part because of the signal attenuation, with a high intensity at the back side of the part reflecting off the interface between the laminate and the surrounding water medium. The various layers are identified in Figure 7, along with the backwall. As shown in Figure 7, the signal weakened due to signal attenuation.

Figure 8 represents a side view of each of the individual lamina, as extracted from the automated algorithm presented in Figure 6 and implemented in the MATLAB software environment. The ROI was taken as a 58.4 mm × 10 mm area with an overall part thickness of 6.23 mm. Of note is the 14th layer of the part that exhibits the highest wrinkle intensity in Figure 8. The top four layers of the part are almost flat, which is in agreement with the visual inspection that was unable to identify any wrinkles from the part surface. Conversely, the wrinkles in the individual layers can be observed to progressively increase from lamina 5 up to the 14th lamina. The 14th lamina is plotted separately in Figure 9, plotted in the MATLAB environment. Notice that the 3D representation of the 14th lamina is reasonably uniform along the axis of the wrinkle, but there are subtle changes along the projection of the wrinkle peak, a detail expanded upon in the next section and highlighting that the present method can capture the spatial variations in the wrinkle intensity.

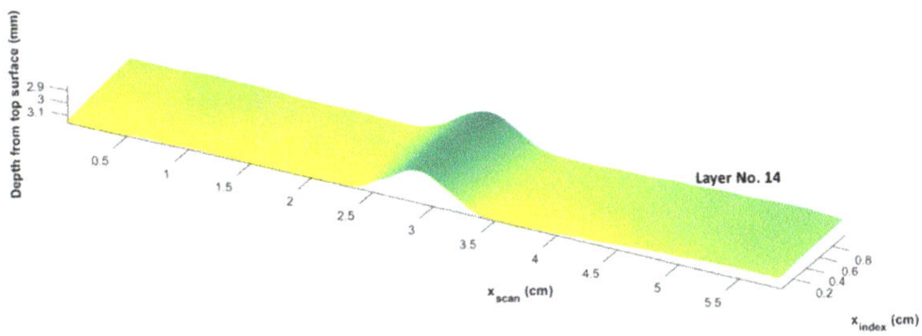

Figure 9. Three-dimensional representation of the 14th lamina of the part showing the subsurface wrinkle.

4. Results

For this study, two sub-surface wrinkle parameters, wrinkle height and wrinkle intensity, were analyzed. As the extracted surface was three-dimensional, a baseline was created by fitting a linear surface to the lamina well outside of the region containing the wrinkle. The wrinkle height H, such as that shown in Figure 10, was measured for each value of x_{index} for each of the individual scans. The base length of the wrinkle was defined as 2L (see Figure 10), where L is the length between the two points where the layer height is one-half of the wrinkle height. The wrinkle intensity is defined as follows (similarly to that in, [15,23,27]):

$$I = \frac{H}{2L} \tag{1}$$

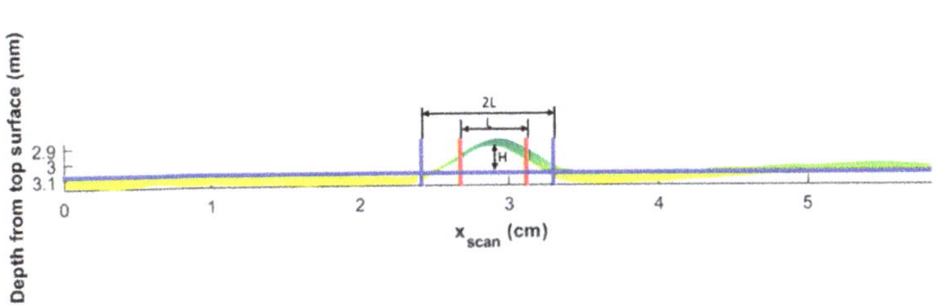

Figure 10. Wrinkle intensity characterization of the 14th extracted lamina.

Figure 11 represents the wrinkle height along x_{index} for each of the 14 laminae. Observe that the height was not constant across the x_{index} direction but it did remain fairly consistent, and an average value is reported to represent the lamina. The wrinkle height generally increased monotonically for each layer as one progressed deeper into the laminate. For example, the wrinkle height was measured as 0.01 mm at layer 5, whereas the maximum wrinkle height was found to be 0.19 mm at layer 14. To report the wrinkle height, a spatial average was computed, and the average wrinkle heights are shown for each layer in Table 1 and Figure 12.

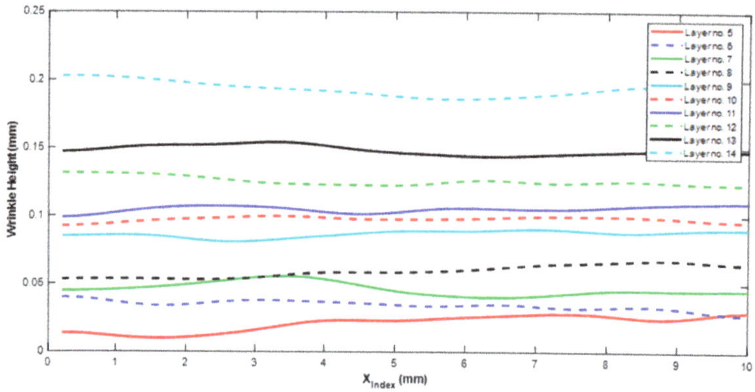

Figure 11. Wrinkle height for each of the individual lamina along the index direction.

Table 1. Wrinkle heights of the individual lamina from the automated non-destructive inspection, where N/A indicates the results are not applicable for analysis.

Layer No.	Wrinkle Height (mm)	Wrinkle Intensity (mm/mm)
1–4	N/A	N/A
5	0.0210	N/A
6	0.0321	N/A
7	0.0435	N/A
8	0.0574	0.0069
9	0.0866	0.0109
10	0.0968	0.0107
11	0.1029	0.0127
12	0.1246	0.0159
13	0.1473	0.0171
14	0.1931	0.0221

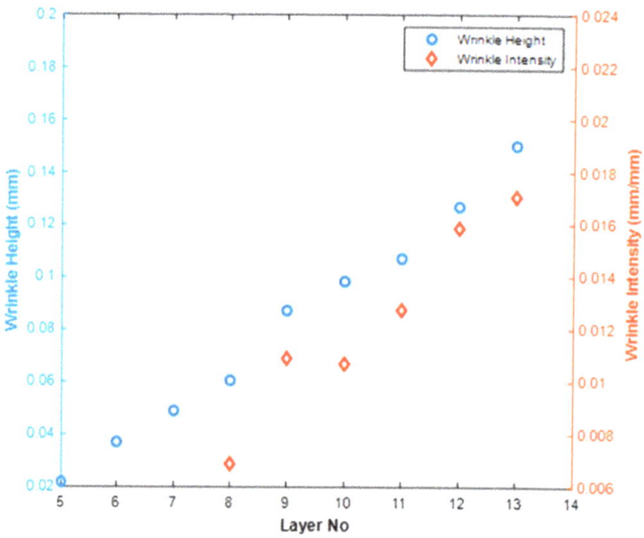

Figure 12. Non-destructively extracted wrinkle height and wrinkle intensity as a function of depth within the investigated laminate.

The top four laminae are not reported, as no detectable wrinkle was identified. Although a wrinkle height could be quantified for the 5th through 7th lamina, the wrinkle width was quite difficult to quantify due to uncertainty, and thus the intensity is not reported for these laminae. For lamina 8 through 14, both the height and intensity are reported and presented in Table 1. The wrinkle intensity was measured as a minimum of 0.0069 in the 8th lamina, and the maximum intensity was found to be 0.0221 on the 14th layer, which was the closet layer to the inserted cross-tows in the part. Looking back to Figure 1, this would suggest that the 14th lamina would have a reduced stiffness of 65% that of the outer most lamina, thus the outer lamina would carry more of the load in any structural application.

5. Conclusions and Discussion

Wrinkles are a common defective feature in a manufactured CFRP composite part, which can significantly compromise the structural performance of a CFRP. The current literature does not provide examples of the extraction, on a lamina-by-lamina basis and in three dimensions, of the wrinkle feature of each individual lamina.

- The present research presents a method for fabricating a laminated composite with a synthetic wrinkle that can be used for inspection methodology development.
- The present research presents both a methodology and results for the extraction of the wrinkled layer surfaces from ultrasonic data.
- The automated code can extract the spatially varying wrinkle geometry and quantify the wrinkle intensity, height, and width as a function of spatial position, for each individual lamina.

In the present study, a part with 28 lamina that were mirrored about the central axis with embedded tows creating an internal wrinkle was studied. The presented results include the 3D surface of each individual lamina, from which the height of the lamina and the wrinkle intensity could be readily characterized. The results presented included a wrinkle with a height as small as 0.01 mm and a wrinkle intensity of only 0.0069 up to a wrinkle height of 0.19 mm and an intensity of 0.022. Future work needs to include validation of the results from sectioned samples characterized using microscopy and potentially a full X-ray CT inspection of the laminate.

Author Contributions: Conceptualization, D.A.J.; methodology, M.A.A. and D.A.J.; software, M.A.A.; validation, M.A.A. and D.A.J.; formal analysis, M.A.A. and D.A.J.; investigation, M.A.A.; resources, D.A.J.; data curation, M.A.A.; writing—original draft preparation, M.A.A. and D.A.J.; writing—review and editing, M.A.A. and D.A.J.; visualization, M.A.A. and D.A.J.; supervision, D.A.J.; project administration, D.A.J.; funding acquisition, D.A.J. All authors have read and agreed to the published version of the manuscript.

Funding: The research was funded in part by Verifi Technologies.

Data Availability Statement: The raw data supporting the conclusions of this article will be made available by the authors for reasonable requests.

Acknowledgments: The authors would like to thank Baylor Materials Testing and Characterization laboratory for contributing to the experimental setup.

Conflicts of Interest: The authors declare no conflicts of interest.

References

1. Ozkan, D.; Gok, M.S.; Karaoglanli, A.C. Carbon Fiber Reinforced Polymer (CFRP) Composite Materials, Their Characteristic Properties, Industrial Application Areas and Their Machinability. In *Engineering Design Applications III: Structures, Materials and Processes*; Öchsner, A., Altenbach, H., Eds.; Springer International Publishing: Cham, Switzerland, 2020; pp. 235–253, ISBN 978-3-030-39062-4.
2. Pawlak, A.M.; Górny, T.; Dopierała, Ł.; Paczos, P. The Use of CFRP for Structural Reinforcement—Literature Review. *Metals* **2022**, *12*, 1470. [CrossRef]

3. Smith, R.A.; Nelson, L.J.; Mienczakowski, M.J.; Challis, R.E. Automated analysis and advanced defect characterisation from ultrasonic scans of composites. *Insight-Non-Destr. Test. Cond. Monit.* **2009**, *51*, 82–87. [CrossRef]
4. Rahul, K.; Jack, D.A.; Smith, D.E. A Statistical Approach for Failure Analysis Involving Uncertainty in Determining Ply Orientation. *Polym. Compos.* **2024**, *45*, 5192–5206. [CrossRef]
5. Nelson, L.J.; Smith, R.A. Fibre direction and stacking sequence measurement in carbon fibre composites using Radon transforms of ultrasonic data. *Compos. Part A Appl. Sci. Manuf.* **2019**, *118*, 1–8. [CrossRef]
6. Koodalil, D.; Rajagopal, P.; Balasubramaniam, K. Quantifying adhesive thickness and adhesion parameters using higher-order SH guided waves. *Ultrasonics* **2021**, *114*, 106429. [CrossRef]
7. Nargis, R.A.; Pulipati, D.; Kokkada, P.; Jack, D.A. Automated Detection of Foreign Object Debris in Woven Carbon Fiber Laminate at Different Environmental Conditions. Available online: https://speautomotive.com/wp-content/uploads/2023/09/Automated-Foreign-Object-Detection-for-Composite-Laminates-Using-High-Resolution-Ultrasound-Testing_Nargis_Rifat.pdf (accessed on 29 February 2024).
8. Katunin, A.; Wronkowicz-Katunin, A.; Dragan, K. Impact Damage Evaluation in Composite Structures Based on Fusion of Results of Ultrasonic Testing and X-ray Computed Tomography. *Sensors* **2020**, *20*, 1867. [CrossRef]
9. Minnie, W.H. Nondestructive Evaluation of Out-of-Plane Wrinkles within Woven Carbon Fiber Reinforced Plastics (CFRP) Using Ultrasonic Detection. Master's Thesis, Baylor University, Waco, TX, USA, 2021. Available online: https://baylor-ir.tdl.org/handle/2104/11585 (accessed on 21 November 2022).
10. Bhat, M.R.; Binoy, M.P.; Surya, N.M.; Murthy, C.R.L.; Engelbart, R.W. Non-destructive evaluation of porosity and its effect on mechanical properties of carbon fiber reinforced polymer composite materials. *AIP Conf. Proc.* **2012**, *1430*, 1080–1087. [CrossRef]
11. De Castro, D.S.V.; Matvieieva, N.; Grosso, M.; Camerini, C.G.; Kotik, H.G.; Heuer, H. Evaluation of Mode II Delamination Area by Non-destructive Techniques: Accuracy and Influence on Fracture Toughness Calculation. *J. Nondestruct. Eval.* **2021**, *40*, 58. [CrossRef]
12. Thor, M.; Kiss, P.; Sause, M.; Hinterhoelzl, R. Strategies for the manufacturing of wrinkle-free composite parts. In Proceedings of the SAMPE Europe Conference 2020, Amsterdam, The Netherlands, 30 September–1 October 2020.
13. Hallander, P.; Akermo, M.; Mattei, C.; Petersson, M.; Nyman, T. An experimental study of mechanisms behind wrinkle development during forming of composite laminates. *Compos. Part A Appl. Sci. Manuf.* **2013**, *50*, 54–64. [CrossRef]
14. Xie, N.; Smith, R.A.; Mukhopadhyay, S.; Hallett, S.R. A numerical study on the influence of composite wrinkle defect geometry on compressive strength. *Mater. Des.* **2018**, *140*, 7–20. [CrossRef]
15. Hsiao, H.M.; Daniel, I.M. Elastic properties of composites with fiber waviness. *Compos. Part A Appl. Sci. Manuf.* **1996**, *27*, 931–941. [CrossRef]
16. Pain, D.; Drinkwater, B.W. Detection of Fibre Waviness Using Ultrasonic Array Scattering Data. *J. Nondestruct. Eval.* **2013**, *32*, 215–227. [CrossRef]
17. Ma, T.; Li, Y.; Zhou, Z.; Meng, J. Wrinkle Detection in Carbon Fiber-Reinforced Polymers Using Linear Phase FIR-Filtered Ultrasonic Array Data. *Aerospace* **2023**, *10*, 181. [CrossRef]
18. Schumacher, D.; Meyendorf, N.; Hakim, I.; Ewert, U. Defect recognition in CFRP components using various NDT methods within a smart manufacturing process. *AIP Conf. Proc.* **2018**, *1949*, 020024. [CrossRef]
19. Zhang, L.; Tham, Z.W.; Chen, Y.F.; Tan, C.Y.; Cui, F.; Mutiargo, B.; Ke, L. Defect imaging in carbon fiber composites by acoustic shearography. *Compos. Sci. Technol.* **2022**, *223*, 109417. [CrossRef]
20. Schmerr, L.W. *Fundamentals of Ultrasonic Nondestructive Evaluation: A Modeling Approach*; Springer: Boston, MA, USA, 1998; ISBN 978-1-4899-0144-6.
21. Sandhu, A.; Dodwell, T.J.; Butler, R. An automated image processing algorithm to determine wrinkle characteristics from B-scans. In Proceedings of the 17th European Conference on Composite Materials, Munich, Germany, 26–30 June 2016; p. 9.
22. Larrañaga-Valsero, B.; Smith, R.A.; Tayong, R.B.; Fernández-López, A.; Güemes, A. Wrinkle measurement in glass-carbon hybrid laminates comparing ultrasonic techniques: A case study. *Compos. Part A Appl. Sci. Manuf.* **2018**, *114*, 225–240. [CrossRef]
23. Larrañaga-Valsero, B.; Smith, R.A.; Tayong, R.B.; Fernández-López, A.; Güemes, A. Wrinkle characterisation from Ultrasonic Scans of Composites. In Proceedings of the 55th Annual Conference of the British Institute of Non-Destructive Testing, Nottingham, UK, 12–14 September 2016; p. 15.
24. Zhang, Z.; Guo, S.; Li, Q.; Cui, F.; Malcolm, A.A.; Su, Z.; Liu, M. Ultrasonic detection and characterization of delamination and rich resin in thick composites with waviness. *Compos. Sci. Technol.* **2020**, *189*, 108016. [CrossRef]
25. An Introduction to Ultrasonic Transducers for Nondestructive Testing | Olympus IMS. Available online: https://www.olympus-ims.com/en/resources/white-papers/intro-ultrasonic-transducers-ndt-testing/ (accessed on 5 April 2023).
26. Blackman, N.J.; Jack, D.A.; Blandford, B.M. Improvement in the Quantification of Foreign Object Defects in Carbon Fiber Laminates Using Immersion Pulse-Echo Ultrasound. *Materials* **2021**, *14*, 2919. [CrossRef] [PubMed]
27. Takeda, T. Micromechanics model for three-dimensional effective elastic properties of composite laminates with ply wrinkles. *Compos. Struct.* **2018**, *189*, 419–427. [CrossRef]

materials

MDPI

Article

Damage Detection in FRP-Reinforced Concrete Elements

Pranit Malla [1,*], Seyed Saman Khedmatgozar Dolati [1], Jesus D. Ortiz [2], Armin B. Mehrabi [1,*], Antonio Nanni [2] and Jiayi Ding [3]

1 Department of Civil and Environmental Engineering, Florida International University, Miami, FL 33174, USA; skhed004@fiu.edu
2 Department of Civil and Architectural Engineering, University of Miami, Coral Gables, FL 33146, USA; jdo72@miami.edu (J.D.O.); nanni@miami.edu (A.N.)
3 AtkinsRéalis, Miami, FL 33126, USA; jiayi.ding@atkinsrealis.com
* Correspondence: pmall011@fiu.edu (P.M.); amehrabi@fiu.edu (A.B.M.)

Abstract: Fiber-Reinforced Polymer (FRP) composites have emerged as a promising alternative to conventional steel reinforcements in concrete structures owing to their benefits of corrosion resistance, higher strength-to-weight ratio, reduced maintenance cost, extended service life, and superior durability. However, there has been limited research on non-destructive testing (NDT) methods applicable for identifying damage in FRP-reinforced concrete (FRP-RC) elements. This knowledge gap has often limited its application in the construction industry. Engineers and owners often lack confidence in utilizing this relatively new construction material due to the challenge of assessing its condition. Thus, the main objective of this study is to determine the applicability of two of the most common NDT methods: the Ground-Penetrating Radar (GPR) and Phased Array Ultrasonic (PAU) methods for the detection of damage in FRP-RC elements. Three slab specimens with variations in FRP type (glass-, carbon- and basalt-FRP, i.e., GFRP, CFRP, and BFRP, respectively), bar diameter, bar depths, and defect types were investigated to determine the limitations and detection capabilities of these two NDT methods. The results show that GPR could detect damage in GFRP bars and CFRP strands, but PAU was limited to damage detection in CFRP strands. The findings of this study show the applicability of conventional NDT methods to FRP-RC and at the same time identify the areas with a need for further research.

Keywords: Fiber-Reinforced Polymer (FRP); ground-penetrating radar (GPR); ultrasonic testing (UT); phased array ultrasonic (PAU); non-destructive testing (NDT); FRP-reinforced concrete (FRP-RC)

1. Introduction

The construction industry predominantly utilizes two structural materials: steel and concrete [1]. However, with the increasing demand for extended service life, reduced maintenance, enhanced resilience, and sustainability, the limitations of traditional construction materials (e.g., steel reinforced/prestressed concrete, structural steel, and timber) have become more evident. In response to these demands, Fiber-Reinforced Polymer (FRP) composites have emerged as a promising alternative, offering improved durability and performance and providing the potential for extended service life and reduced maintenance costs [2]. FRPs are composite materials composed of reinforcing fibers impregnated in a polymeric resin. The reinforcing fibers in the composite are the main load-carrying (reinforcing) elements, while the polymeric matrix or resin helps to form the desired geometry and transfers forces to and between the fibers. In general, the types of FRPs used in the construction industry based on the type of fibers are GFRP (glass-FRP), CFRP (carbon-FRP), BFRP (basalt-FRP), and AFRP (aramid-FRP) composites.

1.1. FRP-Reinforced Concrete (FRP-RC) Elements

Over the past three decades, FRP composites have gained significant popularity in civil engineering, attributed to mainly their increased durability, corrosion resistance, and higher strength-to-weight ratio [3]. They have been used as reinforcement for constructing new structures as well as rehabilitating existing ones. FRPs can be used either in conjunction with concrete elements or as stand-alone structural or non-structural elements in buildings as well as bridge structures. When used in conjunction with concrete elements, FRP application can be divided into two categories: (1) internal application with FRP bars/rods and strands/tendons for new FRP-reinforced/prestressed constructions and (2) external application with FRP laminates/plates/jackets, sheets/fabrics/wraps, and near-surface mounted (NSM) bars for the strengthening, retrofitting, and repair of existing structures. This paper focuses on the internal application of FRP composites, more specifically on the damage detection of FRP rods/strands embedded in concrete elements. For the sake of brevity, FRP bars/rods and strands/tendons are referred to as FRP bars in the following sections of this paper.

1.2. Advantages of FRP-RC Elements

Corrosion is one of the main issues that can compromise the serviceability and safety of conventional steel-reinforced/prestressed concrete structures. A 2002 Federal Highway Administration (FHWA) study conducted in partnership with the National Association of Corrosion Engineers (NACE) International, now known as the Association for Materials Protection and Performance (AMPP), estimated the average annual direct cost of corrosion for US highway bridges to be $8.29 billion [4]. A decade later, in 2013, NACE International estimated an increase in this cost to $13.6 billion per year [5]. Despite these estimates being decades old, the issue of corrosion persists, and it remains a primary cause of bridge deterioration in the US. The latest 2021 American Society of Civil Engineers (ASCE) infrastructure report card scored America's bridges a low grade of C and emphasized the use of innovative materials such as ultra-high-performance concrete (UHPC), corrosion-resistant reinforcement, high-performance steel, composites, and improved coatings to increase the lifespan of the nation's bridges [6].

FRP composites are one of such relatively new construction materials that are resistant to all the factors causing corrosion in steel-reinforced concrete (RC) structures, such as a decrease in concrete pH due to carbonation, chloride penetration, and the diffusion of halides and chemicals [7–10]. Further, FRP composites are not affected by electromagnetic disturbances from sources such as railroads with DC or AC traction, overhead power lines, and unbalanced currents from three-phase power systems, which contribute to the corrosion of metal structures and the deterioration of reinforced concrete [11]. Hence, the use of FRPs as reinforcement in concrete elements is strongly justified for locations where the corrosion of conventional steel reinforcement poses significant economic and safety risks [12].

Additionally, better mechanical performance, superior durability, and the environmental implications of the FRP composites [13,14] offer more flexibility for engineers to build structures that last longer. When compared to steel bars, FRP bars have significantly higher tensile strength [15], about one-fourth of the density of steel, and can achieve a longer service life [16]. Nevertheless, the application of FRP composites is associated with a higher initial cost, which is often quoted as one of the major drawbacks to its implementation. However, in recent years, the initial cost of GFRP bars has benefitted due to price fluctuation in the metal market worldwide since the mid-2020s and has even dropped due to the growth of the GFRP bar industry [17]. Further, despite the fact that FRP bars initially cost more than traditional steel bars, a life cycle cost study shows that they can rather be cost-effective in the long run [18]. Because of these factors, FRP bars are progressively becoming a reliable material in civil engineering. This is evident from a recent example of a coastal bridge fully reinforced with GFRP bars built in 2021 at the 23rd Avenue over Ibis Waterway located in Florida, USA, which is the second of its kind [3,19].

1.3. Damage in FRP-RC Elements

Although FRP bars offer improved durability and performance compared to steel in certain aspects, there are concerns about potential damage and defects in both FRP bars and FRP-reinforced concrete (FRP-RC) elements. Many of the serviceability issues related to conventional RC elements such as cracking, permeability, carbonation, chloride content, and concrete cover may not pose the same concern for FRP-RC elements. FRP bars and FRP-RC elements are prone to a unique set of defects as compared to their steel counterparts. For instance, the bond behavior of the FRP bar–concrete interface differs from that of the steel bar–concrete interface [20]. The bond failure of FRP bars not only occurs in the concrete but also inside the bars, unlike a steel bar [21,22]. Similarly, in a study conducted by Valentine [23], it was found that cracks are the predominant defect reported by the bridge inspectors in the inspection of FRP-reinforced bridge decks, which can be attributed to the low modulus of elasticity of the FRP bars. In this paper, the detectability of three different types of potential damage that might occur in the FRP reinforcements—rupture, debonding, and loss of cross-sectional properties—will be investigated. It should be noted that the term "potential damage" has been used due to the fact that, unlike steel bars, where corrosion is the obvious damage to be expected, there is very limited information on the damage that is possible in FRP bars, a relatively new, corrosion-resistant construction material. Additionally, this paper will also include the detection of damage in concrete such as delamination, cracks, and voids, which would be similar to traditional steel reinforced/prestressed concrete elements.

1.4. Inspection of FRP-RC Elements

The literature on the application of non-destructive testing (NDT) methods for the internal application of FRP is limited and scarce. There is no standard guide available for the inspection of FRP-RC elements [24–26]. This represents a knowledge gap that this research study attempts to address. Hence, although the use of FRP in highway infrastructures has been on the rise [27–30], the absence of reliable condition assessment methods for FRP-RC elements has significantly hindered its extensive application. Bridge engineers are hesitant to use materials that are difficult to detect and assess for maintenance. Therefore, there's a pressing need for research into effective condition assessment techniques for FRP-RC elements, which could greatly encourage the adoption of FRP in future construction projects.

The inspection of FRP-RC elements is limited to detecting the initiation of FRP bars–concrete debonding [31,32] or the initiation of fractures in the FRP [33,34] rather than detecting the damage in the bars themselves. This is in most part because it was believed that FRP bars are undetectable or have low detectability, making it impossible to spot them effectively during an inspection. NDT techniques used for inspecting steel-reinforced concrete rely on identifying differences in specific properties, such as the dielectric constant and acoustic impedance, between steel and concrete. However, FRP reinforcements, unlike steel, exhibit properties similar to concrete that include non-conductivity and comparable density. These similarities introduce complexities in detecting/inspecting FRP, making it a more challenging task.

However, Ékes [35] demonstrated for the first time that ground-penetrating radar (GPR) can detect both CFRP and GFRP bars embedded in concrete and therefore concluded that it is a suitable tool for locating FRP bars on bridge decks. Another study conducted by the authors of this paper showed that the detectability of FRP bars/strands increased with the rise in the antenna center frequency of the GPR device and further showed that phased array ultrasonic (PAU) testing is also effective in detecting GFRP and CFRP strands [36]. PAU is sensitive in detecting air voids and hence it was effective only for FRP strands because of the air voids present within the twisted wires of strands and the uneven surface of the strands, unlike the smooth surfaces of bars. However, these studies do not give any information about the detectability of damage in FRP reinforcements using GPR and PAU.

This paper explores the feasibility of employing commercially available GPR and PAU devices to identify damage in FRP bars embedded in concrete. These methods are selected among various NDTs because they are widely used in inspecting steel RC elements [24]. Further, this paper also aims to determine the detection of damage in the concrete elements reinforced with FRP using GPR and PAU devices. Three small-scale slabs were fabricated with damage simulated in bars and concrete to evaluate the feasibility of the chosen NDT method. The results of this study show that GPR devices can detect damage in FRP bars/strands and concrete. However, it was observed that PAU devices are effective only for detecting damage in CFRP strands along with steel bars and concrete.

The results of this study can be utilized to drive further research on the non-destructive testing of FRP-RC elements and embedded FRP bars. One such prospective field of study in the future could be the use of NDT damage detection methods in conjunction with diagnostic load testing for bridges. Diagnostic load tests are performed to evaluate the integrity and performance of bridges and identify local damage areas based on the variations in measurements of deflections, strains, and vibration responses [19]. Once local damage areas are identified, NDT can be employed to perform a more thorough and refined damage assessment within those areas. When used together, NDT and diagnostic load testing can achieve efficient, comprehensive, and dependable damage detection and assessment of FRP bars embedded in concrete. These will provide owners with inspection options and help them in decision making regarding necessary countermeasures for ensuring the bridge's safety and longer service life.

2. Experimental Program

2.1. Fabrication of FRP-RC Slab Specimens

To determine the capability of GPR and PAU in detecting defects in FRP reinforcements, three concrete slab specimens were fabricated and inspected. These slabs, measuring 36 inches in width, 36 inches in length, and 7 inches in depth, were fabricated with simulated defects in FRP bars and the concrete itself. The concrete mix design used for casting the slabs was determined following the specifications of the Florida Department of Transportation (FDOT) for "Class II 4500 Bridgedeck" concrete. The mixture included Type II cement with a water-to-cement ratio (w/cm) of 0.44, #57 stone as a coarse aggregate, and silica sand as a fine aggregate. Concrete cylinders were tested at 28 days following American Society for Testing and Materials (ASTM C39) standards [37,38] to verify the actual strength, resulting in an average compressive strength of 31.70 MPa with a standard deviation of 0.69 MPa (yielding a coefficient of variation of 2.2%).

The construction of slab specimens aimed to explore the effect of various factors in detectability, including the type of FRP bars/strands (GFRP, CFRP, BFRP), their diameters, the depths of embedment, and the type of defects. Table 1 shows the key characteristics of each slab specimen. Given the prevalent use of GFRP bars compared to other FRP bars in concrete reinforcement, one slab (Slab O) was reinforced only with GFRP bars at varying depths of embedment. According to American Concrete Institute (ACI) CODE-440.11-22 [39], the concrete cover for GFRP-reinforced members ranges between 0.75 inches and 3 inches, which guided the depth variations in the slab specimens to reflect the potential positioning of the top layer of FRP reinforcement. Slab P was designed to include two CFRP strands, a BFRP bar, and a steel bar (which serve as a benchmark for this investigation). This setup enables a direct comparison of the detectability of various FRP bars/strands against that of the steel bar under identical testing conditions. Additionally, the study extended to include Slab Q, which was designed to investigate the detection of simulated defects within the concrete itself.

Each specimen was constructed and marked according to the layout depicted in Figure 1. Every slab is designated by an alphabet (O, P, and Q), and each side is assigned a number (ranging from 1 to 4). The direction of measurement is defined by the starting point and the endpoint numbers of the measurement.

Table 1. Identification of small-scale concrete slab specimens.

Slab Specimen	Slab ID	Bar Diameter	No. of Bars
Slab with damaged GFRP bars	Slab O	#6 GFRP bars	4
Slab with damaged CFRP strands, BFRP bar, and steel bar	Slab P	0.6″ CFRP strands, #4 steel bar, and #6 BFRP bar	4
Slab with damage in concrete	Slab Q	#4 and #6 GFRP bars, #4 and #6 steel bars, #3 CFRP bar, and 0.6″ CFRP strand	6

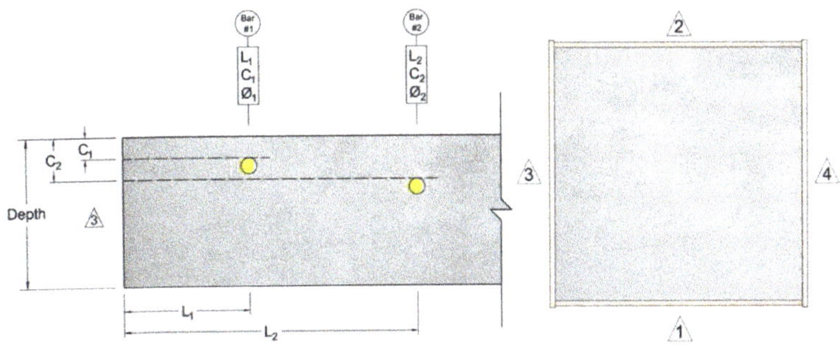

Figure 1. Labeling of slab specimens.

2.2. Simulation of Defects in FRP Bars and Concrete

Slabs O and P contain defects in the bars. These defects correspond to bar rupture, a loss of cross-sectional properties, and debonding. To produce the bar rupture, the cross-section was reduced almost to breakage and then covered with a polyethylene tube to prevent the concrete from penetrating the defect and filling the void, as shown in Figure 2a,b. Cross-sectional property defects were produced by reducing the cross-section by 50% of its initial area and then covering it with polyethylene to prevent the concrete from filling the removed volume, as shown in Figure 2c,d. The loss of cross-sectional "properties" was simulated through the reduction in the cross-sectional "area" only for the experimental purpose of this paper. It should be noted that such an extreme reduction in the cross-sectional "area" of FRP bars may be unlikely, but the reduction in cross-sectional "properties" does occur under harsh environmental conditions. The aim is to explore the capability of NDT methods in detecting the variation in damage levels. Finally, the debonding defect was simulated by covering the bar with bubble wrap to generate a thin layer between the bar and the concrete (see Figure 2e,f). Figure 3 shows the location of the different defects and their dimension details.

In Slab Q, four types of defects in the concrete were simulated to evaluate the feasibility of different NDT methods. Delamination, flexural and split cracks, as well as voids in the concrete were simulated in Slab Q using thin architectural polystyrene foam held in place with the use of epoxy, as shown in Figures 4 and 5. This specimen includes steel, glass, and carbon FRP #4 and #6 bars.

Table 2 provides the geometrical details of the slab specimens, including the distance of the bars from the edge, cover depth, bar diameter, bar material, and slab thickness, following the conventions shown in Figure 1.

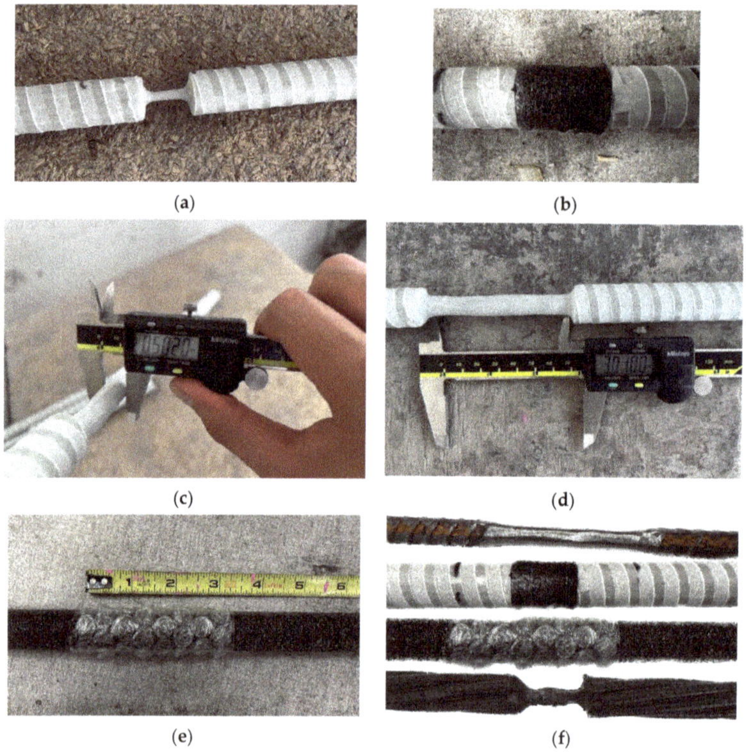

Figure 2. Damage introduced in bars. (**a**) Cross-section rupture, (**b**) covered rupture defect, (**c**) reduction in cross-section, (**d**) 3 inches of cross-section defect, (**e**) 3 inches of debonding defect, (**f**) different defects in bars.

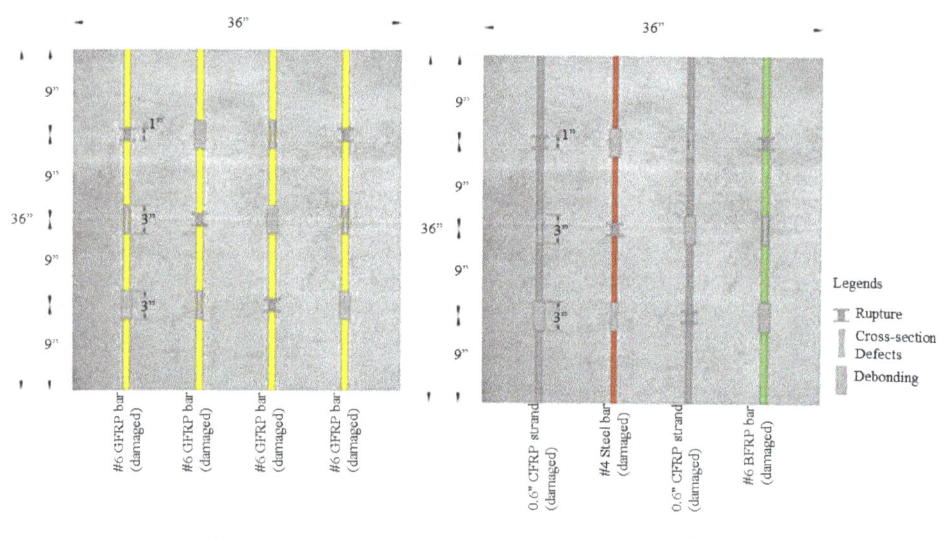

Figure 3. Slab specimen with simulated damage in bars, Slab O (**a**) and Slab P (**b**).

Figure 4. Three-dimensional scheme of Slab Q with simulated damage in the concrete.

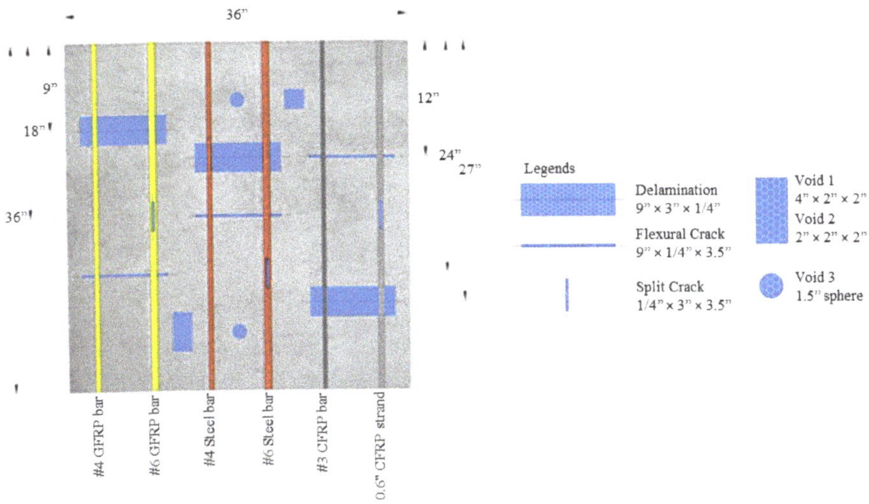

Figure 5. Dimension details of Slab Q with simulated damage in the concrete.

Table 2. Reinforcement/dimension details of slab specimens.

Slab ID	Parameter (Symbol/Units)	Reference Side	Bar 1	Bar 2	Bar 3	Bar 4	Bar 5	Bar 6
O	Distance to edge reference (L/inch)	3	6.0	14.0	22.0	30.0		
	Cover depth (C/inch)		1.0	2.0	3.5	4.5		
	Bar diameter (ϕ/inch)		#6					
	Bar material (T)		Glass					
	Slab thickness (h/inch)		7.0					
P	Distance to edge reference (L/inch)	3	6.0	14.0	22.0	30.0		
	Cover depth (C/inch)		1.0	2.0	3.5	4.5		
	Bar diameter (ϕ/inch)		0.6″	#4	0.6″	#6		
	Bar material (T)		C-Std *	Steel	C-Std *	Basalt		
	Slab thickness (h/inch)		7.0					
Q	Distance to edge reference (L/inch)	3	3.0	9.0	15	21	27	33
	Cover depth (C/inch)		3.3	3.1	3.3	3.1	3.3	3.2
	Bar diameter (ϕ/inch)		#4	#6	#4	#6	#3	0.6′
	Bar material (T)		Glass	Glass	Steel	Steel	Carbon	C-Std *
	Slab thickness (h/inch)		7.0					

* C-Std. (i.e., CFRP strands) labeled as Bars 1 and 3 in Slab P and Bar 6 in Slab Q.

2.3. Non-Destructive Testing (NDT) Methods

As mentioned in the introduction section, this paper aims to determine the detectability of damage in FRP reinforcements and FRP-RC elements using GPR and PAU devices (Figure 6). GPR is a real-time NDT method used to analyze the internal characteristics of civil structures. It operates on the principle that electromagnetic waves are reflected back when they hit a boundary between two materials with different dielectric constants [40,41]. The technique involves transmitting electromagnetic waves into the material under investigation and capturing the waves that are reflected from any irregularities within it. These irregularities could include boundaries between different materials, such as those between concrete and bars, or interfaces created by subsurface anomalies like voids, cracks, and instances of debonding or delamination in concrete [42]. Despite its potential, research on the application of GPR for inspecting FRP-RC elements is limited and its efficacy as a reliable NDT technique for this application remains unexplored. This study aims to investigate the feasibility of using GPR to detect damage in FRP bars.

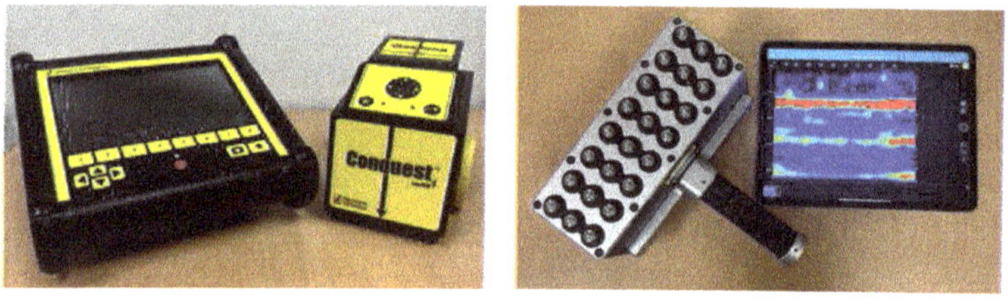

(a) **(b)**

Figure 6. NDT devices used: (**a**) GPR device, (**b**) PAU device.

The GPR system used in this experiment was a Conquest 100 Enhanced GPR (Sensors and Software Inc., Mississauga, ON, Canada) with a monostatic GPR antenna with a center frequency of 1000 MHz. GPR tests were carried out using both individual line scans and comprehensive grid scans. The line scans served as an initial survey to provide a preliminary understanding of the internal structure, including the orientation of reinforcements and the depth of exploration, by producing a cross-sectional image along the scan direction. However, this method proved to be time-consuming and labor-intensive due to the need to interpret multiple line scans. To streamline the process, grid scans were introduced, involving systematic GPR data collection along a predefined grid covering the test area. The grid's line spacing, set at 2 inches for this study, directly influenced the resolution of the collected data, with closer spacing yielding higher-resolution images and easier data interpretation. The data from grid scans produced depth slice images, offering a cross-sectional view parallel to the specimen's surface, facilitating a detailed analysis of the internal features.

PAU is the other NDT method being investigated in this study which is simply an advancement over the conventional ultrasonic testing (UT) method. Similar to GPR, UT is based on the principle that the ultrasonic waves are reflected back upon encountering a boundary between two materials with different acoustic impedances. These ultrasonic waves are generated and received by transducers that convert electrical or optical signals into ultrasonic waves and vice versa. A PAU setup is achieved by arranging multiple transducers in an array (Figure 6) and activating them sequentially with slight delays, allowing the individual waves to interfere constructively and destructively [43]. This arrangement enables the focusing and steering of the ultrasonic waves. PAU offers test results that are easy to interpret, a scanning rate of 5 to 10 times, and better resolution, reliability, portability, and mobility than conventional UT [44,45]. However, one of its major drawbacks is the uncertainty associated with its application as it has not yet been fully tested for inspection of FRP-RC elements.

The PAU device used for this experiment was a Pundit Live Array Pro device (Screening Eagle Technologies, Zurich, Switzerland) with 8 × 3 dry-contact Pundit array transducers. The PAU line scan was performed by moving the array of ultrasonic transducers along a specified line of inspection. Each scan at distinct positions was combined to form a continuous cross-sectional image perpendicular to the surface. For area scans, a stripe scan technique was employed, moving the transducers perpendicular to the line of inspection. Each stripe scan produced a line scan with a width equal to that of the PAU device, and these scans were stitched together to create a depth slice view. This view represents a cross-section parallel to the scanned surface, which can be further developed into a comprehensive 3D iso-surface model. A previous study by the authors [36] can be further explored for in-depth information on the applicability of GPR and PAU techniques for the inspection of structural elements reinforced with FRP.

3. Results

3.1. *Ground-Penetrating Radar (GPR)*

In order to detect damage in FRP bars, it is first important to determine whether the GPR device is able to detect the bars itself. The line scans (B-scans) of the slabs O, P, and Q were obtained for the sole purpose of checking bar detectability before taking the detailed area scan which can be used in the field inspection for the real-time detection of damage. Figure 7 shows the GPR response of a line scan collected perpendicular to the embedded bars in the longitudinal direction (from the reference edge 3 to 4) over Slab O. The top of the hyperbolic shape (i.e., inverted U shape) in the figure indicates the location of the bars.

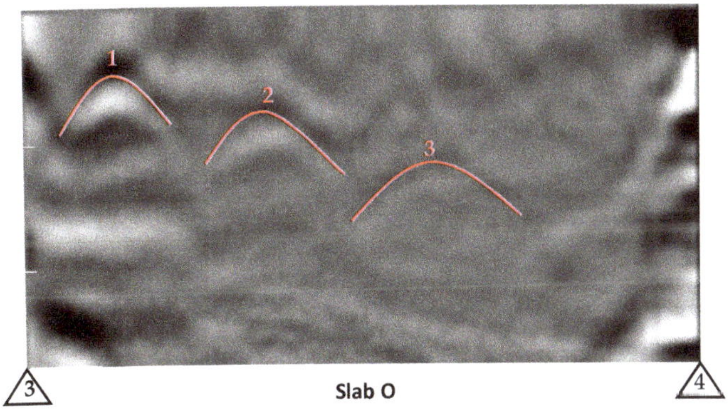

Legends: 1 = GFRP bar 1, 2 = GFRP bar 2, 3 = GFRP bar 3

Figure 7. Line view for bar detection carried out before defect/damage detection (Red markings are superimposed/added on test results to indicate distinctive features).

The GPR line scan of only one of the slabs (Slab O) is presented in this paper for the sake of brevity and Table 3 summarizes the results of the line scan test of all the other slabs. The line scans of Slab P and Q are presented in Appendix A, Figure A1.

Table 3. Summary of bar detectability for damage detection.

Slab ID	Bar 1	Bar 2	Bar 3	Bar 4	Bar 5	Bar 6
O	✓	✓	✓	X	-	-
P	✓	✓	✓	X	-	-
Q	X	X	✓	✓	✓	✓

Note: ✓ = detectable, X = not detectable.

It was observed that GPR could not detect the #6 GFRP bar in Slab O (Bar 4) and the #6 BFRP bar in Slab P (Bar 4), which have a concrete cover of more than 3.5 inches. Hence, it is clear that GPR would not be able to detect bar damage with a larger concrete cover. Slab Q, on the other hand, did not have any damage in the embedded bars but the damage was introduced to the concrete in the vicinity of the bars (in some cases below those bars). In addition to taking line scans perpendicular to the embedded bars in Slab Q, line scans were also taken at the center of the slab parallel to the bars to detect the damage introduced to the concrete. Two such line scans were taken, one from the top surface and the other from the bottom surface (after flipping the slab upside down), which are shown in Figure 8. It can be seen that more damage was visible from the bottom surface than from the top surface because the damage was closer to the bottom. The line scans conducted from the top surface could not detect vertical damage (flexural and split cracks) at all. Moreover, it can be seen that the hyperbolas for vertical damage (damage "c" in Figure 8) are narrower and taller than those for horizontal damage (damage "b" in Figure 8), which are wider and shorter. The damage "a" and "d" in the line scan carried out from the bottom surface (flipped Slab Q) appears to have been located at the surface, which is erroneous. This could be attributed to the use of small polystyrene foam cubes as a base for securing this damage (ping pong balls representing voids) at the bottom of the formwork.

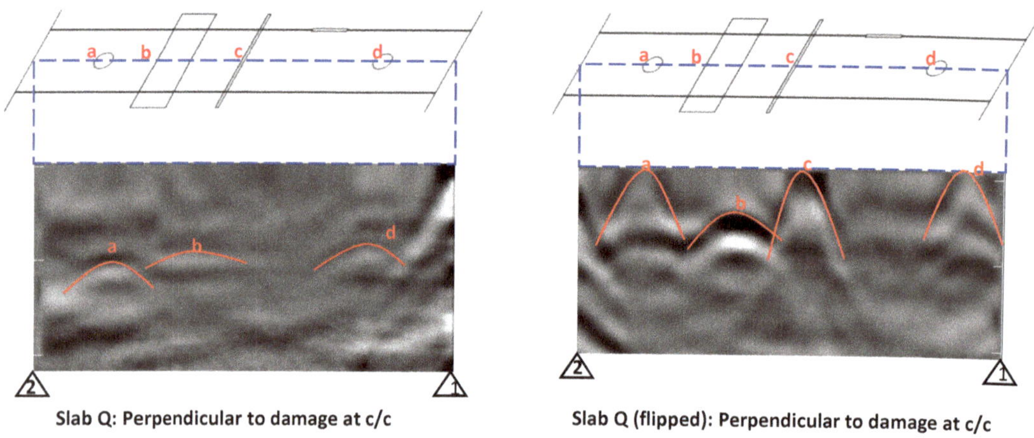

Legends: a/d = Spherical void, b = Concrete delamination, c = Flexural crack

Figure 8. Line views for concrete defects/damage at c/c of Slab Q (red markings are superimposed/added to test results to indicate distinctive features).

Next, depth slices (C-scans) taken with GPR to determine the damage detectability for each of the constructed slabs (Slab O, P, Q) are shown in Figures 9 and 10. To generate the depth slices, data collection was conducted longitudinally from reference edges 3 to 4, with a consistent spacing of 2 inches.

In Slab O with damaged GFRP bars, GPR successfully identified various forms of damage, including rupture, cross-sectional properties reduction, and debonding in Bar 1, as shown in Figure 9. However, the damage in Bar 2 is not quite distinguishable (see Appendix A, Figure A2). The bar itself is not as clearly visible as Bar 1, which is also because the already weak signal from GFRP bars (compared to steel bars) becomes even weaker with increasing depth. Further, while Bar 3 was detectable during the line scan, it was not visible in the depth slice due to its comparatively weaker signals relative to those from Bars 1 and 2. Thus, it was observed that the GPR exhibited limitations in detecting GFRP bars situated at greater depths, rendering it ineffective in identifying damage in these deeper-embedded GFRP bars. Consequently, it can be concluded that GPR's damage detection capability is primarily confined to GFRP bars located at shallower depths.

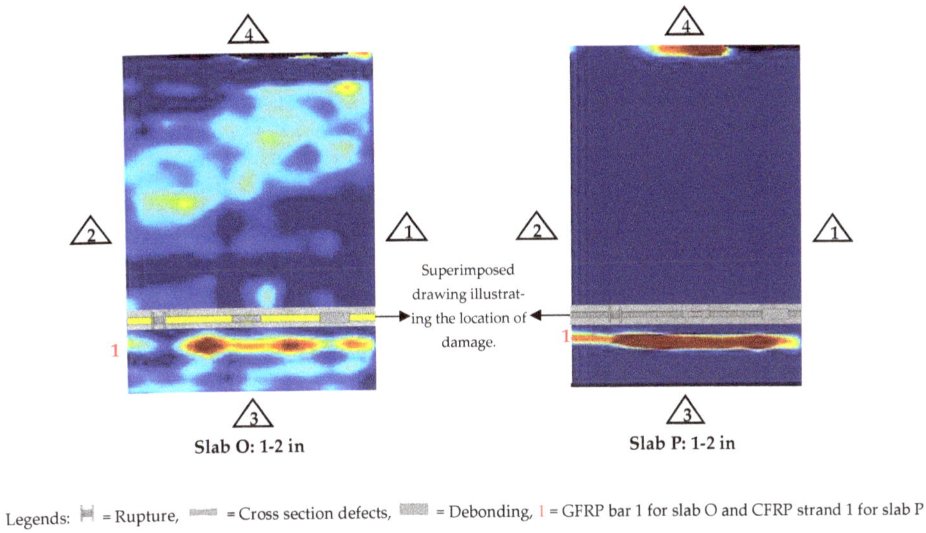

Figure 9. Depth slices of slabs for damage detection in bars.

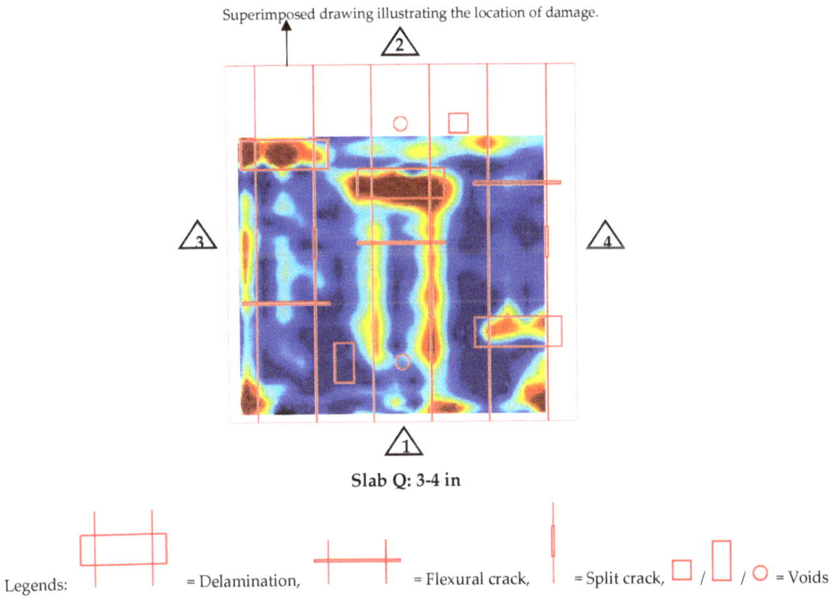

Figure 10. Depth slices of slabs for defect/damage detection in concrete.

The GPR depth slice results for Slab P varied in efficacy for detecting damage across different reinforcement materials. Specifically, for the CFRP strand located near the surface, GPR failed to identify any damage (see Figure 9). However, for the CFRP strand positioned at a depth of 3.5 inches, there were faded indications of damage (see Appendix A, Figure A2). In contrast, GPR exhibited comprehensive damage detection capabilities for the steel bar, identifying all present damage (see Appendix A, Figure A2). The BFRP bar presented a unique challenge; GPR was unable to detect it, particularly in the presence of dominant signals from the steel bar and CFRP strands. Consequently, damage in the BFRP bar remained undetected. However, it is worth noting that if the slab was only reinforced with

BFRP bars, it is possible that GPR might have been able to detect damage in BFRP bars to a certain degree.

The depth slice of slab Q shown in Figure 10 illustrates GPR's capability to clearly identify delamination within the concrete slab. However, its performance was limited when it came to the detection of deeper vertical cracks. In contrast, when the GPR test was conducted over the bottom surface of the slab, it proved adept at identifying shallower vertical cracks and voids (see Appendix A, Figure A3). These findings underscore the nuanced performance of GPR in detecting different types of defects in concrete, highlighting its selective sensitivity to various forms of damage depending on their nature and depth within the elements.

3.2. Phased Array Ultrasonic (PAU)

The PAU line scans were taken with a similar approach used for GPR line scans for first detecting the bars as shown in Figure 11. For Slab O, no GFRP bars were detectable, and for Slab P, the steel bar (Bar 2) and the CFRP strand (Bar 3) were clearly detectable. Similarly, for Slab Q, the steel bars (Bars 3 and 4) were clearly detectable. Further, there were also some indications of vertical cracks (labeled a and b in Figure 11).

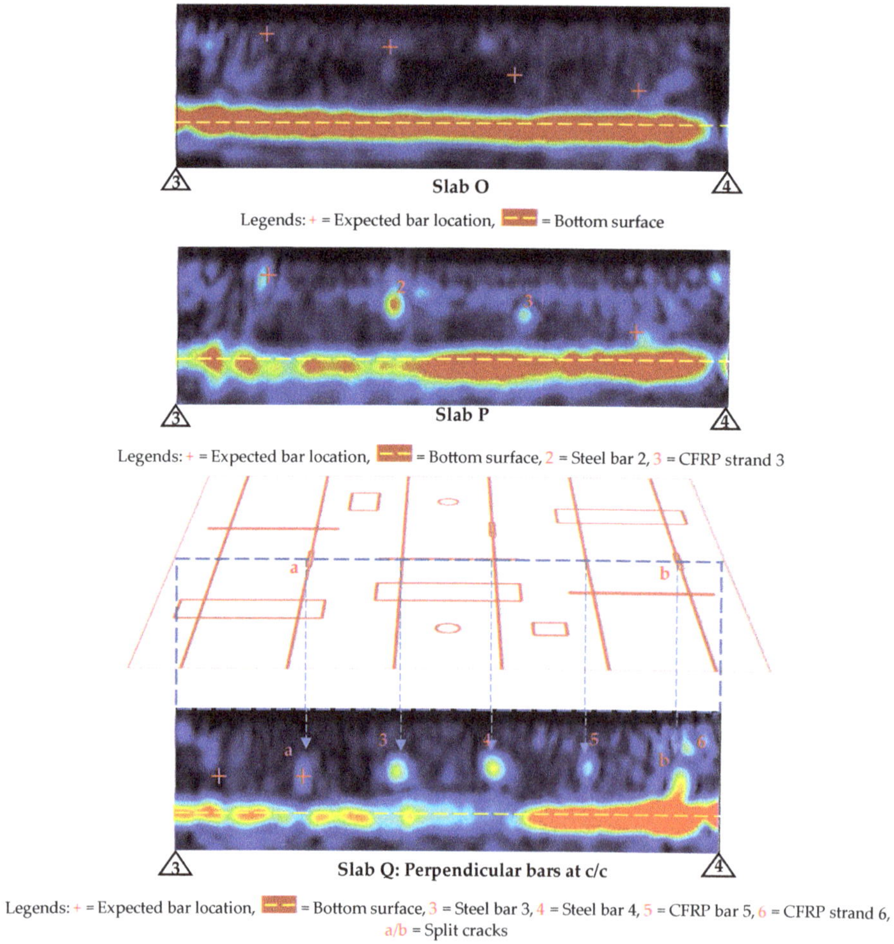

Figure 11. PAU line views for bar detection before damage detection.

In addition to performing line scans perpendicular to the embedded bars in Slab Q, line scans were also conducted at the center of the slab parallel to the bars to detect the damage introduced in the concrete, as shown in Figure 12. It can be seen that the horizontal delamination (labeled as b in the figure) was distinctly visible but the vertical cracks (labeled as c in the figure) were only visible as a dot at the top point of the crack where it initiates (instead of being visible as a vertical line). Similarly, the voids (labeled as a and d in the figure) were also visible but not as distinct as the delamination, which could be because they were smaller in size.

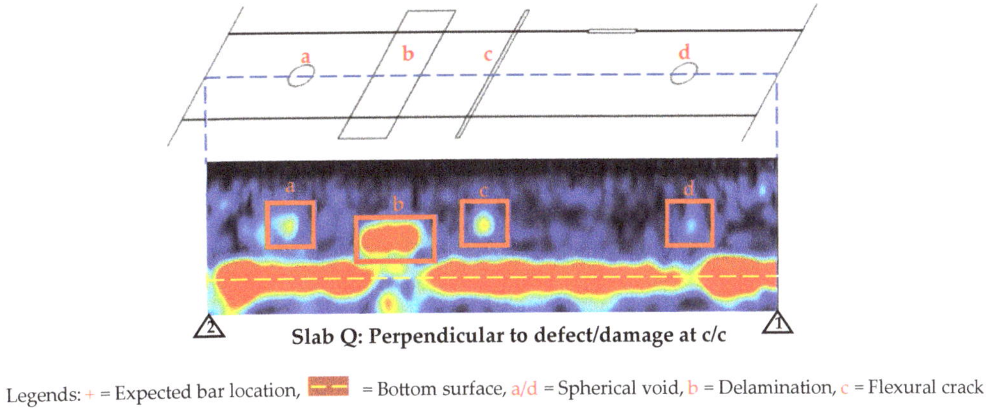

Figure 12. Line view for concrete defects/damage at c/c of Slab Q using PAU.

As expected, since GFRP bars could not be detected using PAU (as mentioned earlier in the PAU line scan results), the area scans over Slab O did not yield any damage detection. However, the depth slices and 3D view of PAU tests for Slabs P and Q (shown in Figures 13 and 14) produced some promising results.

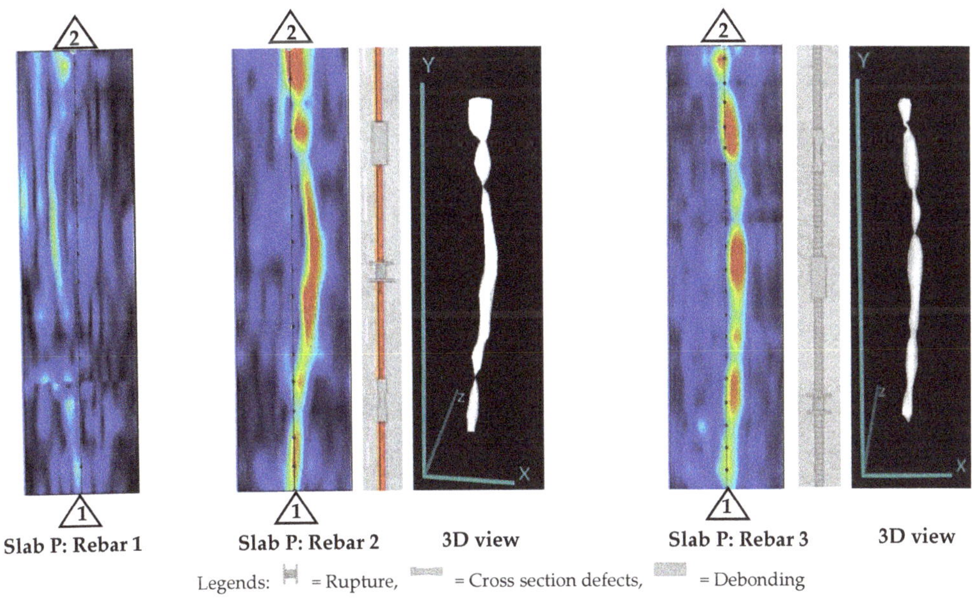

Figure 13. Depth slices of slabs for damage detection in bars using PAU.

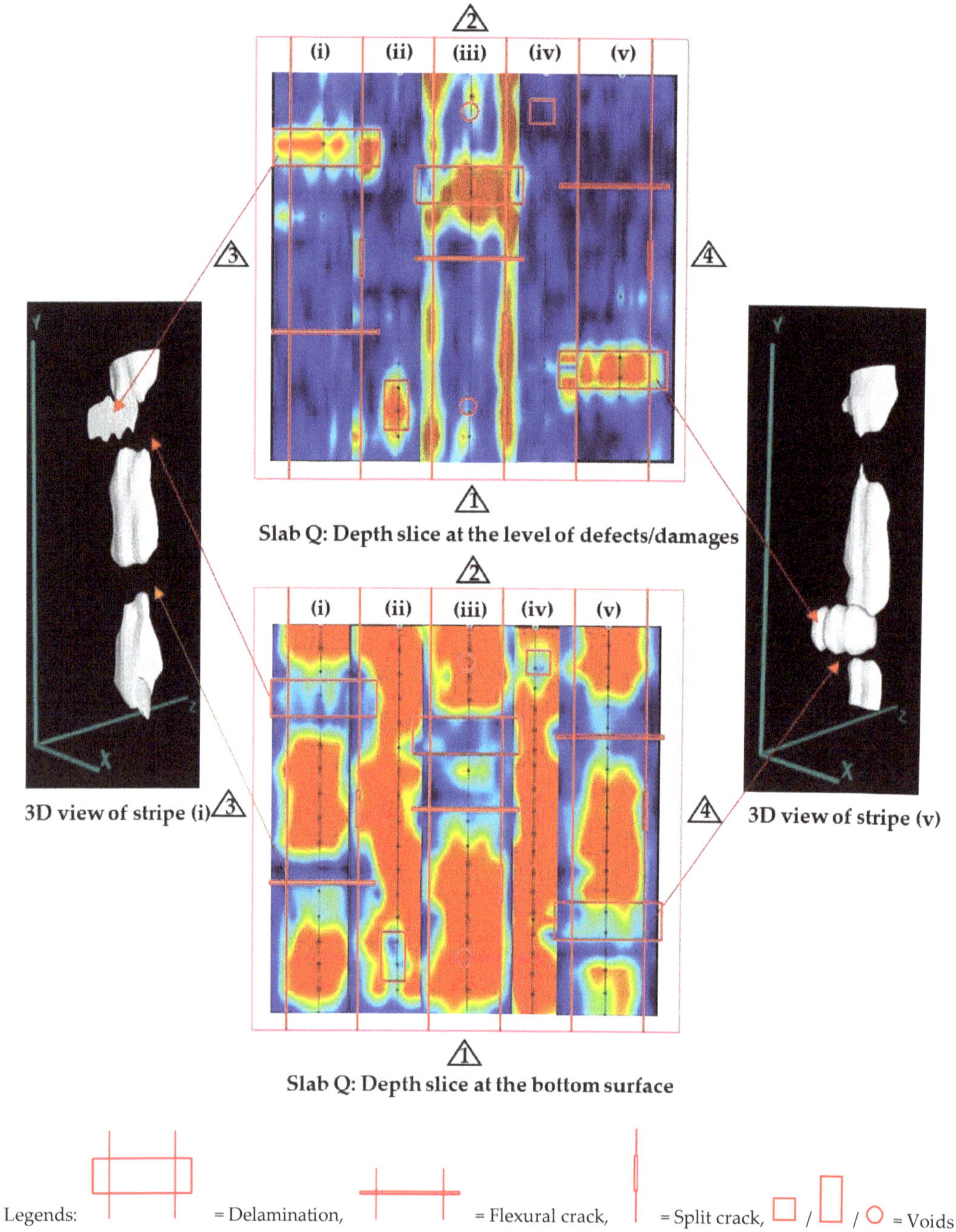

Figure 14. Depth slices of slabs for damage in concrete detectability tests using PAU.

It can be observed that Bar 2 and Bar 3 in Slab P do not appear to be straight. This is because Bar 2 (steel bar) had a kink because of a rupture at its mid-length resulting in a shift during casting, making it look like a bent bar. Similarly, in the case of Bar 3 (CFRP strand), the detected shape appears to be curved since these strands were flexible and curved outwards and downwards (sag) during the concrete pouring process. The CFRP strand had the tendency to bend to its transported coiled shape. It should also be noted that although it seemed the CFRP strand closer to the top surface (Bar 1) was visible in the line view (Figure 11), upon conducting the area scan over the strand, the detection proved not distinct enough for the entire strand (Figure 13). Only a portion of the strand at the mid-span was visible, which could be due to the fact that it sagged at the center with the weight of the concrete during casting, therefore increasing its depth from the top surface and making it free of surface reflections. These nuances of the PAU test results obtained from Slab P further illustrate the precision of PAU devices in detecting the geometrical orientation of the steel bar and CFRP strands embedded in concrete.

Similarly for Slab Q, while the CFRP bar (Bar 5) and CFRP strand (Bar 6) seem to be visible in the line scan (Figure 11), they were not visible on the depth slices (Figure 14), which could be because the signals from the damage in concrete (delamination) dominated over the weaker signal from the CFRP bar/strand.

In the context of damage detection in FRP bars, from the results of Slab O, it is evident that the PAU test cannot detect embedded GFRP bars at any depth and hence it also cannot detect any damage in GFRP bars. In slab P, the PAU was unable to detect the CFRP strand situated closer to the top surface. Additionally, its capability was limited when it came to the BFRP bar, rendering it ineffective in identifying damage within these BFRP bars. In contrast, PAU displayed comprehensive detection capabilities for the steel bar, with all damage, including the rupture at the bent position, being clearly identifiable. This clarity in damage detection was further enhanced when viewed in a 3D perspective. However, the damage in the CFRP strand at an intermediate depth, specifically Bar 3, was not as readily discernible as those in the steel bar. The presence of damage in this strand could be interpreted from the non-uniform color scale along the detection path, with red being the stronger signal and green to yellow being the weaker signal associated with defective/damaged areas.

Finally, for Slab Q, the PAU was adept at clearly identifying horizontal delamination and voids within the slab. Regarding the detection of vertical cracks in concrete (flexural and split cracks), although the direct detection proved challenging, these cracks could be inferred from the observed discontinuities or gaps present at the slab's bottom reflection [46], shown at the bottom center image and the 3D views shown in Figure 14. The bottom surface of the slab specimen is not visible below delamination and cracks because these discontinuities prevent the propagation of ultrasonic waves below them and may even trap the waves to bounce back and forth between them and the top surface, resulting in multiple equally spaced reflections [47].

4. Discussion

The test results are summarized in a comprehensive test result matrix shown in Table 4. From the table, it can be seen that except for the case of the damaged BFRP bar, there is no other case where both GPR and PAU collectively have the label "ND" (not detectable). Therefore, the application of these two NDT methods together will not miss the detection of damage in internal FRP bars/strands and concrete. Hence, all the test parameters are found to be either detectable, "D", or to have limited detectability, "LD" (limitation based on depth), by at least one of the GPR or PAU devices. In other words, in case a parameter is not detectable by GPR, then the detectability or limited detectability can still be ensured using PAU and vice versa. Thus, using both the GPR and PAU would be the best option to inspect FRP-RC elements. Otherwise, detectability can still be achieved using just one of these devices for the parameters mentioned as "LD" and "D" in Table 4.

Table 4. Effectiveness of GPR and PAU methods for the inspection of FRP-RC elements [1].

Slab	Parameters		Selected NDTs	
			GPR	PAU
O	Damaged GFRP bars	Rupture	LD	ND
		Cross-sectional property loss	LD	ND
		Debonding	LD	ND
P	Damaged CFRP strands	Rupture	LD	LD
		Cross-sectional property loss	LD	LD
		Debonding	LD	LD
	Damaged BFRP bar	Rupture	ND	ND
		Cross-sectional property loss	ND	ND
		Debonding	ND	ND
	Damaged steel bar	Rupture	D	D
		Cross-sectional property loss	D	D
		Debonding	D	D
Q	Defects/damage in concrete	Horizontal delamination	D	D
		Vertical Cracks	LD	LD
		Voids	LD	LD

Note: D = detectable; LD = limited detectability (based on depth); ND = not detectable. [1] The results were obtained from slab specimens with a maximum thickness of 7 inches and a maximum reinforcement depth of up to 4.5 inches.

However, there are some limitations related to the research conducted in this paper that warrant further investigation and consideration in future research studies concerning nondestructive testing techniques for the inspection of FRP-RC elements. It should be noted that the results shown in Table 4 have been obtained under laboratory conditions and the range of parameters used for the test specimens and test methods may not be applicable generally. For example, certain techniques were used to simulate the damage in FRP, e.g., the reduction in cross-section by grinding out the material and taping the bar to simulate debonding. Nevertheless, because the specimen and test conditions aimed to be as practical as possible, the results can provide credible guides for the use of NDT methods for FRP-RC elements. Regarding the inability to detect damage in the BFRP bar in Slab P, further experiments need to be conducted on slab specimens only reinforced with BFRP bars (similar to Slab O) before coming to a conclusion about the detectability of damage in BFRP bars. The low frequency 1 GHz GPR device used in this research was not able to detect the BFRP bar but a previous study by the authors [36] has proved that it can be detected using higher frequency GPR devices. Future research on detecting damage in BFRP bars should focus on using high-frequency GPR devices to detect several types of simulated damage with variations in parameters such as depth, extent, and type of damage. Since BFRP bars have gained popularity in recent years, it is justifiable to dedicate future research to the detectability of damage in BFRP bars to gain confidence among engineers in its use as a reinforcing material. Additionally, the scope of this study was limited to the real-time test results obtained from the proprietary software that comes with the commercially available GPR and PAU devices used in this study. This method of data collection represents the actual field conditions faced by the inspectors, who do not have access to complicated post-processing software at the inspection site location. Future research on refining these NDT techniques could even explore the integration of artificial intelligence and machine learning [48,49] for improved data analysis on remotely sensed data. Similarly, this study is limited to only three specimens with limited variation in depth of bars up to 4.5 inches, which can be overcome in future studies by conducting experimental verification on several specimens with a wider range of test parameters to collect more data for a statistically sound validation. One such extension could be testing the ability of the NDT devices to detect damage in multiple layers of reinforcements, as this study is limited only to the investigation of the first layer of reinforcement at the selected cover depth.

5. Conclusions

This study investigated the application of GPR and PAU in detecting damage in the FRP-RC elements. Three slab specimens with variations in several parameters, such as FRP type (GFRP, CFRP, BFRP), bar diameter, bar depth, and defect types were fabricated to determine the limitations and detection capabilities of these two NDT devices. Damage in the FRPs was simulated by a reduction in the cross-section to represent changes in cross-sectional properties and by wrapping the bars with tape to represent debonding. Foam pieces, flat and solid, were used to simulate the damage expected in the concrete. The findings of this study generally conclude that the combined use of GPR and PAU can detect potential internal defects associated with FRP-RC elements, as well as delamination, cracks, and voids in concrete.

The findings of this study have contributed significantly to the field of non-destructive testing (NDT) for FRP-RC elements. The successful demonstration of the combined use of GPR and PAU methods in detecting a variety of damage in FRP-RC elements under laboratory conditions lays the groundwork for future research and practical applications. The practical implications of these findings could involve the development of guidelines for the application of GPR and PAU methods in the inspection of FRP-RC elements in real-world scenarios, potentially improving the safety, maintenance, and durability of such structures. The specific conclusions of this paper include:

- Damage associated with the FRP-RC elements can be categorized as damage in the FRP reinforcements and those in the concrete. The potential types of damage in the FRP bars were identified as ruptures, loss of cross-sectional properties, and debonding, while the internal damage in concrete includes delamination, cracks, and voids.
- GPR could detect damage in GFRP bars, CFRP strands, steel bars, and all the internal damage introduced in concrete. It was not able to detect damage in BFRP bars in the experimental setup considered in this study, but there is a possibility that a higher frequency GPR device may be able to detect damage in BFRP bars, which is to be investigated in future studies.
- PAU showed limitations in its capability to detect damage in GFRP and BFRP bars but performed well in detecting damage in CFRP strands, steel bars, and concrete.
- Using GPR and PAU testing together would be the best option to inspect FRP-RC elements as the damage missed by one method would be detectable by the other. However, detectability (except for BFRP bars) can still be achieved using just one of these devices but with some limitations on the depth of the FRP.
- There may be some limitations related to the research conducted in this paper that may warrant further investigation. The experiments were under laboratory conditions and for a range of parameters for specimens and test methods and may not be applicable generally. Nevertheless, because the specimen and test conditions were chosen to be as practical as possible, e.g., FRP type, sizes, and concrete cover similar to actual values, the results can provide a credible guide for the use of NDT methods for FRP-RC elements.

Author Contributions: Conceptualization, P.M., S.S.K.D. and A.B.M.; methodology, P.M., S.S.K.D. and J.D.O.; software, P.M. and S.S.K.D.; validation, A.B.M. and A.N.; formal analysis, A.B.M. and A.N.; investigation, P.M., S.S.K.D. and J.D.O.; resources, A.B.M. and A.N.; data curation, P.M., S.S.K.D. and J.D.O.; writing—original draft preparation, P.M.; writing—review and editing, A.B.M., A.N., S.S.K.D., J.D.O. and J.D.; visualization, P.M., S.S.K.D. and J.D.O.; supervision, A.B.M. and A.N.; project administration, A.B.M. and A.N.; funding acquisition, A.B.M. and A.N. All authors have read and agreed to the published version of the manuscript.

Funding: This research received no external funding.

Institutional Review Board Statement: Not applicable.

Informed Consent Statement: Not applicable.

Data Availability Statement: The data presented in this study are available by request from the corresponding author.

Acknowledgments: The authors greatly acknowledge the support of the Department of Civil and Environmental Engineering at Florida International University, the Florida International University Doctoral Evidence Acquisition Fellowship program, and the Department of Civil and Architectural Engineering at the University of Miami. The contents of this paper reflect the views of the authors, who are responsible for the facts and the accuracy of the information presented herein.

Conflicts of Interest: Author Jiayi Ding was employed by the company AtkinsRéalis. The remaining authors declare that the research was conducted in the absence of any commercial or financial relationships that could be construed as a potential conflict of interest.

Appendix A

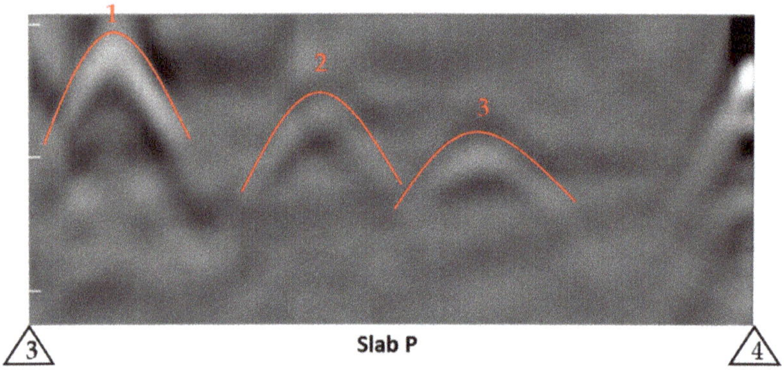

Legends: 1 = CFRP strand 1, 2 = Steel bar 2, 3 = CFRP strand 3

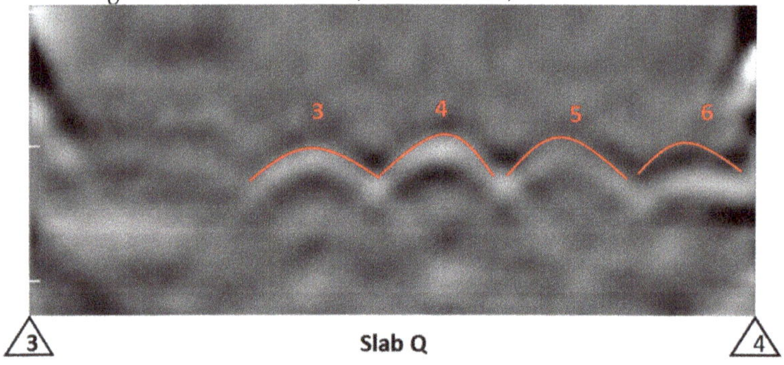

Legends: 3 = Steel bar 3, 4 = Steel bar 4, 5 = CFRP bar 5, 6 = CFRP strand 6

Figure A1. Line views for bar detection carried out before damage detection (for Slabs P and Q).

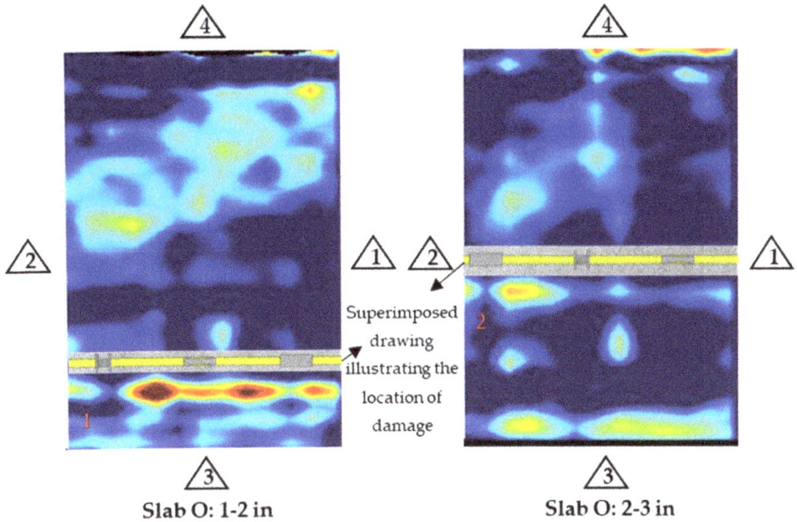

Legends: = Rupture, = Cross section defects, = Debonding, 1 = GFRP bar 1, 2 = GFRP bar 2

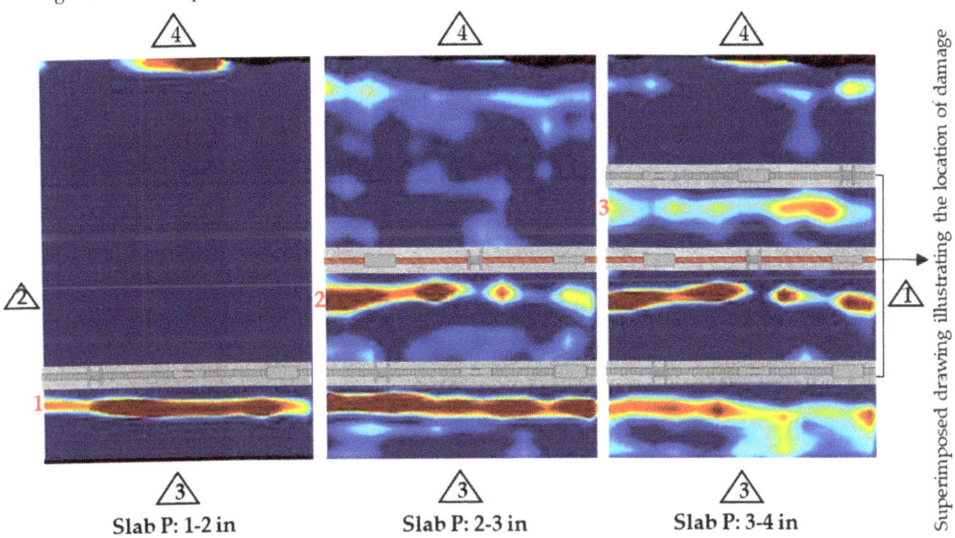

Legends: = Rupture, = Cross section defects, = Debonding, 1 = CFRP strand 1, 2 = Steel bar 2, 3 = CFRP strand

Figure A2. Depth slices of slabs O and P for damage detection in all the bars.

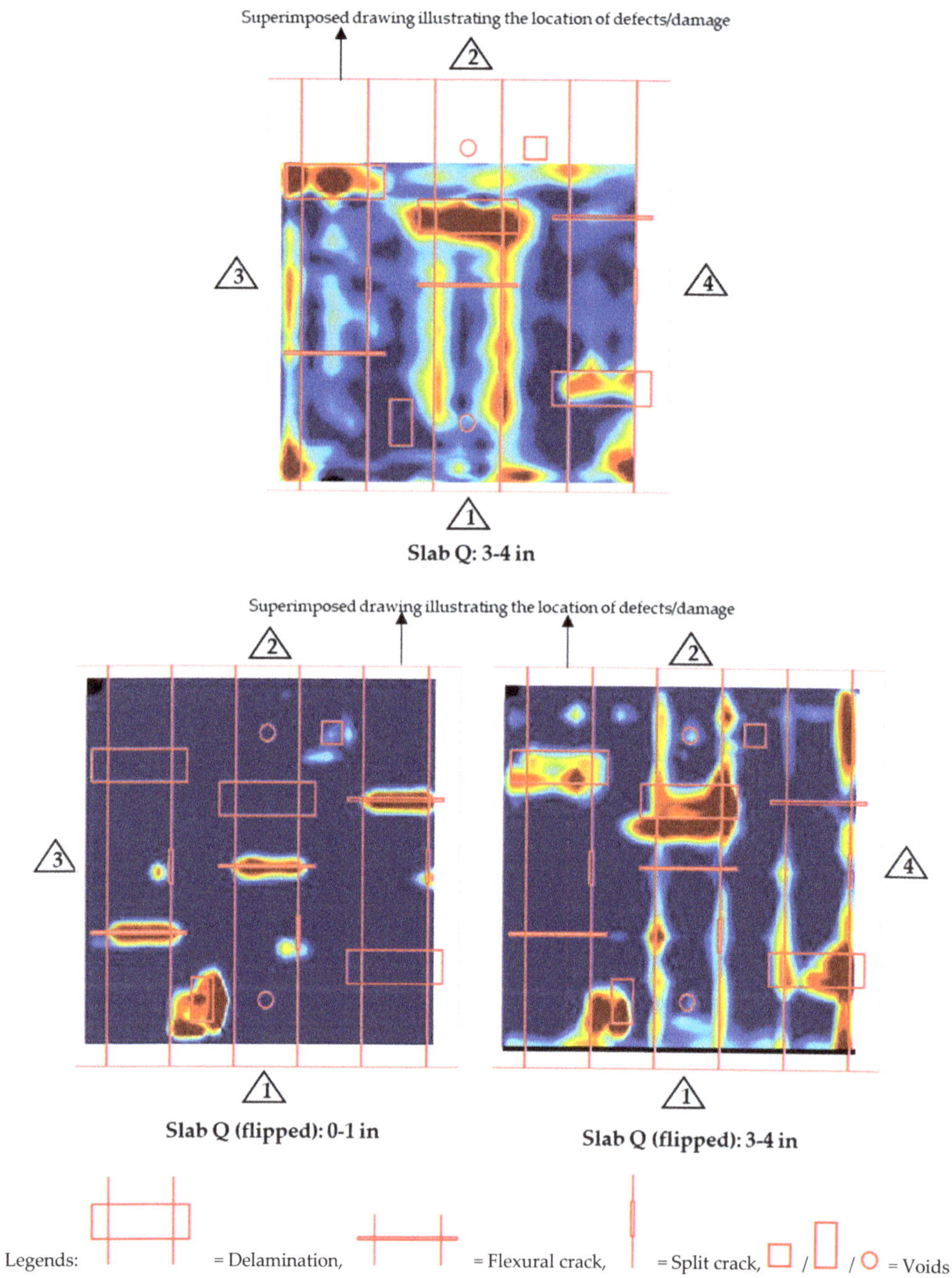

Figure A3. Depth slices of slab Q for damage detection in concrete (from top and bottom surface).

References

1. Brózda, K.; Selejdak, J. Analysis of FRP Bars Used as Reinforcement in Concrete Structures. *Prod. Eng. Arch.* **2016**, *12*, 2–4. [CrossRef]
2. Nolan, S.; Rossini, M.; Knight, C.; Nanni, A. New Directions for Reinforced Concrete Coastal Structures. *J. Infrastruct. Preserv. Resil.* **2021**, *2*, 1. [CrossRef]
3. Ekenel, M.; y Basalo, F.D.C.; Nanni, A. Fiber-Reinforced Polymer Reinforcement for Concrete Members. *Concr. Int.* **2021**, *43*, 18–22.
4. Koch, G.H.; Brongers, M.P.H.; Thompson, N.G.; Virmani, Y.P.; Payer, J.H. *Corrosion Cost and Preventive Strategies in the United States*; Elsevier: Amsterdam, The Netherlands, 2002.
5. NACE. Highways and Bridges. Available online: https://www.ampp.org/technical-research/what-is-corrosion/corrosion-reference-library/highways-bridges (accessed on 28 February 2024).
6. ASCE. *Report Card for America's Infrastructure*; ASCE: Reston, VA, USA, 2021.
7. Mohamed, H.M.; Benmokrane, B. Design and Performance of Reinforced Concrete Water Chlorination Tank Totally Reinforced with GFRP Bars: Case Study. *J. Compos. Constr.* **2014**, *18*, 05013001. [CrossRef]
8. Benmokrane, B.; Mohamed, H.M.; Mousa, S.; Elsafty, A.; Nolan, S. Design, Construction, Testing, and Behavior of Driven Precast Concrete Piles Reinforced with GFRP Bars and Spirals. *J. Bridge Eng.* **2021**, *26*, 04021050. [CrossRef]
9. Lau, K.; Permeh, S.; Faridmarandi, S.; Azizinamini, A. Review of Mitigation of Carbonation-Induced Corrosion with UHPC. In Proceedings of the AMPP Annual Conference on Corrosion, Denver, CO, USA, 19 March 2023; p. AMPP-2023.
10. Lau, K.; Samanbar, P.; Ghahfarokhi, M.; Faridmarandi, S.; Azizinamini, A. *Establishing Inspection Periods and Preliminary Recommendations for Preventive Measures for Mid-Rise Buildings near Coastal and Inland South Florida Environments*; Florida International University (FIU): Miami, FL, USA, 2023.
11. Lingvay, I.; Bors, A.M.; Lingvay, D.; Radermacher, L.; Neagu, V. Electromagnetic Pollution of the Environment and Its Effects on the Materials from the Built up Media. *Rev. Chim.* **2018**, *69*, 3593–3599. [CrossRef]
12. Rossini, M.; Saqan, E.; Nanni, A. Prediction of the Creep Rupture Strength of GFRP Bars. *Constr. Build. Mater.* **2019**, *227*, 116620. [CrossRef]
13. Siddika, A.; Al Mamun, M.A.; Ferdous, W.; Alyousef, R. Performances, Challenges and Opportunities in Strengthening Reinforced Concrete Structures by Using FRPs—A State-of-the-Art Review. *Eng. Fail. Anal.* **2020**, *111*, 104480. [CrossRef]
14. Gharde, S.; Kandasubramanian, B. Mechanothermal and Chemical Recycling Methodologies for the Fibre Reinforced Plastic (FRP). *Environ. Technol. Innov.* **2019**, *14*, 100311. [CrossRef]
15. Emparanza, A.R.; Kampmann, R.; De Caso y Basalo, F. State-of-the-Practice of Global Manufacturing of FRP Rebar and Specifications. *ACI Fall Conv.* **2017**, *327*, 45.1–45.14.
16. Goyal, A.; Pouya, H.S.; Ganjian, E.; Claisse, P. A Review of Corrosion and Protection of Steel in Concrete. *Arab. J. Sci. Eng.* **2018**, *43*, 5035–5055. [CrossRef]
17. Emergen Research. *Fiber Reinforced Polymer Rebar Market*; Emergen Research: Vancouver, BC, Canada, 2022.
18. Malnati, P. A Hidden Revolution: FRP Rebar Gains New Strength. *Compos. Technol.* **2011**, *17*, 24–29.
19. Kiani, N.; Abedin, M.; Steputat, C.C.; Mehrabi, A.B.; Nanni, A. Structural Health Monitoring of FRP-Reinforced Concrete Bridges Using Vibration Responses. In Proceedings of the European Workshop on Structural Health Monitoring; Springer: Berlin/Heidelberg, Germany, 2022; pp. 735–744.
20. Zhao, D.; Zhou, Y.; Xing, F.; Sui, L.; Ye, Z.; Fu, H. Bond Behavior and Failure Mechanism of Fiber-Reinforced Polymer Bar–Engineered Cementitious Composite Interface. *Eng. Struct.* **2021**, *243*, 112520. [CrossRef]
21. Hossain, K.M.A.; Ametrano, D.; Lachemi, M. Bond Strength of Standard and High-Modulus GFRP Bars in High-Strength Concrete. *J. Mater. Civ. Eng.* **2014**, *26*, 449–456. [CrossRef]
22. Kim, B.; Lee, J.-Y. Resistance of Interfacial Debonding Failure of GFRP Bars Embedded in Concrete Reinforced with Structural Fibers under Cycling Loads. *Compos. Part B Eng.* **2019**, *156*, 201–211. [CrossRef]
23. Valentine, S. *Expected Service Life of Michigan Department of Transportation Reinforced Concrete Bridge Decks*; University of Michigan: Ann Arbor, MI, USA, 2015.
24. Khedmatgozar Dolati, S.S.; Malla, P.; Ortiz, J.D.; Mehrabi, A.; Nanni, A. Non-Destructive Testing Applications for in-Service FRP Reinforced/Strengthened Concrete Bridge Elements. In Proceedings of the Nondestructive Characterization and Monitoring of Advanced Materials, Aerospace, Civil Infrastructure, and Transportation XVI; SPIE: Bellingham, WA, USA, 2022; Volume 12047, pp. 59–74.
25. Ortiz, J.D.; Khedmatgozar Dolati, S.S.; Malla, P.; Nanni, A.; Mehrabi, A. FRP-Reinforced/Strengthened Concrete: State-of-the-Art Review on Durability and Mechanical Effects. *Materials* **2023**, *16*, 1990. [CrossRef]
26. Malla, P.; Khedmatgozar Dolati, S.S.; Ortiz, J.D.; Mehrabi, A.; Nanni, A. Damages and Defects in FRP Reinforced and FRP Strengthened Concrete Elements. *J. Compos. Constr.* **2023**, *287*, 116155.
27. Kim, Y.J. *Use of Fiber-Reinforced Polymers in Highway Infrastructure*; The National Academies Press: Washington, DC, USA, 2017; ISBN 9780309390040.
28. Karbhari, V.M.; Chin, J.W.; Hunston, D.; Benmokrane, B.; Juska, T.; Morgan, R.; Lesko, J.J.; Sorathia, U.; Reynaud, D. Durability Gap Analysis for Fiber-Reinforced Polymer Composites in Civil Infrastructure. *J. Compos. Constr.* **2003**, *7*, 238–247. [CrossRef]

29. Mohammedameen, A.; Gülşan, M.E.; Alzeebaree, R.; Çevik, A.; Niş, A. Mechanical and Durability Performance of FRP Confined and Unconfined Strain Hardening Cementitious Composites Exposed to Sulfate Attack. *Constr. Build. Mater.* **2019**, *207*, 158–173. [CrossRef]
30. Benmokrane, B.; El-Salakawy, E.; El-Ragaby, A.; El-Gamal, S. Performance Evaluation of Innovative Concrete Bridge Deck Slabs Reinforced with Fibre-Reinforced-Polymer Bars. *Can. J. Civ. Eng.* **2007**, *34*, 298–310. [CrossRef]
31. Li, W.; Ho, S.C.M.; Patil, D.; Song, G. Acoustic Emission Monitoring and Finite Element Analysis of Debonding in Fiber-Reinforced Polymer Rebar Reinforced Concrete. *Struct. Health Monit.* **2017**, *16*, 674–681. [CrossRef]
32. Xu, K.; Ren, C.; Deng, Q.; Jin, Q.; Chen, X. Real-Time Monitoring of Bond Slip between GFRP Bar and Concrete Structure Using Piezoceramic Transducer-Enabled Active Sensing. *Sensors* **2018**, *18*, 2653. [CrossRef] [PubMed]
33. Ghaib, M. Detection and Localization of Damage in Fiber Reinforced Polymer Bars Using Acoustic Emission, Micro Computed Tomography, and Scanning Electron Microscopy Techniques. Ph.D. Thesis, University of Manitoba, Winnipeg, MB, Canada, 2018; p. 197.
34. Sharma, G.; Sharma, S.; Sharma, S.K. Fracture Monitoring of Steel and GFRP Reinforced Concrete Beams Using Acoustic Emission and Digital Image Correlation Techniques. *Struct. Concr.* **2021**, *22*, 1962–1976. [CrossRef]
35. Ékes, C. GPR: A New Tool for Structural Health Monitoring of Infrastructure. In Proceedings of the 3rd International Conference on Structural Health Monitoring of Intelligent Infrastructure, Vancouver, BC, Canada, 13–16 November 2007.
36. Malla, P.; Khedmatgozar Dolati, S.S.; Ortiz, J.D.; Mehrabi, A.; Nanni, A.; Dinh, K. Feasibility of Conventional Non-Destructive Testing Methods in Detecting Embedded FRP Reinforcements. *Appl. Sci.* **2023**, *13*, 4399. [CrossRef]
37. *ASTM C39*; Standard Test Method for Compressive Strength of Cylindrical Concrete Specimens. ASTM International: West Conshohocken, PA, USA, 2010.
38. Hameed, A.; Rasool, A.M.; Ibrahim, Y.E.; Afzal, M.F.U.D.; Qazi, A.U.; Hameed, I. Utilization of Fly Ash as a Viscosity-Modifying Agent to Produce Cost-Effective, Self-Compacting Concrete: A Sustainable Solution. *Sustainability* **2022**, *14*, 11559. [CrossRef]
39. *ACI CODE-440.11-22*; Building Code Requirements for Structural Concrete Reinforced with Glass Fiber—Reinforced Polymer (GFRP) Bars—Code and Commentary. American Concrete Institute: Farmington Hills, MI, USA, 2022.
40. Dutta, S.S. *Nondestructive Evaluation of FRP Wrapped Concrete Cylinders Using Infrared Thermography and Groud Penetrating Radar*; West Virginia University: Morgantown, WV, USA, 2006.
41. ACI Committee. *228 ACI 228.2R-13*; Report on Nondestructive Test Methods for Evaluation of Concrete in Structures. American Concrete Institute: Farmington Hills, MI, USA, 2013.
42. Javed, A.; Sadeghnejad, A.; Rehmat, S.; Yakel, A.; Azizinamini, A.; Florida International University, M.; Florida Department of Transportation. *Magnetic Flux Leakage Method for Damage Detection in Internal Post-Tensioning Tendons*; Florida Department of Transportation: Miami, FL, USA, 2021; 166p.
43. Shull, P.J. *Nondestructive Evaluation Theory, Techniques, and Applications*; Marcel Dekker, Inc.: New York, NY, USA, 2001; ISBN 0824788729.
44. Boychuk, A.S.; Generalov, A.S.; Stepanov, A.V. CFRP Structural Health Monitoring by Ultrasonic Phased Array Technique. In Proceedings of the EWSHM—7th European Workshop on Structural Health Monitoring, IFF-STTAR, Nantes, France, 11 July 2014; pp. 2206–2211.
45. Ryan, T.W.; EricMann, J.; Chill, Z.M.; Ott, B.T. *Bridge Inspector's Reference Manual*; BIRM 1; Federal Highway Administration: Washington, DC, USA, 2012; p. 1020.
46. Shokouhi, P.; Wolf, J.; Wiggenhauser, H. Detection of Delamination in Concrete Bridge Decks by Joint Amplitude and Phase Analysis of Ultrasonic Array Measurements. *J. Bridge Eng.* **2014**, *19*, 04013005. [CrossRef]
47. FHWA. Pavements—Ultrasonic Tomography (UST). Available online: https://infotechnology.fhwa.dot.gov/ultrasonic-tomography-ust/ (accessed on 28 February 2024).
48. Mahara, A.; Rishe, N. Integrating Location Information as Geohash Codes in Convolutional Neural Network-Based Satellite Image Classification. *IPSI Trans. Internet Res.* **2023**, *19*, 24–30. [CrossRef]
49. Amiri Margavi, A.; Babaee, H. Rank-Adaptive Reduced-Order Modeling of 2D Incompressible Linearized Navier Stokes Equations with Time-Dependent Bases. In Proceedings of the 76th Annual Meeting of the Division of Fluid Dynamics, Washington, DC, USA, 19–21 November 2023.

materials

MDPI

Article

Detecting the Sigma Phase in Duplex Stainless Steel by Magnetic Noise and First Harmonic Analysis

João Silva [1], Edgard Silva [1], Augusto Sampaio [1], Rayssa Lins [1], Josinaldo Leite [2], Victor Albuquerque Silva [3] and João Manuel R. S. Tavares [4,*]

1 Academic Unit of Industry, Federal Institute of Education, Science and Technology of Paraíba, Av. Primeiro de Maio, 720—Jaguaribe, João Pessoa 58015-435, Brazil; silvajbo@ifpb.edu.br (J.S.); edgard@ifpb.edu.br (E.S.); augusto.sampaio@academico.ifpb.edu.br (A.S.); rayssasatlins@gmail.com (R.L.)

2 Department of Mechanical Engineering, Federal University of Paraíba, Jardim Universitário, s/n—Castelo Branco, João Pessoa 58051-900, Brazil; josinaldo@ct.ufpb.br

3 Department of Teleinformatics Engineering, Federal University of Ceará (UFC), Fortaleza 60455-970, Brazil; victor.albuquerque@ieee.org

4 Instituto de Ciência e Inovação em Engenharia Mecânica e Engenharia Industrial, Departamento de Engenharia Mecânica, Faculdade de Engenharia, Universidade do Porto, Rua Dr. Roberto Frias, S/N, 4200-465 Porto, Portugal

* Correspondence: tavares@fe.up.pt

Abstract: Non-destructive electromagnetic tests based on magnetic noise analysis have been developed to study, among others, residual stress, heat treatment outcomes, and harmful microstructures in terms of toughness. When subjected to thermal cycles above 550 °C, duplex stainless steels form an extremely hard and chromium-rich constituent that, if it is superior to 5%, compromises the steel's corrosion resistance and toughness. In the present work, a study was carried out concerning the interaction of excitation waves with duplex stainless steel. Hence, by analyzing the magnetic noise and variations in the amplitude of the first harmonic of the excitation waves, the detection of the deleterious sigma phase in SAF 2205 steel is studied. To simplify the test, a Hall effect sensor replaced the pick-up coil placed on the opposite surface of the excitation coil. Sinusoidal excitation waves of 5 Hz and 25 Hz with amplitudes ranging from 0.25 V to 9 V were applied to samples with different amounts of the sigma phase, and the microstructures were characterized by scanning electron microscopy. The results show that the best testing condition consists of applying waves with amplitudes from 1 V to 2 V and using the first harmonic amplitude. Thus, the test proved effective for detecting the formation of the deleterious sigma phase and can follow the ability to absorb energy by impact and, thus, the material embrittlement.

Keywords: magnetic noise; non-destructive test; duplex stainless steel

1. Introduction

Duplex stainless steels are characterized by presenting equal volumetric percentages of ferrite and austenite constituents in their structure [1–3]. However, when thermal cycles above 550 °C are allowed, a harmful phase called sigma is formed, with a hardness of around 900 HV and rich in chromium. The high hardness compromises the energy absorption capacity by impact, i.e., toughness, and the high chromium content reduces the corrosion resistance [4–7]. The paramagnetic sigma phase arises from the ferrite structure, which is ferromagnetic, changing the material permeability. However, it does not occur in the austenite phase, which is paramagnetic [1,3,4,6].

Low quantities of the sigma phase promote a considerable decrease in toughness without a notable influence on the hardness. For instance, the precipitation of 1.3% of this phase decreased the impact toughness from 320 J (solute-treated) to 24 J (aged samples at 800 °C for 10 min) [1,2,4,5], and, as the precipitation of the sigma phase increases, cracks occur preferentially in the contours of the particles of this phase. Also, as the sigma phase is rich in chromium, it deflects the adjacent region of Cr, reducing corrosion resistance [1,5].

Changes in magnetic permeability characterize the formation of new constituents by phase transformations in ferromagnetic materials. This makes non-destructive electromagnetic tests interesting, such as the ones based on eddy currents, magnetic Barkhausen noise (MBN), and magnetic permeability measurements in the reversibility region of the movement of magnetic domain walls. These tests have been applied, for example, to studies of heat treatments, residual stress, fatigue, the presence of second phases, the integrity of welded joints, and embrittlement [2,3,8–19].

The imposition of an electromagnetic wave on ferromagnetic materials leads to the movement of the magnetic domain walls and an interaction of these with the material's microstructure. Factors such as grain boundaries, grain size, precipitates, dislocations, second-phase particles, and residual stresses act as anchorage points for the movement of the walls, and when the external magnetic field strength is sufficient for them to overcome the obstacles of the microstructure, the generation of magnetic Barkhausen noise occurs. This noise analysis brings information about the changes in the material, which is valuable in non-destructive electromagnetic tests [20–35].

Electromagnetic tests based on MBN are characterized by two coils: the excitation, which applies the signal, and the pick-up, which acquires the interaction signal. In these, there are three configurations of pick-up coils: one on the same surface as the excitation, another on the opposite surface, and the last surrounding the material. From these, the most common is the layout on the same surface of the excitation coil. A pick-up coil wound closely around the sample has also been used in studies of stress measurements in thin film samples, where the thicknesses are relatively small [16]. These tests have various configurations for the devices used, mainly in terms of excitation coils and receivers, the shape of the cores of the coils, and the characteristics of the signal applied to them, such as in terms of amplitude, frequency, and type of wave. Positioning the receiver coil on the opposite surface favors studying along the thickness of the material, thus sweeping a larger volume and allowing the evaluation of areas thermally affected by processes such as welding. Positioning the excitation coil on the same surface is limited in this regard, and the surrounding coil requires different configurations depending on the thickness of the material to be analyzed [10,13,15].

When a magnetic field is applied to a ferromagnetic material, the resulting shape of the magnetic induction is distorted due to magnetic hysteresis, and the non-linearity of the material's permeability, i.e., a sinusoidal shape wave being applied induces a non-sinusoidal one. This distorted magnetic induction waveform contains components at the harmonic frequencies of the applied magnetic field [23,24,26]. Harmonic analysis has been applied to investigate material failures, detect corrosion degradation, perform non-destructive evaluations of steels subjected to heat treatment, and monitor harmful microstructures in stainless steels [23–28].

Electromagnetic tests in the reversibility region of the magnetic domain walls' movement have used Hall effect sensors to detect the interaction of magnetic flux density and the material and, thus, analyze microstructural variations, magnetic anisotropy, and residual stress, among other applications. These sensors are sensitive to small variations in the intensity of the magnetic flux passing through a material, cost only a few dollars, and are simple to use. There are initial studies for replacing the receiver coil in tests in the irreversibility region. However, a deeper study of the applied wave intensity and a better transmission wave frequency is lacking in the literature [2,3,8,10].

Magnetic permeability measurements using a Hall effect sensor have proven effective for studying the α' phase in duplex stainless steels. The results confirm that Hall voltage measurements are affected by the phase transformations that occur in SAF 2205 duplex stainless steel at 425 °C and 475 °C and are suitable for tracking the formation of the α' phase in a non-destructive way. This was confirmed by correlation with the X-ray diffraction technique, an already well-consolidated inspection technique [3]. Similar results were obtained by authors who observed a decrease in magnetic susceptibility measurement in the same temperature range [29]. The microstructure formed prevents the movement of the magnetic domain walls, and, therefore, the magnetic susceptibility is decreased.

Permeability measurements have also been used to detect the formation of the sigma phase in the temperature range from 700 °C to 1000 °C through measurements carried out with Hall effect sensors and applying magnetic field strengths of up to 300 A/m in a pick-up coil. An ideal external magnetic field should be applied to obtain the best amplitude for the phase assessment. A value of 211.5 A/m has been suggested as the ideal field. The paramagnetic sigma phase reduces the induced magnetic field even when the phase is in reduced quantities of 2% [2]. The sigma phase was also studied in SAF 2205 duplex stainless steels at temperatures of 800 °C and 900 °C by analyzing the MBN with emitting and receiving coils positioned on the same surface and applying 10 Hz sinusoidal waves. Root mean square (RMS) values were correlated with the amount of the sigma phase. The MBN intensity was significantly reduced with the increased heat treatment time, indicating fewer ferromagnetic phases [4,7].

MBN noise has been applied to analyze several types of materials. The effect of quenching embrittlement in supermartensitic steels was studied in samples treated at temperatures of 620 °C and 640 °C cooled in water and in an oven after tempering, and those cooled slowly showed a lower toughness and a higher volumetric fraction of austenite [10]. Another study analyzed the stress profile generated in a 1070 steel sample subjected to a three-point bending test. The results suggested that the technique can detect the applied voltage profiles. Furthermore, the variation in microstructure in carbon steel welded joints was monitored by MBN. The results permitted the identification of the welded joint's heat-affected zone (HAZ) using MBN signals [30].

Non-destructive inspection techniques based on MBN have also been applied to detect non-homogeneous regions in carbon steel sheets. The non-homogeneous or damaged regions were produced by plastic deformations in rolled and annealed sheets of SAE 1060 and 1070 steels. The behavior of the mean square root of the Barkhausen magnetic noise signal was correlated with the position of the uneven regions detected in the samples. The results showed that in all the studied cases, it was possible to detect the position of the damage through the variation in the magnetic field [31]. Another application based on MBN is detecting early-stage fatigue, which is associated with plastic deformation in ferromagnetic metallic structures. Experimental results demonstrated that MBN is promising for this characterization [32]. The presentation of the behavior of the interaction between excitation waves and different types of anchoring points, presented previously, serves as a foundation for understanding the influence of the microstructure on magnetic noise.

In the present work, a study was carried out of the best excitation wave to be applied in duplex stainless steel (DSS) to detect the formation of the sigma phase in a configuration where the receiving coil, positioned on the opposite surface to the excitation, is replaced by a Hall effect sensor to simplify the test. In this, the magnetic noise and the harmonic of the emitting wave were studied, as well as their ability to follow the formation of the harmful sigma phase. The methodology presented was applied to duplex stainless steel because, under the studied condition, only one harmful constituent is formed from the decomposition of ferrite. This would facilitate the interpretation of the signals resulting from the interaction between the excitation waves and the material, as well as the analysis of the methodology. Furthermore, the applied field strength was carried out in the region of magnetic reversibility of the movement of the magnetic domain walls [2,3] and the magnetic noise analyzed, which, according to the literature, will not be termed Barkhausen, as it is not in the irreversibility region.

2. Materials and Methods

The material studied in this work was SAF 2205 hot rolled steel supplied in plate format with a thickness of 8 mm. To carry out the experiments, four duplex stainless steel SAF 2205 samples were used, which were machined by electro-erosion in a circular shape with a diameter of 24 mm, one without aging, and three underwent thermal treatments at a temperature of 850 °C for ¼ h, 1 h, and 2 h in a resistance oven and were then cooled in water. The treatment at 850 °C for ¼ h can generate 5% of the sigma phase in the material, which is enough for the steel under study to become embrittled, affecting its microstructure and compromising its mechanical properties [1–3]. The other treatment times are sufficient to consolidate the formation of the sigma phase and serve to follow the deleterious sigma phase.

The plate was received in the hot rolled state, and all the measurements were performed in the rolled direction, corresponding to the easy magnetic direction. The direction of easy magnetization of the received material was determined by the non-destructive technique for measuring magnetic anisotropy developed by [8]. The procedure used to study microstructural anisotropy was a non-destructive test carried out in the region of magnetic reversibility using direct current. It consisted of applying a magnetic flux intensity in the center of a circular sample 24 mm in diameter and 8 mm in thickness to determine the magnetic flux density resulting from the interaction through a Hall effect sensor applying a frequency of 5 Hz and 10 V. The sample was rotated from 0° to 360° in 22.5° steps.

Duplex stainless steel was chosen for this study because it is widely used in industry due to its toughness and corrosion resistance characteristics in the as-received condition. However, when subjected to thermal cycles, such as in the welding process, the sigma phase may form from the decomposition of the ferromagnetic ferrite phase. A mere 5% of the sigma phase is enough to compromise its mechanical properties and corrosion resistance, as this phase has a hardness of around 1000 HV and is rich in chromium, thus impoverishing its matrix and reducing its resistance to corrosion.

Manufacturing processes of the studied steel, such as welding, require non-destructive testing to evaluate their quality by detecting the formation of the sigma phase and predicting corrective interventions. The apparatus used in this study can be applied to detect the presence of this harmful constituent. Furthermore, in the studied temperature range, the steel under analysis presents only the formation of a harmful constituent to its properties, making it easier to predict the origin of the changes in the used signals.

The samples were prepared for optical and scanning electron microscopy analysis, the first to detect the amount of the sigma phase and the second for the microstructural analysis. The samples were maintained under chemical attack with an electrolytic solution of 10% KOH using a voltage of 3.7 V and a current of 0.75 A for ¼ h, which preferably reveals the sigma phase and facilitates its visualization and identification, i.e., segmentation. It is possible that some χ phase was also formed and quantified as σ phase. However, it was not the objective of this work to separate these two phases because they have similar effects on steel's properties. The χ phase forms itself before the precipitation of the σ phase and disappears once the σ phase starts to precipitate [1]. Twenty images of each condition were acquired and segmented. The amount of the constituent sigma phase was determined according to a 95% confidence interval.

Ferrite microhardness measurements in the as-received and treated for ¼ h and 2 h conditions were carried out to analyze the presence of the sigma phase. Ten measurements were taken for each condition, and the confidence interval was obtained at 95%. A Shimadzu microhardness tester model HMV-G 20 S was used with an applied load of 250 g.

The X-ray diffraction tests were carried out using copper Kα radiation, with a voltage of 40 kV and a current of 30 mA; a step of 0.02°, with a time per step equal to 9.6 s; and adopting a viewing angle (2θ) ranging from 41° to 53°.

The workbench consists of two modules, one for excitation and the other for acquisition. The scheme of the experimental setup is shown in Figure 1.

The excitation module comprises a function generator (Minipa MFG 4205B model, São Paulo, Brazil) and a transmitter coil. The function generator transmits waves of different formats to the pick-up coil. Here, a sinusoidal wave was chosen since it presents the least interference from the harmonics of the excitation wave. The coil was positioned in the center of one of the sample faces, inducing a magnetic flux density in the material. The studied waves were applied to the center of the circular samples, and the signal resulting from the interaction was acquired on the opposite surface. The surfaces of the samples were sanded with 320-grit sandpaper to eliminate possible surface oxidation during cooling after treatment.

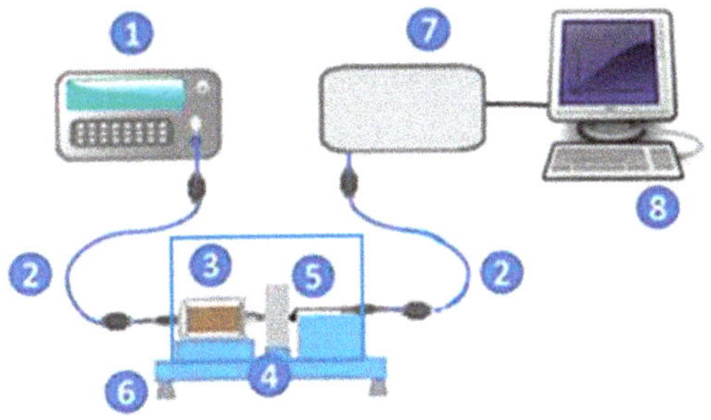

Figure 1. Used experimental setup: (1) signal generator, (2) shielded cables, (3) excitation coil, (4) material sample, (5) Hall effect sensor, (6) test bench with Faraday cage, (7) acquisition board, and (8) computer.

The acquisition module comprises a Hall effect sensor, an acquisition board, and a computer. The sensor is positioned in the center of the other face of the sample to detect the interaction field of the interaction between the excitation wave and the material. The acquisition board connects to the sensor and the computer via USB cables. The computer performs automatic data acquisition using software developed in-house. The chosen SS495A model is a Hall linear effect sensor with a sensitivity of 3.125 mV/gauss and an input voltage between 0 V and 10 V, it being supplied here with a continuous voltage of 5 V. This sensor has a working range of up to 700 gauss, and the measurements were carried out at around 550 gauss, remaining below its saturation region.

To detect the characteristics of the excitation waves that can follow the sigma phase formation, the as-received and treated at 850 °C for ¼ h samples were analyzed. Thus, the characteristics of the wave station that can have the sensitivity to detect the presence of 5% of the sigma phase from the thermal treatment were determined. The best condition of this analysis was then used to follow the formation of this harmful phase in samples with different amounts of sigma phase, mainly corresponding to ¼ h, 1 h, and 2 h of treatment times, as well as in the initial condition.

The signal analysis was divided into two stages. In the first one, high-pass filters for magnetic noise analysis were applied. Thus, using a fast Fourier-transform (FFT) algorithm, the signals were transformed to the frequency domain and then processed by 50 Hz and 250 Hz digital high-pass filters. These cutoff frequencies were used to analyze the noise region. The second stage involved analyzing the harmonics of the 5 Hz and 25 Hz excitation waves by identifying their peaks based on the signal's root mean square (RMS).

Thus, based on the signal's RMS parameter analysis, the best condition was used to follow the sigma phase formation in the SAF 2205 material. The excitation voltages of the waves used were 0.25 V, 0.5 V, 0.75 V, 1 V, 1.25 V, 1.5 V, 1.75 V, 2 V, 3 V, 5 V, 7 V, and 9 V. These values refer to the magnetic field strength applied to the samples of 1.4 Oe, 2.3 Oe, 3.4 Oe, 3.9 Oe, 4.6 Oe, 5.2 Oe, 5, 6 Oe, 6.7 Oe, 9.9 Oe, 15.7 Oe, 22.4 Oe, and 28.2 Oe, respectively. For each condition, 50 measurements were taken, and the 95% confidence interval was determined.

The test power was used to analyze the ability of the studied experiment to differentiate conditions with and without the sigma phase. An accurate estimation of the test power can predict the probability that a statistically significant difference will be detected based on a finite sample size under a real alternative hypothesis. If the power is too low, there is little chance of a significant difference being detected, and non-significant results are likely, even if there are, in fact, real differences. In Figure 2, $(1 - \beta)$ corresponds to the test power.

Values of (1 − β) greater than or equal to 1 (one) indicate a 100% or more difference and that the measurements will always differ. Values below 1 (one) indicate the probability of overlapping results. The value (1 − β) indicates the measurement system's accuracy and corresponds to the test power in Figure 2.

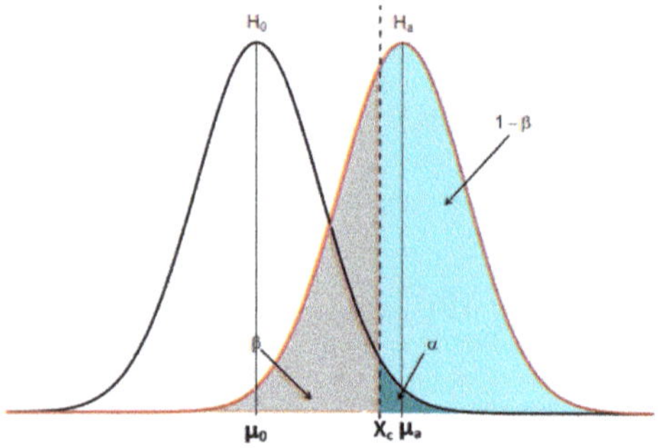

Figure 2. Representation of the histogram for two situations, H_0 and H_a, where one seeks to know with an error α how precisely the two measurements differ.

In Figure 2, α is the error if the null hypothesis (H_0) is rejected in favor of the alternative hypothesis (H_a) when H_0 is true. β is the error if H_0 is not rejected when H_a is true. (1 − β) is the power of a hypothesis test, i.e., the probability of rejecting H_0 when H_a is the true hypothesis. When (1 − β) is greater than or equal to 1 (one), it indicates a 100% or more difference and that the measurements will always differ.

The following equations are used to determine the test power:

$$\overline{x}_c = \mu_0 + t_c \times \frac{S_0}{\sqrt{n}}, \quad (1)$$

where $\overline{x}_c$ is the critical value of the H_0, S_0 is the H_0 standard error, μ_0 is the H_0 mean, n the number of measures, and t_c the t Student for (n − 1) and α = 0.05, and

$$Z_\beta = \frac{(\overline{x}_c - \mu_a) \times \sqrt{n}}{S_a}, \quad (2)$$

where Z_β is the standard score, which represents how far from the mean a data point is, μ_a is the H_a mean, and S_a is the H_a standard error.

The Z_β value is then used to find the area of the region of (1 − β) shown in Figure 2, which represents how different H_0 and H_a are, using a Z score table.

3. Results and Discussions

Figure 3a shows the optical microscopy of the as-received material, and Figure 3b depicts the polar plot of the measured magnetic flux density for the related stainless steel sample. Figure 1 shows that the easy magnetization direction is lamination (180°). This angle has the highest value of magnetic flux density, i.e., this direction corresponds to the one with the lowest magnetic losses. The present work was carried out with direct current and a fixed solenoid pole; therefore, under the experimental conditions, the polar graphs show the maximum at an angle of 0°. The plate was received in the hot rolled state, and all the measurements were performed in the rolled direction, corresponding to the easy magnetic direction.

In scanning electron microscopy, the angle formed between the X-ray source and the detector, i.e., the angle between the transmitted beam and the reflected beam, is usually designated as 2θ. In an experiment, the crystallographic plane cannot be observed; however, the transmitted and the reflected beam can be observed and, therefore, the 2θ can be experimentally assessed.

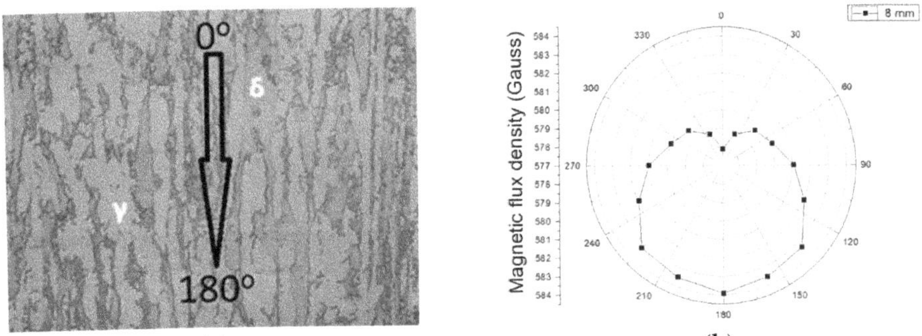

Figure 3. (**a**) The optical microscopy of the DSS structure in the as-received condition, where δ is the ferrite phase and γ the austenite, and (**b**) the magnetic flux density variation as a function of the rotation angle.

Figure 4a,b show the material's scanning electron microscopy analysis results in the as-received condition and its X-ray diffraction. In Figure 4a, one can observe the presence of austenite islands, clear regions on a ferrite matrix, and dark regions, showing no formation of precipitates. The presence of these constituents is confirmed in the diffractogram of Figure 4b, where only peaks of austenite and ferrite can be seen. The ferrite constituent is ferromagnetic and the austenite paramagnetic. In the as-received condition, SAF 2205 steel has no ferrite matrix transformation. However, when heated at temperatures above 550 °C, precipitates are formed, including the sigma phase, which has a hardness of around 900 HV and, as it is rich in chromium, impoverishes the matrix of this element and compromises both the corrosion resistance and the tenacity of the material [1–4,8].

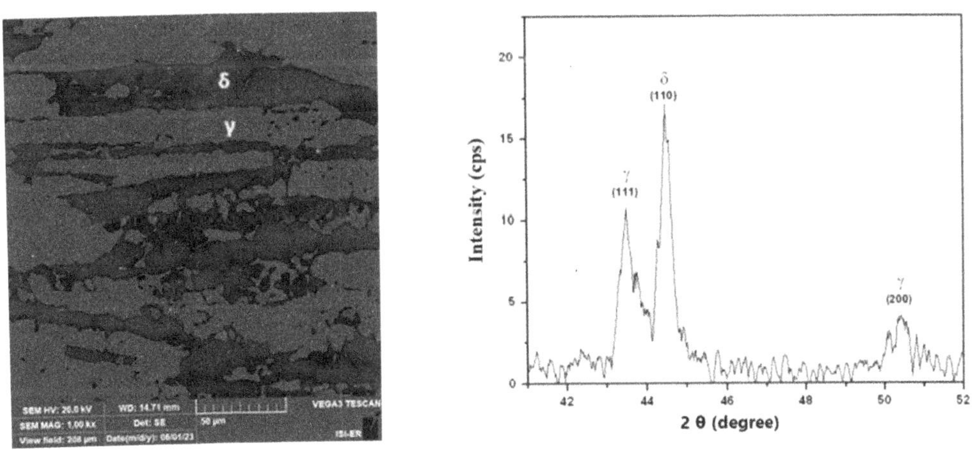

Figure 4. (**a**) The scanning electron microscopy image (10% KOH attack, 1000x of magnification) and (**b**) the X-ray diffractogram of the as-received condition sample (δ—ferrite and γ—austenite).

Figure 5 presents the microscopy of the sample treated at 850 °C for ¼ h and its X-ray diffractogram. The presence of the sigma phase in the ferritic region can be observed as discontinuous precipitates formed from the ferrite grain boundaries. The austenite of duplex stainless steel does not transform, as the diffusion of atoms in it is about 100 times greater than in ferrite, which leads to concentrations occurring in this phase [1–3,10].

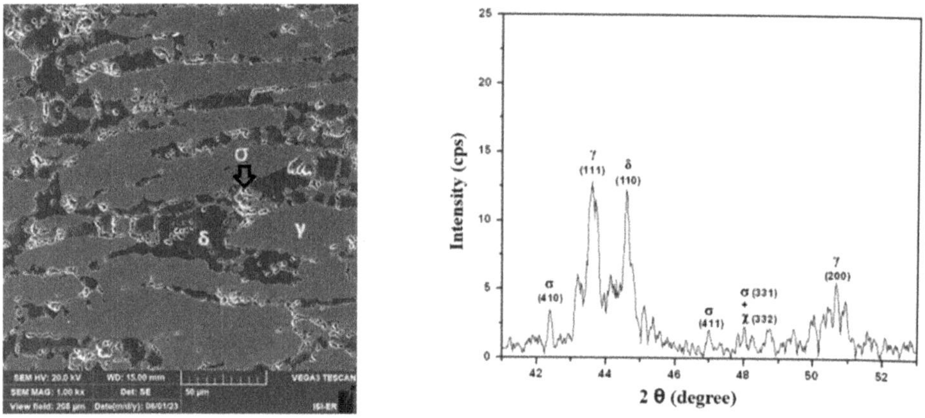

(a) (b)

Figure 5. (**a**) The scanning electron microscopy image (10% KOH attack, 1000X of magnification) and (**b**) the X-ray diffractogram of the sample treated at 850 °C for ¼ h (δ—ferrite, γ—austenite, and σ—sigma).

Figure 6 presents the microscopy of the sample treated at 850 °C for 2 h and its X-ray diffractogram. A greater presence of the sigma constituent can be observed in this case. The amount of the sigma phase was measured as being 5%, 17%, and 18% for the treatment times of ¼ h, 1 h, and 2 h, respectively. As the difference between the sample treated for 1 h and the one for 2 h was only 1%, Figure 6 presents only the scanning electron microscopy image and the X-ray diffractogram of the sample treated for 2 h.

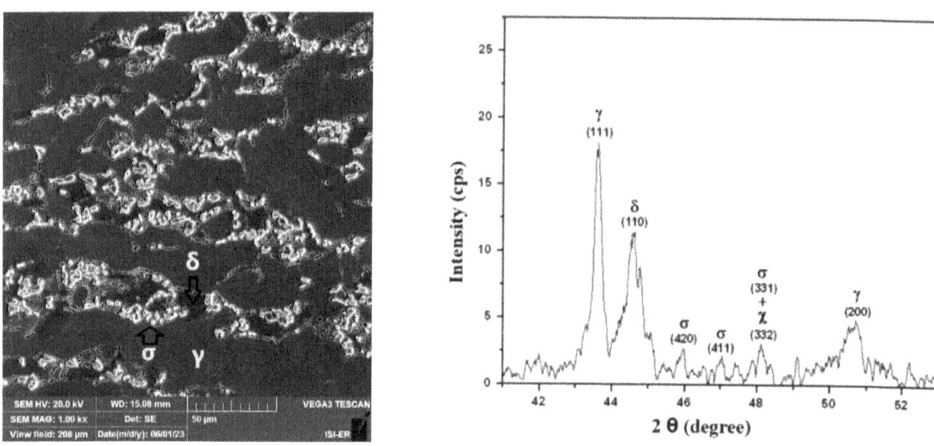

(a) (b)

Figure 6. (**a**) The scanning electron microscopy image (10% KOH attack, 1000X of magnification) and (**b**) the X-ray diffractogram of the sample treated at 850 °C for 2 h (δ—ferrite, γ—austenite, and σ—sigma).

The treatments with times of 1/4 h and 2 h show the presence of the chi phase. However, the χ phase forms before the precipitation of the σ phase and disappears once the σ phase starts to precipitate. The precipitation of the σ and χ phases decreases the magnetic properties of the steel because ferrite is ferromagnetic, and the χ and σ phases are paramagnetic phases. Electrolytic etching with 10% KOH solution preferably reveals the σ phase [1,33].

Ferrite microhardness measurements in the as-received, treated for ¼ h, and for 2 h conditions were carried out, being 215.15 (±13.4) for the as-received condition, 254.1 (±10.79) for the treated for ¼ h, and 292.5 (±10.4) for 2 h. Ten measurements were taken for each condition, and the confidence interval was obtained at 95%. The increase in hardness shows the formation of the sigma phase during treatment. The results of X-ray diffraction and scanning electron microscopy, combined with the microhardness, reinforce the understanding that there was no sigma formation before annealing, or if it did occur, the quantity was insufficient for detection with the techniques used. This indicates that the 5% sigma obtained is already enough to influence the results and that the effect of ferrite softening, if any, will not be a main factor in the present work.

Figure 7a,b show the magnetic noise of the conditions without the sigma phase and with 5% of it, respectively, after applying the 50 Hz high-pass filter for an excitation wave of 5 Hz frequency and 1 V amplitude. The presence of sigma precipitates acts as anchor points for the movement of the magnetic domain walls and reduces the magnetic flux density value [2,3]. Also, there is the contribution of the paramagnetism of the sigma phase [1,5]. The RMS values for the conditions without and with the sigma phase were determined as equal to 0.04825 and 0.04437 gauss, respectively. The test power was calculated to be around 6, indicating a more than 100% certainty in differentiating between the two cases despite only having a 10% difference. Although the values obtained are small in magnitude, measurements in different conditions can differentiate the two cases without overlapping measurements, thus confirming accuracy.

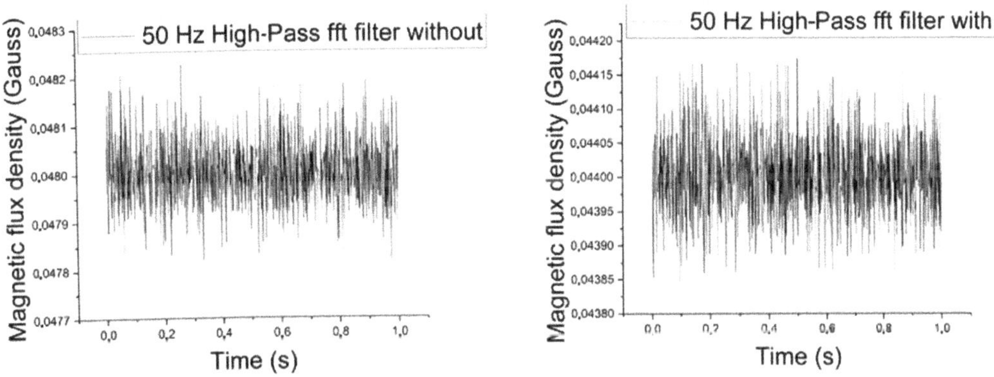

(a) (b)

Figure 7. (**a**) Magnetic noise, measured in the magnetic flux density, of the conditions without the sigma phase, as a function of the time, after applying the 50 Hz high-pass filter for an excitation wave of 5 Hz and 1 V. (**b**) Magnetic noise of the conditions with 5% of the sigma phase, as a function of time, after applying the 50 Hz high-pass filter, for an excitation wave of 5 Hz and 1 V.

Next, this study concerned itself with the characteristics of the excitation wave able to detect the presence of the sigma phase by magnetic noise analysis in the samples under study, with the application of excitation waves with a frequency of 5 Hz and cutoff frequencies of 50 Hz and 250 Hz. Figure 8 shows the RMS variation in the signal acquired by the Hall sensor as a function of the amplitude of the excitation wave applied to the samples without precipitate and treated at a temperature of 850 °C for ¼ h, for an excitation frequency of 5 Hz and amplitudes from 0.25 V to 9 V, and a cutoff frequency of 50 Hz.

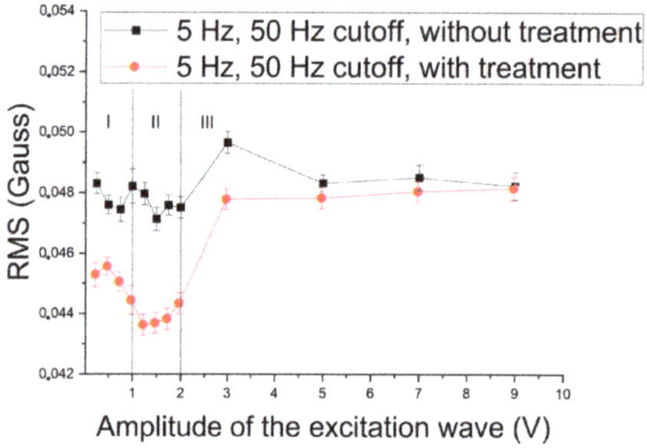

Figure 8. RMS of the signal acquired by the Hall sensor as a function of the amplitude of a 5 Hz excitation wave applied in samples with and without treatment and a cutoff of 50 Hz.

Analyzing Figure 8, one can perceive the existence of three distinct regions, named I, II, and III, one for 0.25 V to 1 V amplitudes, another for 1 V to 2 V amplitudes, and the last one for 3 V to 9 V amplitudes. This figure depicts the RMS variation as a function of the amplitude of the excitation wave for the samples as received and with 5% of the paramagnetic sigma constituent. The curve for the sample with the presence of the sigma constituent is shifted downwards due to the reduction in its ferromagnetism, with the transformation of part of the ferrite constituent into sigma precipitates. In region I, there is also a reduction in RMS values due to blocking the movement of the walls of the magnetic domains due to the presence of sigma precipitates. The increased magnetic flux in region II leads to a greater detection of these precipitates and their contribution to the blockage. However, in the third region (III), as there is a large increase in magnetic flux, there is an increase in the noise generated by the movement of the magnetic domain walls, which try to overcome the blocking effect of the sigma constituent. Similar behavior was observed in [2,3,10]. The logarithmic scale was chosen for the amplitude to facilitate the visualization of the regions in the presented graphs.

A point was chosen in each region of Figure 8 to analyze the test power of the measurements, that is, their ability to differentiate the two conditions. The test power was applied to the waves with 0.25 V, 1.25 V, and 7 V of excitation amplitude, and results of 4.04, 6.26, and 0.6, respectively, were obtained. The value of 0.6 indicates that in the third region for this amplitude, there is a 60% overlap of the results of the two conditions. However, for the first and second regions, the values were above 1 (one), indicating with total certainty that the technique accurately differentiates the two conditions. The values of 4.04 and 6.26 indicate the sensitivity of the measurement setup.

Figure 9 shows the RMS variation of the signal acquired by the Hall sensor as a function of the amplitude for an excitation frequency of 5 Hz and a cutoff frequency of 250 Hz, applied in the samples with and without treatment. The same regions identified in the previous case can be observed in Figure 9. It can also be noted that the RMS values of the treated sample signal are lower than those of the untreated sample. This can occur due to the paramagnetism of the sigma phase, which reduces the material's permeability [2,3], or to the movement of magnetic domain walls' blocking because of this newly formed structure [2,13,34].

To identify the best working region for the 5 Hz excitation wave, a graph was built concerning the module of the difference in the RMS of the two cases, without and with the presence of the sigma phase in a percentage, as shown in Figure 10. It is noted that the greater differences between the two cases in the central region of the graph are the most evident for the performed test in detecting the sigma phase. This region corresponds to excitation wave amplitudes ranging from 1 V to 2 V.

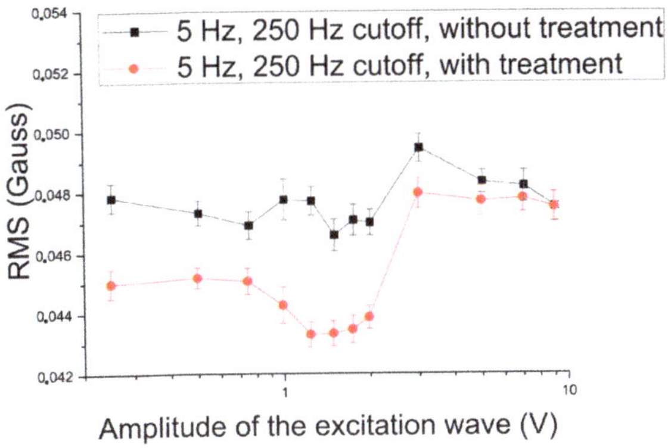

Figure 9. RMS of the signal acquired by the Hall sensor as a function of the amplitude of the 5 Hz excitation wave applied in samples with and without treatment and a cutoff frequency of 250 Hz.

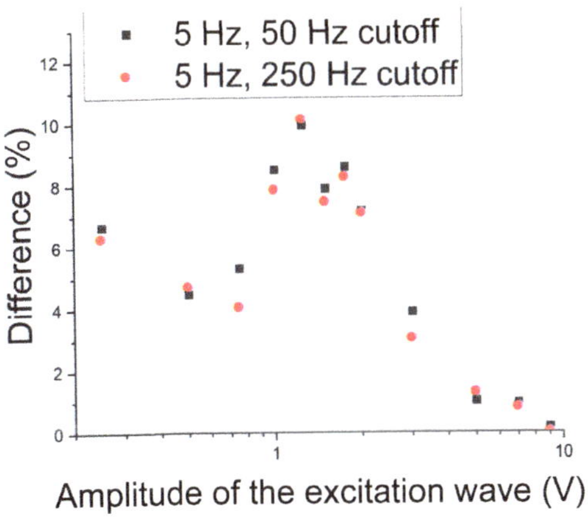

Figure 10. Module of the difference in RMS as a function of the amplitude of the 5 Hz excitation wave and the two tested cutoff frequencies.

Next, the influence of applying 25 Hz sinusoidal waves was studied. Figures 11 and 12 show the RMS values as a function of the amplitude of the excitation wave with 50 Hz and 250 Hz cutoffs. Figure 11 shows the influence of the 50 Hz cutoff, where the three regions detected when applying the 5 Hz wave can be observed. The first stage presents behavior in the conditions of with and without precipitate in a parallel way, followed by a fall, but with a gradual increase from the second region. This means that the harmonics of the excitation wave begin to influence the RMS values in the noise region, indicating that the increase in frequency and the application of cutoffs in the lower-frequency region lead the harmonics of the excitation wave to influence the noise region and increase its amplitude.

In Figure 12, one can see that the first region is still quite well-defined. The second stage goes from 1 V to 2 V, and a gradual increase is observed due to the increase in frequency, leading the harmonics to interference in this region. However, for these, the behavior was like that of the 5 Hz excitation waves.

Figure 13 shows the module of the difference in the RMS as a function of the wave amplitude to determine which regions lead to the best values that differentiate the percentage as a function of the amplitude of the 25 Hz excitation wave. One can perceive that in the region from 1 V to 2 V, the best results were found for the cutoff frequency of 25 Hz. For the 50 Hz cutoff, there appear points with greater values for the amplitudes of waves of 7 V and 9 V, where one can perceive that as the cutoff is reduced to 50 Hz and the frequency is increased, these points are also more distinguished. These are due to the harmonic of the main wave interfering in this region.

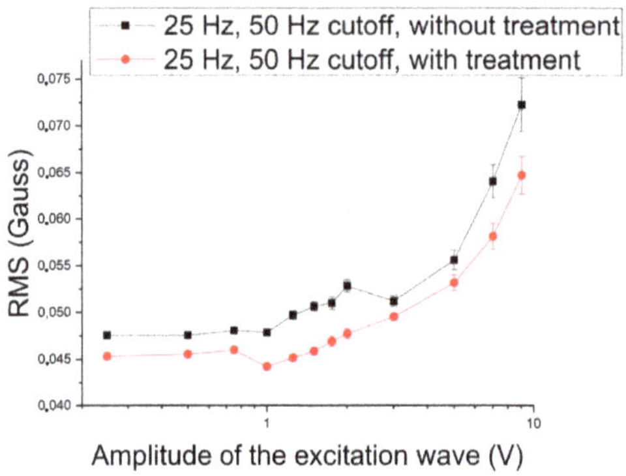

Figure 11. RMS of the signal acquired by the Hall sensor as a function of the amplitude of the excitation wave applied to the samples with and without treatment for an excitation frequency of 25 Hz and a cutoff of 50 Hz.

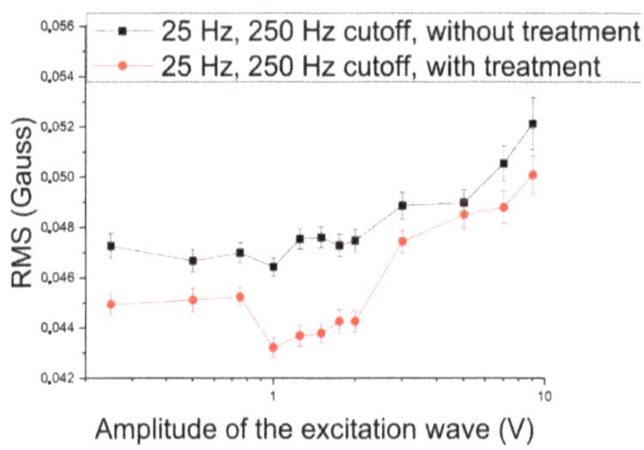

Figure 12. RMS of the signal acquired by the Hall sensor as a function of the amplitude of the excitation wave applied to the 8 mm samples with and without treatment with an excitation frequency of 25 Hz and a cutoff of 250 Hz.

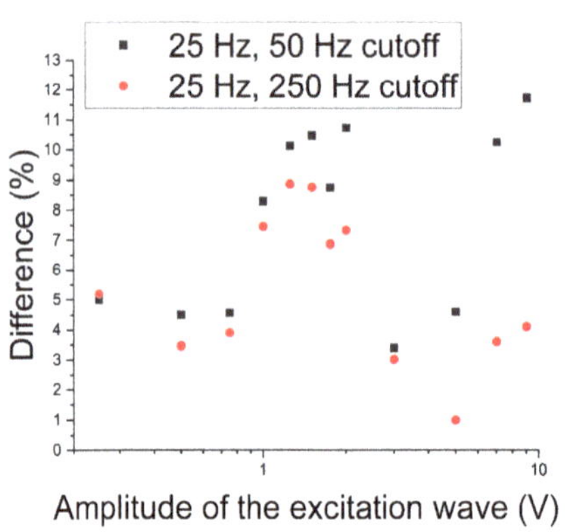

Figure 13. RMS difference module as a function of the wave amplitude of the applied 25 Hz transmitter wave and the two cutoff frequencies.

Figure 14 shows the RMS variation as a function of the excitation amplitude for waves of 5 Hz and 1 V regarding the amount of the sigma constituent. The excitation wave was chosen to be in the region with the best results. Thus, the magnetic noise obtained by the methodology can follow the formation of the sigma phase.

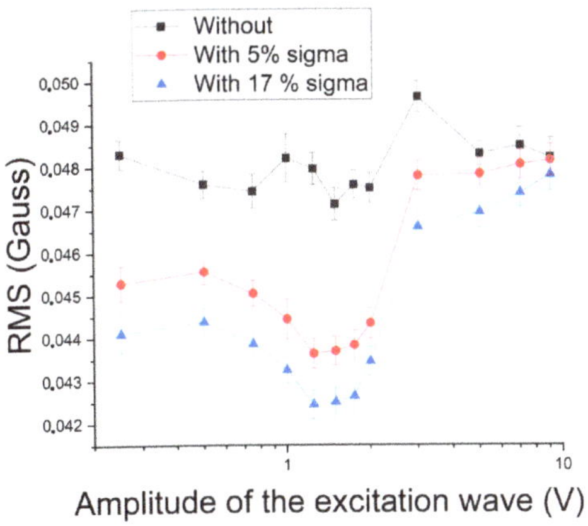

Figure 14. RMS of the signal acquired by the Hall sensor as a function of the amplitude of the 5 Hz excitation wave applied to 8 mm samples for conditions with different amounts of the sigma phase and a cutoff of 50 Hz.

Next, the ability to detect the constituent sigma's presence was studied by analyzing the excitation wave's harmonics [18–20]. Figure 15 shows the variation in the RMS of the first harmonic of 5 Hz and 25 Hz excitation waves as a function of their amplitude. A similar behavior could be observed between the studied frequencies: an increase to the amplitude of 1 V, followed by a decrease to 2 V, and a linear growth again. In addition, the signals with precipitates present lower values than those without the sigma constituent. The lower values, obtained with precipitates, must combine the paramagnetism of the sigma constituent, formed from the ferromagnetic ferrite, and the blocking of the magnetic domain walls' movement by forming coarse precipitates. This type of behavior has been observed in the literature [2,3]. However, the drop in values from 2 V, as occurred for both conditions, indicates that the increase in amplitude leads to the detection of the contribution of paramagnetic austenite with the increase in magnetic flux [10]. It can also be noted in Figure 15 that the RMS values fall with increasing frequency. This was expected, as the penetration depth has an inverse behavior to the increase in frequency due to the surface effect [12]. Because the error range of the measurements is too small to be easily readable, a magnification showing the range detail for the sample with thermal treatment and the 25 Hz and 3 V excitation wave is visible in Figure 15.

To determine the best working region, a graph was built of the difference between the conditions without and with precipitate in a percentage as a function of the amplitude of the 5 Hz and 25 Hz excitation waves (Figure 16). One can note that the best region continues to be from 1 V to 2 V. Still, the range of difference in this region with magnetic noise analysis was around 7% to 10% before, and now it becomes from 25% to 32% when the amplitude of the first harmonic is analyzed. Thus, the first harmonic was applied to monitor the formation of the sigma constituent.

Next, a study concerning the follow-up of the formation of the sigma constituent was carried out. Considering that in the range from 1 V to 2 V, the increase in the magnetic flux density begins to detect the paramagnetic austenite, the study was conducted for 1 V, which takes values still in the region of better results for detecting the sigma phase. Figure 17 shows the RMS variation of the first harmonic of the excitation wave as a function of the treatment time at 850 °C for an amplitude of 1 V and frequencies of 5 Hz and 25 Hz.

Figure 17 shows a drop in RMS values to a plateau after 1 (one) hour of treatment. The test was shown to be sensitive to the detection of 5% of the sigma phase, which is harmful to the tenacity of the material and occurs after 15 min of treatment. The plateau is due to the amount of formed constituent being close to the treatment times of 1 h and 2 h.

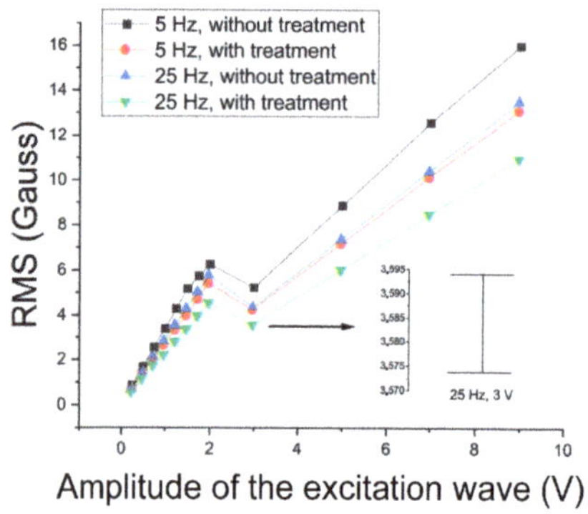

Figure 15. RMS variation of the first harmonic as a function of the amplitude of the 5 Hz and 25 Hz excitation waves and samples with and without the sigma phase. The 25 Hz and 3 V excitation wave error range is shown.

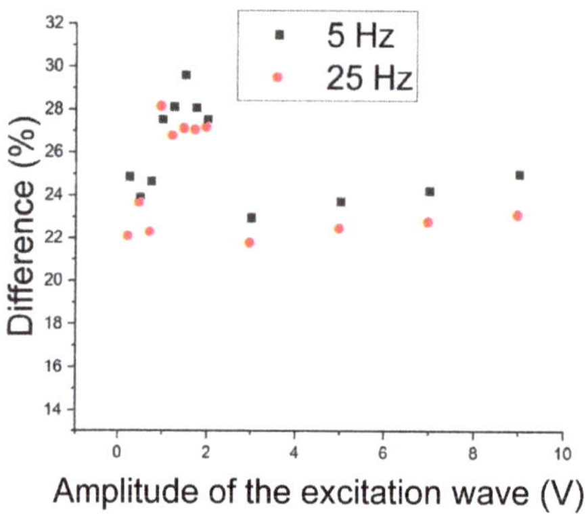

Figure 16. Differences in percentage between conditions without and with the sigma phase as a function of the amplitude of the 5 Hz and 25 Hz excitation waves.

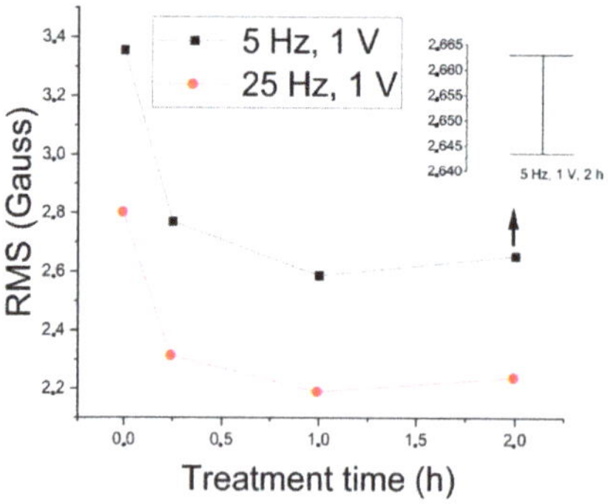

Figure 17. RMS variation of the first harmonic as a function of the treatment time for the 5 Hz and 25 Hz excitation waves. The 5 Hz and 1 V excitation wave error range is shown.

A correlation between the absorbed energy per impact and the amount of the sigma phase for the same studied plate showed an absorbed energy of 76.67 J for the as-received condition, 13.87 J for the amount of 5% of the sigma phase, and 10.2 J for the amounts of 17% and 18% of the sigma phase [2]; that is, there is a quick drop to 5% and then a stabilization around 10 J. The values of 5%, 17%, and 18% correspond to the amounts of the sigma phase for the treatment times of ¼ h, 1 h, and 2 h, respectively. This indicates that the variation in the first harmonic RMS directly correlates with the variation in energy absorbed by impact and serves as a parameter to follow the formation of the harmful sigma phase in the studied stainless steel. Correlations between the amount of the sigma phase and the energy absorbed per impact have shown the same behavior in the literature [1,35].

In the present work, a study of detecting the formation of sigma precipitates was carried out through an electromagnetic test with the replacement of the pick-up coil by a magnetic field sensor. The magnetic noise and the first harmonic amplitude variation for detecting harmful sigma precipitate were analyzed. One can note that both parameters detect the constituent studied, and the best results were obtained with the harmonic analysis. It was also verified that the best test amplitude of the excitation wave to detect the presence of the sigma phase, without interference from paramagnetic austenite, is 1 V for frequencies of 5 Hz and 25 Hz.

4. Conclusions

In this work, a study was carried out on applying magnetic noise and the amplitude of the first harmonic of the excitation wave to detect the formation of the harmful constituent sigma in duplex stainless steel. An electromagnetic test was applied with the replacement of the pick-up coil by a magnetic field sensor. The following conclusions were obtained:

- The applied electromagnetic test detected the presence of the sigma phase, with the replacement of the pick-up coil by the Hall effect sensor.
- The RMS values of the samples with the presence of the sigma phase were lower than those of the condition without precipitate, and the graphs of the RMS values as a function of the amplitude of the excitation wave showed three regions with different behaviors.
- The reduction in the RMS values in the first region occurred probably due to the movement of magnetic domain walls being blocked by the presence of the sigma phase. On the other hand, the increase in the magnetic flux in region II led to a greater detection of the phase precipitates and their contribution to the blockage. However, for region III, there was an increase in noise generated by the movement of the walls, which tried to overcome the blocking effect of the sigma constituent.
- Applying waves with the studied frequencies detected the presence of the sigma constituent, with the best results being for excitation waves with amplitudes from 1 V to 2 V, both by magnetic noise and the amplitude of the first harmonic analysis.
- The analysis, both by magnetic noise and the amplitude of the first harmonic, showed the presence of a plateau between 1 V and 2 V, which was attributed to the increase in the magnetic flux becoming influenced by the presence of paramagnetic austenite.
- The best results for detecting the sigma phase were obtained by analyzing the amplitude of the first harmonic, which was used to follow the constituent sigma formation and proved its effectiveness.

Author Contributions: Conceptualization, J.S., E.S., A.S., R.L., J.L., V.A.S., and J.M.R.S.T.; methodology, J.S., E.S., A.S., R.L., and J.L.; formal analysis, J.S., E.S., A.S., R.L., V.A.S., and J.M.R.S.T.; investigation, J.S., E.S., A.S., and R.L.; resources, J.S., A.S., and R.L.; data curation, J.S., A.S., and R.L.; writing—original draft preparation, J.S., E.S., A.S., and R.L.; writing—review and editing, E.S., J.L., V.A.S., and J.M.R.S.T.; visualization, J.S., E.S., V.A.S., and J.M.R.S.T.; supervision, E.S. and J.M.R.S.T.; project administration, E.S. and J.L.; funding acquisition, J.S., E.S., and J.L. All authors have read and agreed to the published version of the manuscript.

Funding: This research was funded by IFPB, UFPB, CAPES, and CNPq.

Institutional Review Board Statement: Not applicable.

Informed Consent Statement: Not applicable.

Data Availability Statement: The data supporting this study's findings are available from the first author, João Silva, upon reasonable request.

Acknowledgments: The authors gratefully acknowledge the financial support for this research work provided by the Brazilian Science Foundations CNPq and CAPES, and by the IFPB and UFPB institutions.

Conflicts of Interest: The authors declare no conflicts of interest.

References

1. Tavares, S.S.M.; Pardal, J.M.; Guerreiro, J.L.; Gomes, A.M.; Da Silva, M.R. Magnetic detection of sigma phase in duplex stainless steel UNSS 31803. *J. Magn. Magn. Mater.* **2010**, *322*, 29–33. [CrossRef]
2. Chen, T.H.; Weng, K.L.; Yang, J.R. The effect of high temperature exposure on the microstructural stability and toughness property in a 2205 duplex stainless steel. *Mater. Sci. Eng.* **2002**, *A338*, 259–270. [CrossRef]
3. Silva, E.M.; Leite, J.P.; Neto, F.A.F.; Leite, J.P.; Fialho, W.M.L.; Albuquerque, V.H.C.; Tavares, J.M.R.S. Evaluation of the Magnetic Permeability for the Microstructural Characterization of a Duplex Stainless Steel. *J. Test. Eval.* **2016**, *44*, 1106–1111. [CrossRef]
4. Normando, P.G.; Moura, E.P.; Souza, J.; Tavares, S.S.M.; Padovese, L.R. Ultrasound, eddy current and magnetic Barkhausen noise as tools for sigma phase detection on a UNS S31803 duplex stainless steel. *Mater. Sci. Eng. A* **2010**, *527*, 2886–2891. [CrossRef]
5. Silva, E.M.; Leite, J.P.; Leite, J.P.; Fialho, W.M.L.; de Albuquerque, V.H.C.; Tavares, J.M.R. Induced Magnetic Field Used to Detect the Sigma Phase of a 2205 Duplex Stainless Steel. *J. Nondestruct. Eval.* **2016**, *35*, 28. [CrossRef]
6. Silva, E.M.; de Albuquerque, V.H.C.; Leite, J.P.; Varela, A.C.G.; Moura, E.P.; Tavares, J.M.R.S. Phase transformations evaluation on a UNS S31803 duplex stainless steel based on nondestructive testing. *Mater. Sci. Eng. A* **2009**, *516*, 126–130. [CrossRef]
7. Huallpa, E.A.; Monlevade, E.F.; Capó Sánchez, J.; Campos, M.A.; Padovese, L.; Goldenstein, H. Use of Magnetic Barkhausen Noise (MBN) to Follow Up the Formation of Sigma Phase in SAF 2205 (UNS S31803) Duplex Stainless Steel. *Mater. Res.* **2016**, *19*, 1008–1016. [CrossRef]
8. Silva, E.M.; Paula, A.M.R.; Leite, J.P.; Andrade, L.S.S.; Albuquerque, V.H.C.; Tavares, J.M.R.S. Detection of the Magnetic Easy Direction in Steels Using Induced Magnetic Fields. *Metals-Open Access Metall. J.* **2016**, *6*, 317–328. [CrossRef]
9. Ortiz, P.M.; Benitez, J.A.P.; Hernandez, J.H.E.; Caleyo, F.; Mehboob, N.; Grossinger, R.; Hallen, J.M. Influence of the maximum applied magnetic field on the angular dependence of Magnetic Barkhausen Noise in API5L steels. *J. Magn. Magn. Mater.* **2016**, *401*, 108–115. [CrossRef]
10. Tavares, S.S.M.; Noris, L.F.; Pardal, J.M.; Silva, M.R. Temper embrittlement of supermartensitic stainless steel and non-destructive inspection by magnetic Barkhausen noise. *Eng. Fail. Anal.* **2019**, *100*, 322–328. [CrossRef]
11. Ducharne, B.Y.A.; Tene Deffo, P.; Tsafack, S.H.; Nguedjang, K. Directional magnetic Barkhausen noise measurement using the magnetic needle probe method. *J. Magn. Magn. Mater.* **2021**, *519*, 167453. [CrossRef]
12. He, Y.; Mehdi, M.; Hilinskiedrisy, E.J.A. Through-process characterization of local anisotropy of Non-oriented electrical steel using magnetic Barkhausen noise. *J. Magn. Magn. Mater.* **2018**, *453*, 149–162. [CrossRef]
13. Kahrobaee, S.; Hejazi, T.H.A. RSM-based predictive model to characterize heat treating parameters of D2 steel using combined Barkhausen noise and hysteresis loop methods. *J. Magn. Magn. Mater.* **2017**, *433*, 131–140. [CrossRef]
14. Stupakov, A.; Pereverton, A.; Neslusan, M. Reading depth of the magnetic Barkhausen noise. I. One-phase semi-hard ribbons. *J. Magn. Magn. Mater.* **2020**, *513*, 167086. [CrossRef]
15. Lin, Q.; Jiang, S.; Tian, H.; Ding, H.; Wang, W.; Guo, J.; Liu, Q. Study on non-destructive testing of rail rolling contact fatigue crack based on magnetic barkhausen noise. *Wear* **2023**, *528–529*, 204965. [CrossRef]
16. Omae, K.; Yamazaki, T.; Oka, C.; Sakurai, J.; Hata, S. Stress measurement based on magnetic Barkhausen noise for thin films. *Microelect. Eng.* **2023**, *279*, 112057. [CrossRef]
17. Nebair, H.; Helifa, B.; Bensaid, S.; Zidelmel, S.; Lefkaier, I.K. Martensite morphology and volume fraction evaluation of dual-phase X70 steel using magnetic Barkhausen noise technique. *J. Magn. Magn. Mater.* **2022**, *555*, 169397. [CrossRef]
18. Astudillo, M.R.N.; Núñez, N.M.; Pumarega, M.I.L.; Ferrari, G.; Ruzzante, J.; Gómez, M. Study of martensite induced by deformation with Magnetic Barkhausen Noise technique. *J. Magn. Magn. Mater.* **2022**, *556*, 169454. [CrossRef]
19. Fagan, P.; Ducharne, B.; Daniel, L.; Skarlatos, A.; Domenjoud, M.; Reboud, C. Effect of stress on the magnetic Barkhausen noise energy cycles: A route for stress evaluation in ferromagnetic materials. *Mater. Sci. Eng. B* **2022**, *278*, 115650. [CrossRef]
20. Qian, Z.; Zeng, H.; Liu, H.; Ge, Y.; Cheng, H.; Huang, H. Effect of Hard Particles on Magnetic Barkhausen Noise in Metal Matrix Composite Coatings: Modeling and Application in Hardness Evaluation. *IEEE Trans. Magn.* **2022**, *58*, 2101412. [CrossRef]
21. Lewis, N.J.; Anderson, P.I.; Gao, Y.; Robinson, F. Development and application of measurement techniques for evaluating localised magnetic properties in electrical steel. *J. Magn. Magn. Mater.* **2018**, *451*, 495–501. [CrossRef]
22. Silva, R.; Kugelmeier, C.L.; Vacchi, G.S.; Martins Junior, C.B.; Dainezi, I.; Afonso, C.R.M.; Mendes Filho, A.A.; Rover, C.A.D. A comprehensive study of the pitting corrosion mechanism of lean duplex stainless steel grade 2404 aged at 475 °C. *Corro. Sci.* **2021**, *191*, 109738. [CrossRef]

23. İspirli, M.M.; Yilmaz, A.E. An investigation on characteristics of tracking failure in epoxy resin with harmonic and fractal dimension analysis. *Turkish J. Electr. Eng. Comput. Sci.* **2018**, *26*, 245–256. [CrossRef]
24. Ryu, K.S.; Park, J.S.; Nahm, S.H.; Yu, K.M.; Kim, Y.B.; Son, D. Nondestructive evaluation of aged 1Cr-1Mo-0.25 V steel by harmonic analysis of induced voltage. *J. Magn. Magn. Mater.* **2001**, *231*, 294–298. [CrossRef]
25. Kaleti, T.; Gur, C.H. Determination of surface residual stresses in carburised AISI 8620 steel by the magnetic Barkhausen noise method. *Bri. Ins. Non-Destr. Test.* **2020**, *62*, 416–421.
26. Matlack, K.H.; Kim, J.Y.; Jacobs, L.J.; Qu, J. Review of second harmonic generation measurement techniques for material state determination in metals. *J. Nondestruct. Eval.* **2015**, *34*, 273. [CrossRef]
27. Vedalakshmi, R.; Manoharan, S.P.; Song, H.; Palaniswamy, N. Application of harmonic analysis in measuring the corrosion rate of rebar in concrete. *Corros. Sci.* **2009**, *51*, 2777–2789. [CrossRef]
28. Liu, S.; Best, S.; Neild, S.; Croxford, A.; Zhou, Z. Measuring bulk material nonlinearity using harmonic generation. *NDT E Inter.* **2012**, *48*, 46–53. [CrossRef]
29. Lo, K.H.; Lai, J.K.L. Microstructural characterization and change in a.c. magnetic susceptibility of duplex stainless steel during spinodal decomposition. *J. Nucl. Mater.* **2011**, *401*, 143–148. [CrossRef]
30. Franco, F.A.; Padovese, L.R. Non-destructive scanning for applied stress by the continuous magnetic Barkhausen noise method. *J. Magn. Magn. Mater.* **2018**, *446*, 231–238. [CrossRef]
31. Serna-Giraldo, C.P.; Padovese, L.R. Monitoring of carbon steels welded joints by magnetic Barkhausen noise. *Rev. Chil. Ingen.* **2015**, *23*, 413–420. [CrossRef]
32. Noris, L.F.; Padovese, L.R.; Tavares, S.S.M.; Pardal, J.M. Continuous scanning technique with Barkhausen magnetic noise for carbon steel sheets. *Mater. Resear.* **2019**, *22*, e20180366. [CrossRef]
33. Heping, L.; Hu'er, S.; Bin, L.; Dazhao, L.I.; Fenger, S.; Yongtao, Z. Analysis of χ Phase Precipitation in Aged 2205 Duplex Stainless Steel. *J. Wuhan Univ. Technol.-Mater. Sci. Ed.* **2015**, *30*, 591–595.
34. Deng, Y.; Li, Z.; Chen, J.; Qi, X. The effects of the structure characteristics on Magnetic Barkhausen noise in commercial steels. *J. Magn. Magn. Mater.* **2018**, *451*, 276–282. [CrossRef]
35. Martins, M.; Forti, L.R.N. Effect of aging on impact properties of ASTMA890 Grade 1C super duplex stainless steel. *Mater. Charact.* **2008**, *59*, 162–166. [CrossRef]

materials

MDPI

Article

Generation of Mechanical Characteristics in Workpiece Subsurface Layers through Milling

Michael Storchak [1,2,*], Larysa Hlembotska [2] and Oleksandr Melnyk [2]

[1] Institute for Machine Tools, University of Stuttgart, Holzgartenstraße 17, 70174 Stuttgart, Germany
[2] Department of Mechanical Engineering, Zhytomyr Polytechnic State University, Chudnivska Str. 103, 10005 Zhytomyr, Ukraine; zid_gle@ztu.edu.ua (L.H.); o.l.melnyk@ztu.edu.ua (O.M.)
* Correspondence: michael.storchak@ifw.uni-stuttgart.de; Tel.: +49-711-685-83831

Abstract: The generation of mechanical characteristics in workpiece subsurface layers as a result of the cutting process has a predominant influence on the performance properties of machined parts. The effect of the end milling process on the mechanical characteristics of the machined subsurface layers was evaluated using nondestructive methods: instrumented nanoindentation and sclerometry (scratching). In this paper, the influence of one of the common processes of materials processing by cutting—the process of end tool milling—on the generation of mechanical characteristics of workpiece machined subsurface layers is studied. The effect of the end milling process on the character of mechanical property formation was evaluated through the coincidence of the cutting process energy characteristics with the mechanical characteristics of the machined subsurface layers. The total cutting power and cutting work in the tertiary cutting zone area were used as energy characteristics of the end milling process. The modes of the end milling process are considered as the main parameters affecting these energy characteristics. The mechanical characteristics of the workpiece machined subsurface layers were the microhardness of the subsurface layers and the total work of indenter penetration, determined by instrumental nanoindentation, and the maximum depth of indenter penetration, determined by sclerometry. Titanium alloy Ti10V2Fe3Al (Ti-1023) was used as the machining material. Based on the evaluation of the coincidence of the cutting process energy characteristics with the specified mechanical characteristics of the machined subsurface layers, the milling mode effect of the studied titanium alloy, in particular the cutter feed and cutting speed, on the generated mechanical characteristics was established.

Keywords: cutting; milling; subsurface layers; nanoindentation; sclerometry; indenter penetration work; indenter penetration depth

1. Introduction

The most common methods of ensuring the required service properties of parts for various machines and mechanisms are various chip formation–cutting processes. As a result of the tool's thermomechanical impact on the machined workpiece, certain mechanical properties are generated in the workpiece's subsurface layers [1,2]. These properties, evaluated in terms of the hardness parameters (including microhardness) [3], residual stresses [4,5], material structure [6,7] and others, have a decisive influence on the service properties of machined parts and their durability [8]. The formation patterns of these mechanical properties are essentially determined by the machining process conditions, which in turn are regulated by the used cutting modes. Numerous publications have been devoted to the study of the relationship between the formation regularities of mechanical properties of subsurface layers and the machining process conditions. In the past few decades, this research focus has also been supported by numerical modeling [9]. This significantly expands the possibilities and application area of this research field, as well as providing a significant reduction in the cost of experimental research. In particular, it is relevant in the study of mechanical property generation patterns in subsurface layers of complex-profile parts [10,11] with the use of spatial cutting processes. Spatial cutting processes are characterized by a significant variability in the contact conditions between the tool and the machined material, of which end milling is a striking representative, generating a significant gradient of mechanical properties in the subsurface layers, which is all the more aggravated in the case of hard-to-machine metals and alloys [12]. These circumstances cause significant difficulties both in evaluating the above mechanical properties and in establishing their relationship with the cutting process conditions. Significant support in the evaluation of mechanical characteristics is provided by micro- and nanometer methods, in particular, instrumental nanoindentation [13,14] and the sclerometry (scratching) of surfaces [15,16], ensuring the determination of integral mechanical characteristics, which are quite closely related to the conditions of their generation, in this case, the cutting processes [17].

The present study is devoted to the investigation of the relationship between the mechanical characteristics of subsurface layers generated during the end milling of ß-phase titanium alloy Ti10V2Fe3Al (Ti-1023) and the cutting process conditions.

2. Brief Description of the State of the Art on the Determination of Mechanical Characteristics from Machined Subsurface Layers

The most commonly measured mechanical characteristics of machined subsurface parts include microhardness, residual stress magnitude, and the microstructure of the machined material. These characteristics are also used to evaluate the mechanical properties of the workpiece subsurface layers of hard-to-machine metals and alloys subjected to milling [18]. Dai et al. [19] studied the hardening process of machined subsurface layers from the Inconel 718 workpiece. They established the effect of the cooling method in the cutting zone on work hardening. Analyzing the effect of cutting speed and tool feed on the work hardening of the same machined material was the focus of a study by Ren and Liu [20]. As a result, the optimal cutting modes were determined, providing the required hardening of machined subsurface layers. Investigating the cutting process of Inconel 718 at different tool rake angles and the machined material structure, Xu and colleagues [21] determined the relationship between the geometric parameters of the tool and the mechanical characteristics of the workpiece's machined subsurface layers. The deformation value of the subsurface layers and their hardening value were used as the studied mechanical characteristics. The study of the microhardness formation process in the subsurface layers during the micro milling of nickel alloy Inconel 718 depending on the cutting modes, in particular, cutting speed, cutter feed, and axial cutting depth, was undertaken by Lu et al. [22]. Xavior and colleagues [23] examined the formation mechanism of Inconel 718 hardening, as well as the formation of residual stresses in it and its microstructure under different tool materials and cutting conditions. The influence of alternative machining processes on surface integrity and the regularities of residual stress formation in the machined subsurface layers of alloy 718 are the subject of a study by Suárez et al. [24]. The formation of machined surface microhardness of titanium alloy Ti-6Al-4V is the subject of studies by Hou with Li [25] as well as Mathoho and coworkers [26]. Monka and colleagues [27] considered the influence of cutting modes and tool geometric parameters on the microhardness of the machined surface in orthogonal and oblique cutting. They obtained response surfaces and correlation dependences of microhardness on the above-studied parameters. The microhardness prediction of the milling machined subsurface layers is the subject of a study by Wang [28]. He used different methods of regression analysis in carrying out this process. The residual stress formation in the subsurface layers of such difficult-to-machine materials as duplex steel and titanium alloy Ti-6Al-4V as a result of the cutting process was studied by dos Santos and colleagues [29] and Rangasamy et al. [30]. A considerable amount of research on the mechanical characteristics of surface layers machined by cutting is devoted to the analysis of the machined material microstructure and the cutting process conditions' influence on it, as found in [31]. At the same time, the microstructure of the machined material was evaluated based on microhardness distribution. Thus Mendas et al. [32] and Ameri with colleagues [33] determined dislocation density by measuring microhardness distribution. Among other mechanical characteristics of the subsurface layers, Alijani et al. [34] studied the microstructure of titanium-nickel alloy after milling. As a result, they determined the effect of machining process conditions on the formation of the studied microstructure. The study by Chen and coworkers investigates the effect of cutting depth and the corresponding magnitude of plastic deformation of Inconel 690 nickel alloy subsurface layers as a result of milling [35]. Haddag and colleagues [36] studied the influence of cutting modes during the machining of titanium alloy Ti-6Al-4V, mainly the cutting speed and tool feed, on the formation of the machined subsurface layer structure. The need to determine the optimal cutting modes for the milling of nickel alloy Inconel 625, providing the necessary microstructure of the machined material, is reported in the study of da Silva et al. [37]. At the same time, the form parameters of the relative movement of the tool and the workpiece, which have the greatest influence on the microstructure of the machined material, were determined. Rajguru and Vasudevan [38] studied the effect of the Inconel 625 milling process on the microhardness of subsurface layers when machining without coolant with coated milling cutters. The influence of both the strain hardening and thermal softening of the machined surfaces was considered. It should be noted that in addition to evaluating the mechanical characteristics of the workpiece's subsurface layers, a significant part of the studies on the surface integrity of difficult-to-machine materials is devoted to investigating the microtopography and, consequently, the microgeometry of the machined surface and the influence of cutting modes and conditions on microtopography parameters (see, for example, [39]).

Further development of the methodological and instrumental base for evaluation methods of the physical and mechanical characteristics of surfaces contributed to the creation of nondestructive testing methods for various surfaces and, in particular, for machined subsurface layers of parts. In this way, micro- and nanometric methods were established, in particular, instrumented nanoindentation [40–42], and sclerometry (scratch test) [43,44]. The research and development carried out by Atkins and Tabor [45] served as a prototype for the creation of an instrumented nanoindentation method. Further improvement of the method, using the continuous penetration of the indenter into the test material through the development of new devices and algorithms for evaluating the measurement results [46,47], ensured the creation of the currently widely used method of instrumented nanoindentation (see, e.g., [48,49]). Almost simultaneously with the instrumented nanoindentation method, the sclerometry method was established [15]. This method ensures the qualitative evaluation of coatings [16,50], such as coatings of carbide inserts used as cutting elements of various tool types, by determining the contact adhesion of the coating with the substrate [51,52]. A part of the uncertainty in the estimation of the indenter load at the moment of the studied surface fracture [53] has been recently compensated for through the use of the multi-pass scratching method [54,55]. The improvement and further development of the instrumented nanoindentation method and sclerometry depend predominantly on the instruments and devices that enable the realization of these methods. Therefore, a significant part of the studies devoted to these nondestructive testing methods is related to the creation and improvement of devices. In this regard, it is necessary to mention the study of Li et al. [56], devoted to the improvement of the calibration method of the device for instrumented nanoindentation through the use of an optical interferometer; the study of Peng et al. [57], devoted to the creation of a device for the realization of nanoindentation on subsurface layers, providing a significant increase in the accuracy of measurements; the work of Ding and colleagues [58], devoted to the study of different calibration methods of measured values; and the study of Fritz and Kiener [59], evaluating the influence of the environment of the device on the measurement results. An important role in the process of the further development of instrumented nanoindentation and sclerometry methods is played by research aimed at improving the methodology for evaluating the determined mechanical properties of materials. Such studies include works devoted to improving the stability of measurement results [60], the selection of an optimization method for determining mechanical properties [61], the analysis of contact stiffness fluctuations in the studied materials [62], the influence of the indenter contact conditions with the tested material on the measurement results, and many others. In this connection, it is necessary to point out the study of Harsono et al. [63], devoted to the investigation of the friction effect in the specified contact; the study of Wang, on establishing the relationship between the indenter penetration modes and the friction value in the contact with the measured mechanical properties [64]; the work of Sivaram et al., devoted to the study of the relationship between pile-up or sink-in effects and the strain hardening of the tested material [65]; and the investigation of friction phenomena in contact with synthesized materials through instrumental nanoindentation and sclerometry by Farayibi and colleagues [66].

The development of the instrumented nanoindentation method was mainly aimed at determining the microhardness of subsurface layers (see, e.g., [67–69]), the deformation degree of variously formed parts (see, e.g., [50,58,70]), the microstructure of materials (see, e.g., [71–73]), and the residual stresses in the subsurface layers of specimens generated by their previous formation (see, e.g., [74–76]). The determination of the material mechanical properties using the sclerometry method was also used somewhat later to investigate the microstructure and hardening of materials by estimating the indenter penetration depth into the test material (see, e.g., [77–79]). Recently, the tool indentation method and the sclerometry method have been used not only to determine the above-mentioned individual mechanical characteristics of the parts' subsurface layers subjected to various machining processes but also to evaluate the integral (energy) mechanical characteristics [17,80]. In this regard, it is necessary to note the study of Bezyazychnyy et al. [81], devoted to the establishment of the relationship between cutting modes and physical and mechanical characteristics of the machined material, and the specific accumulated strain energy of this material along the workpiece depth. The relationship between the elastic and plastic components, as well as the total indentation energy and microhardness of the investigated specimens was the topic of the study of Yamamoto et al. [82]. The relationship between the mechanical properties of carbide metals and alloys and the cutting process conditions has been studied by Wang and coworkers [83] and Ren and Liu [20].

The analysis of the possibilities of the instrumented nanoindentation method and the sclerometry method shows the perspectives of these methods' application to the determination of the mechanical characteristics of specimens' subsurface layers, especially specimens from hard-to-machine materials, subjected to machining by cutting with essentially changing conditions of contact between the tool and the workpiece.

Simultaneously with the formation of a specified macro- and microgeometry of the machined part, the cutting process generates certain physical and mechanical characteristics in the machined subsurface layers of the workpiece. Under all other equal conditions, the value and distribution of these characteristics in the machined workpiece volume are mainly determined by the thermomechanical conditions of contact between the tool and the workpiece. In this regard, there are significant difficulties in determining these contact conditions in real three-dimensional cutting processes and, in particular, when machining difficult-to-machine materials. At the same time, such conditions are attracting the most interest from the industry. Taking into account this interest, a study of the influence of the milling process with end milling cutters on the formation of physical and mechanical characteristics of the workpiece subsurface layers was carried out, and titanium alloy was used as the machined material. This study is a continuation and methodological development of a previously published paper [17]. In the previous study [17], a coincidence of cutting power and cutting work with the mechanical characteristics of subsurface layers formed during the orthogonal cutting of structural steel was established. The cutting power and cutting work in the orthogonal cutting process were determined for different cutting speeds and tool rake angles. The mechanical characteristics of the machined subsurface layers were characterized by the total work of indenter penetration during tool nanoindentation and the maximum depth of indenter penetration during sclerometry.

3. Materials and Methods

The methodology for performing the present study to determine the physical and mechanical characteristics of the subsurface layers generated by the end milling process is explained by the scheme presented in Figure 1.

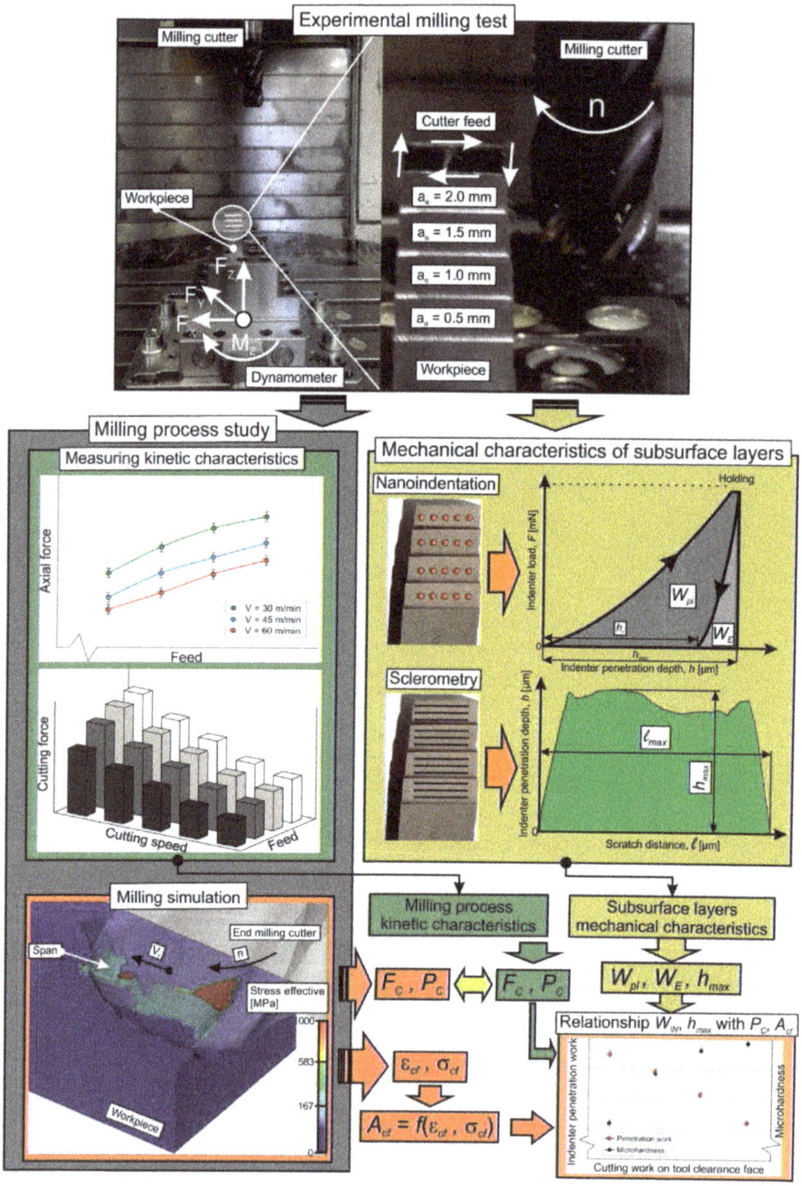

Figure 1. Methodology scheme for the determination of physical and mechanical characteristics of subsurface layers machined by milling.

The initial stage of determining the physical and mechanical characteristics of the workpiece subsurface layers machined by the end cutter is to carry out experimental studies of the cutting process characteristics. These characteristics include the set of characteristics by which the cutting process expresses itself externally. Typically, these include kinetic characteristics, such as the cutting force components and their power, thermal characteristics, such as cutting temperature, heat flows in the workpiece and tool, stresses and strains in the tool-machined material contact pair, etc. In the present study, the kinetic characteristics of cutting were determined, namely, the resultant cutting force F_C and the total cutting power P_C. The next step is devoted to measuring the mechanical characteristics of the milled subsurface layers of the workpiece. These characteristics include the microhardness of the machined layers, the indenter penetration total work, and the maximum depth of indenter penetration [17,55]. The first two mechanical characteristics are measured using instrumented nanoindentation, and the last characteristic is determined by the sclerometry of the workpiece machined surfaces. In parallel with the first two steps, a simulation of the end milling process is performed through a numerical cutting model. The adequacy of the milling numerical model is verified by comparing the measured and simulated values of cutting force and cutting power. After that, a simulation of the studied machining process is performed, as a result of which the stresses σ_{cf} and strains ε_{cf} of the machined material in the area of the tertiary cutting zone are determined. These stresses and strains are subsequently used to calculate the thermomechanical effect of the tool on the machined material A_{cf} in the specified cutting zone. This cutting work in the tertiary cutting zone A_{cf} mainly determines the physical and mechanical characteristics of the machined subsurface layers of the workpiece. The subsequent and last stage of the research methodology for determining the physical and mechanical characteristics of the subsurface layers generated by the end milling process is devoted to analyzing the coincidence of the mechanical characteristics measured at the previous stage with the energy characteristics of the milling process: the total cutting power P_C and the work of thermomechanical impact of the tool on the machined material A_{cf} in the tertiary cutting zone. The presented methodology of the integral mechanical characteristics evaluation in the workpiece subsurface layers formed as a result of the milling process and the study of the coincidence between these characteristics and the energy characteristics of the cutting process influencing them will provide a deeper understanding of the physical processes that generate the mechanical properties of the machined surfaces of parts [17].

3.1. Materials

The machining of the test material by end milling was carried out at the machining center UWF 1202 H by Hermle—Figure 1. Titanium alloy Ti10V2Fe3Al (Ti-1023) was used as the tested material. The machined material was vacuum-annealed before cutting. The chemical composition of titanium alloy Ti-1023 is specified in Table 1, and its mechanical and thermal properties are given in Table 2. The initial characteristics of the machined titanium alloy Ti-1023 specimen are shown in Figure 2. The initial metallographic microstructure of the machined specimen is shown in Figure 2a. Figure 2b illustrates the initial surface topography of the machined specimen. The arithmetic mean profile height of the specimen initial surface was R_a = 0.69 µm, and the maximum height of profile was R_z = 3.74 µm.

The description of the experimental setup, the measurement equipment for the cutting force and torque components, the used tool (carbide end cutter) and its geometrical parameters, and the initial geometry of the machined workpiece (tested specimens) are described in a previously published study [84]. Table 2 shows the mechanical and thermal properties of the carbide end cutter. The milling was performed on a rectangular track along the cross-section of the workpiece in four levels, with a varying radial depth of cut a_e of 0.5 mm, 1.0 mm, 1.5 mm, and 2.0 mm—Figure 1. The feed of the end milling cutter was changed from 0.06 mm/tooth to 0.12 mm/tooth in steps of 0.02 mm/tooth. The cutting speed V_C was varied in five levels: 30 m/min, 45 m/min, 60 m/min, 90 m/min, and 120 m/min. The reliability of experimental values was ensured by repeating each set of cutting modes at least 5 times. The confidence interval was chosen to be equal to 0.9. The choice of confidence interval was based on analyzing the scatter of the individual experimental values of the cutting forces. Since there were no significant differences between the individual measured cutting forces, the average value was used as a representative value of the measured data. The maximum uncertainty in measuring the cutting forces was no more than 11%.

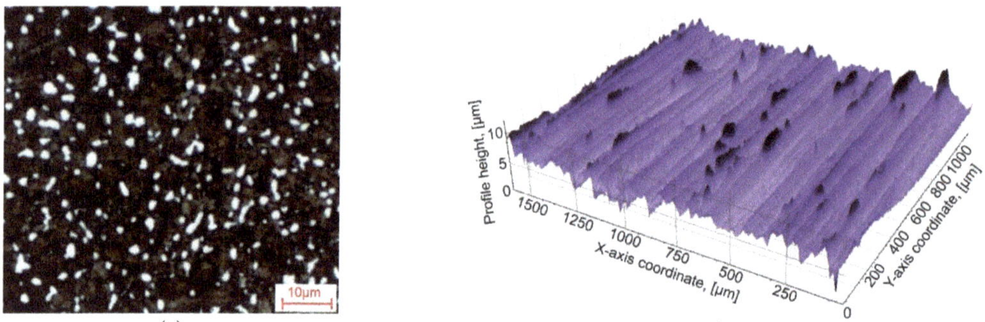

(a) (b)

Figure 2. Initial characteristics of the machined titanium alloy Ti-1023 specimen: (**a**) initial metallographic microstructure and (**b**) initial surface topography of the machined specimen.

Table 1. Chemical composition of titanium alloy Ti10V2Fe3Al [85,86].

Material	Ti	Al	V	Fe	C	N	H	O	Other
Ti10V2Fe3Al	82.86–86.8%	2.6–3.4	9.0–11%	1.6–2.2%	<0.05%	<0.05%	<0.015%	<0.13%	≤0.3%

Table 2. Mechanical and thermal properties of titanium alloy Ti10V2Fe3Al and the milling cutter [85,87].

Material	Strength (MPa)		Elastic Modulus (GPa)	Elongation (%)	Hardness	Poisson's Ratio	Specific Heat (J/kg·K)	Thermal Expansion (µm/m·°C)	Thermal Conductivity (W/m·K)
	Tensile	Yield							
Ti10V2Fe3Al	1282	1220	110	4–10	HV 430	0.35	527	9.7	7.0
Milling cutter	-	-	650	-	HV 1550	0.25	251	-	59

The influence of the milling process on the mechanical characteristics of the machined workpieces' subsurface layers was evaluated through the instrumented nanoindentation and sclerometry of the machined surfaces. The mechanical properties of the milled subsurface layers of the workpiece were evaluated through instrumented indentation using a Fischer Picodentor HM500 measuring system—Figure 3a. The instrumented indentation was performed on the milled surfaces of the specimen with different radial depths of cut a_e—Figure 3b. The measurements were performed with a Berkovich indenter at a maximum force on the indenter equal to 450 mN. The load applied to the indenter was recorded using a force sensor. The accuracy of the indenter load was 0.02 mN, and the accuracy of the indenter depth measurement was 5 nm. At the same time, the depth of indenter penetration into the specimen was recorded by the position-measuring system. The change rate of the indenter load was 20 mN/s. Once the maximum indenter load was reached, the indenter was maintained at this load and then unloaded (see diagram in Figure 3b). The delay time was 5 s. The instrumented nanoindentation process was repeated at least 10 times for each set value of cutting speed V_C, cutter feed f, and radial depth of cut a_e. According to the instrumented nanoindentation diagram obtained as a test result, the Vickers HV microhardness of the subsurface layers, the indenter elastic penetration work W_E, its plastic penetration work $W_{pl,}$ and total penetration work W_{IN} were determined (see the diagram in Figure 3b). The determined values of the total indenter penetration energy were averaged over the tests performed. In this case, the measurement's largest error was no more than 10%.

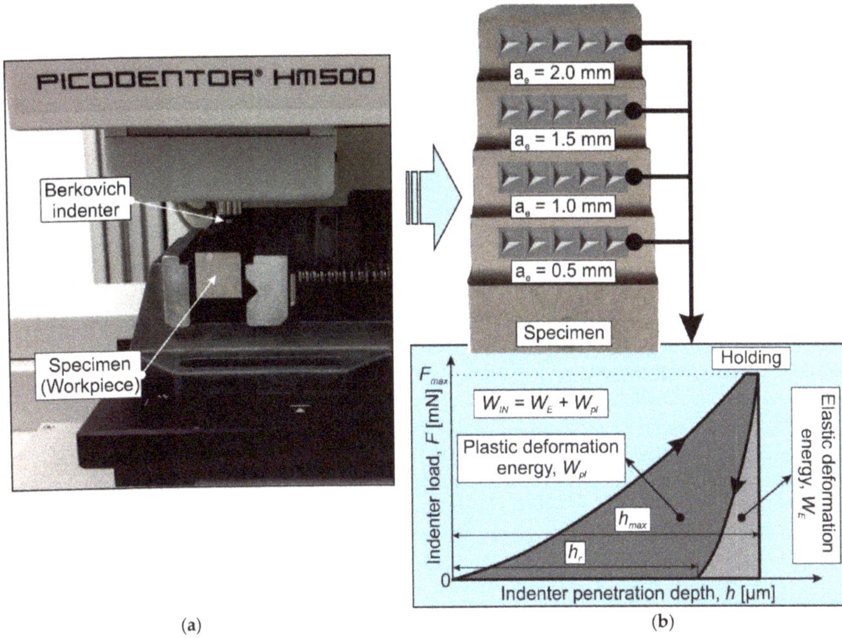

Figure 3. Experimental setup for measurements via instrumental nanoindentation and measurement scheme: (**a**) experimental setup for nanoindentation; (**b**) specimen image and instrumented indentation diagram.

The mechanical property evaluation of the milled subsurface layers was performed through sclerometry using the "Micron-gamma" device [17,55] at a Berkovich indenter load of 100 mN—Figure 4.

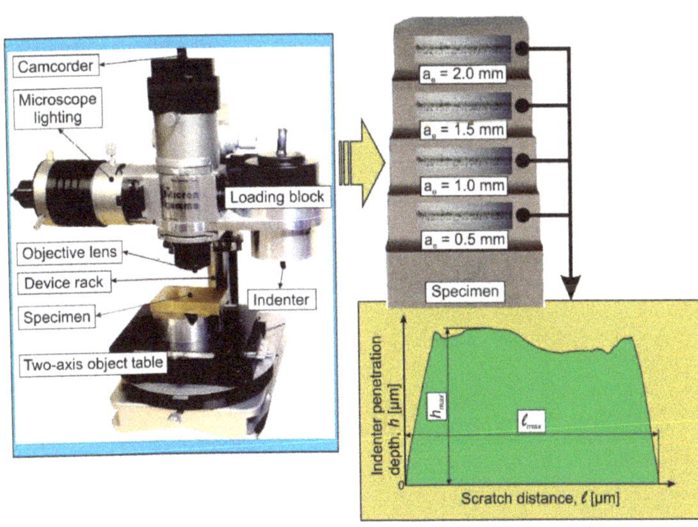

Figure 4. Experimental setup for sclerometry and its measurement scheme: (**a**) experimental setup for sclerometry of machined surfaces; (**b**) specimen image and sclerometry diagram.

In this case, the indenter displacement velocity was 20 μm/s and the scratch length was about 630 μm. The Berkovich indenter was set so that the projection of one of its edges was parallel to the velocity vector of the indenter movement. Sclerometry tests were repeated at least 8 times for each set value of cutting speed V_C, cutter feed f, and radial depth of cut a_e. The error of the averaged values of the maximum indenter penetration depth along the scratch length (see the sclerometry diagram in Figure 4b) did not exceed 11%. To perform scratch analysis on the "Micron-gamma" device, an optical profilometer "Micron-alpha" was used [17,55]. The vertical resolution of the microtopography images of the machined surfaces obtained on this device was about 2 nm.

3.2. Methods

In the first part of the study on the mechanical property generation of the machined subsurface layers [17], the relationship between the conditions of the machining process and the mechanical characteristics of the above-mentioned subsurface layers is shown. Similar to the previously studied process of orthogonal cutting (see [17,55]), the milling process generates certain physical and mechanical characteristics in the subsurface layers of the workpiece due to the elastic–plastic interaction of the end cutter with the machined material and its subsequent separation into chips and the machined surface of the workpiece with the fracture of this material. Unlike orthogonal cutting, in which the stress–strain state of the machined material corresponds to a plane (two-dimensional), in the milling process, as in other real cutting processes, the machined material is in a spatial (three-dimensional) stress–strain state [88–90]. However, both in the spatial process stress–strain state (for the considered case in the milling process), and in the two-dimensional stress–strain state, the mechanical properties of the machined subsurface layers are determined to a significant degree by the cutting process conditions. In this case, either the adiabatic hardening of the machined material or its isothermal softening is realized in the cutting zones [91]. These conditions are also determined by the thermomechanical interaction (contact) between the end cutter and the machined material in the cutting zones. At the same time, the influence of contact conditions was evaluated by the well-proven energy characteristics of the cutting process for this purpose [17]. Taking into account the complexity and multiplicity of physical processes in the cutting zones during the spatial process of milling, the contact conditions of the end cutter with the machined material were estimated using the total cutting power P_C and the plastic deformation work of the machined material in the tertiary cutting zone A_{cf}. Thus, the mechanical characteristics of the machined subsurface layers generated during end milling, namely, microhardness and indenter penetration work, determined using instrumented nanoindentation, and maximum indenter penetration depth, determined using sclerometry, were considered according to the total cutting power P_C and the plastic deformation work of the machined material in the tertiary cutting zone A_{cf}. This is postulated by the following two statements:

- *The thermomechanical interaction of the end cutter with the machined workpiece is evaluated using the total milling power and is proportional to the indenter penetration work in the workpiece machined surface, determined using the instrumented nanoindentation of the machined subsurface layers, and proportional to the maximum depth of the indenter penetration in the subsurface layers, determined using the sclerometry of the machined subsurface layers:*

$$\underset{\Re}{\forall} S_C \in \Re \overset{S_{Cn}}{\underset{S_{C1}}{\exists}} P_C \propto W_{IN} \vee P_C \propto h_{\max}, \tag{1}$$

  where $\Re$ is the existence space of cutting process states (conditions); S_C is the cutting process state; P_C is the total cutting power; W_{IN} is the total indenter penetration work through the instrumented nanoindentation of the milled surface; and h_{max} is the maximum indenter penetration depth during the sclerometry of the milled surface.
- *The thermomechanical interaction of the end cutter with the machined workpiece is evaluated through the plastic deformation work of the machined material in the tertiary cutting zone*

during milling and is proportional to the indenter penetration work in the workpiece machined surface, determined using the instrumented nanoindentation of the machined subsurface layers, and proportional to the maximum depth of the indenter penetration in the subsurface layers, determined using the sclerometry of the machined subsurface layers:

$$\underset{\Re}{\forall} S_C \in \Re \overset{S_{Cn}}{\underset{S_{C1}}{\exists}} A_{cf} \propto W_{IN} \vee A_{cf} \propto h_{\max}. \quad (2)$$

The total cutting power P_C was determined according to the dependence known from cutting theory [88,89,92]:

$$P_C = F_C \cdot V_C, \quad (3)$$

where F_C is the total cutting force, and V_C is the cutting speed.

The cutting force F_C was calculated as the resultant of the cutting force components measured during milling, as is presented in Figure 1 and Section 3.1. The plastic deformation work of the machined material in the tertiary cutting zone A_{cf} was determined as a dependence function of the equivalent stresses σ_{cf} acting in the machined material in the tertiary cutting zone on the strains ε_{cf} of the machined material in this zone:

$$A_{cf} = f(\sigma_{cf}, \varepsilon_{cf}) = V_m \cdot \int_{t_s}^{t_e} \sigma_{cf} d\varepsilon_{cf}, \quad (4)$$

where V_m is the material removal volume, and t_s and t_e are the simulation start and end times, respectively.

A determination of the dependence function (4) was performed by simulating the milling process of the titanium alloy Ti-1023 workpiece (see Section 3.2) with an end cutter using the previously developed numerical model of the milling process [84,86]. The cutting tool in these models was modeled as a perfectly rigid body and the workpiece material as an isotropic material defined by the Johnson–Cook constitutive equation [93,94]. The contact conditions between the tool and chip and between the tool and workpiece were specified via the Coulomb model [95]. Friction coefficients were determined according to a previously developed methodology [96]. In this case, the friction coefficient in the plastic area of the secondary cutting zone was F_{RFp} = 0.786, the friction coefficient in the elastic area of the secondary cutting zone was F_{RFe} = 0.405, and the friction coefficient in the tertiary cutting zone was F_{CF} = 0.623. The fracture mechanism of the machined material [97] was realized using the Cockcroft and Latham model [98]. The critical stress value of the Cockcroft and Latham model as well as the parameters of the constitutive equation were found through sensitivity analysis by DOE (Design of Experiment) [84,86].

A determination of stresses and strains of the machined material in the region of the tertiary cutting zone through a simulation of the milling process with the end cutter was performed using tracking points. The layout of tracking points is shown in Figure 5. The five tracking points (P_1, P_2, P_3, P_4, and P_5) located in the workpiece material in the tertiary cutting zone region were used. Before the plastic deformation work of the machined material in the tertiary cutting zone A_{cf} was calculated, the stresses and strains determined at the indicated points were averaged over all five points.

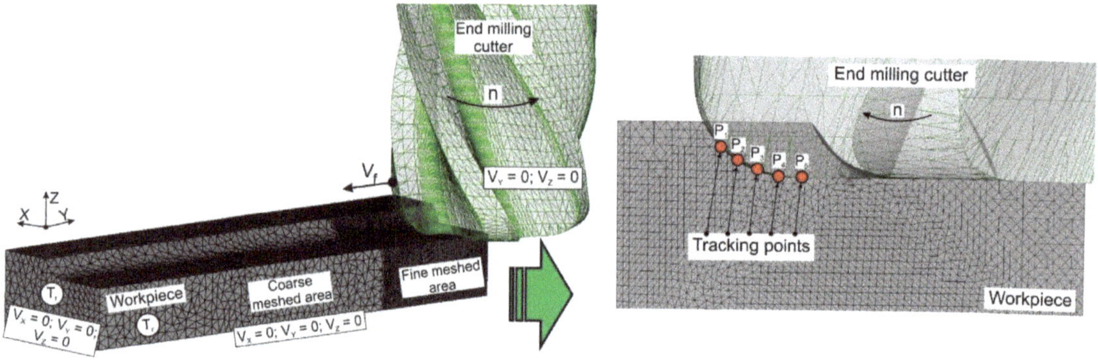

(**a**) (**b**)

Figure 5. Layout scheme of tracking points: (**a**) initial geometric model of milling with a mesh and boundary conditions; (**b**) location of tracking points.

4. Results and Discussion

The influence of the milling process on the mechanical characteristics of the machined material subsurface layers was evaluated using the value of the resulting cutting force F_C and cutting power P_C, and the thermomechanical effect of the end cutter on the machined material A_{cf} in the tertiary cutting zone (see Section 3.2). The dependence of the resultant milling force F_C calculated from the measured values of the cutting force components and cutting power P_C on the cutter feed and cutting speed is shown in Figure 6. The resulting cutting force increases proportionally as the cutter feed increases—Figure 6a.

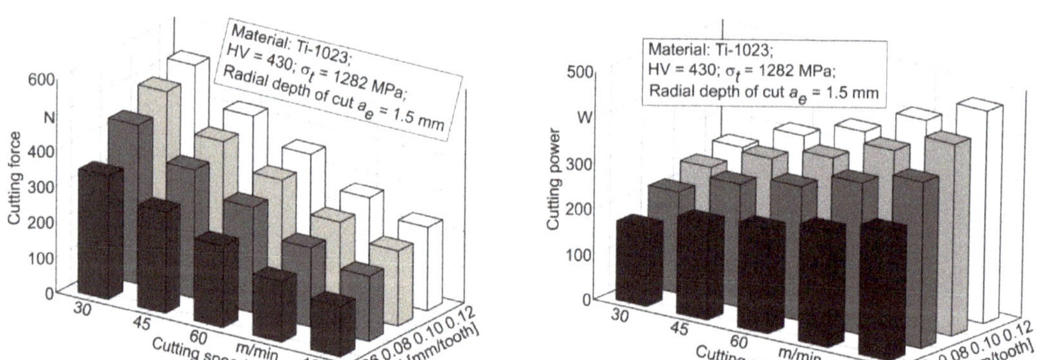

(**a**) (**b**)

Figure 6. Dependence of the resultant cutting force and cutting power on cutting speed and cutter feed: (**a**) cutting mode influence on the resultant cutting force; (**b**) cutting mode influence on cutting power.

The change in cutting force is explained with a proportional increase in the volume of removed material in chip form. At the same time, increasing the cutting speed leads to a significant reduction in the resultant cutting force—Figure 6a. This effect of cutting speed is explained by the predominant influence of machined material softening over its strain hardening [90,99,100]; in other words, the increase in cutting temperature caused by an increase in cutting speed softens the machined material to a greater extent than this material is hardened due to the speed factor. An increase in cutter feed also causes a corresponding increase in cutting power, as is presented in Figure 6b, due to the increased volume of workpiece material removed. In contrast to the effect on the resulting cutting force, an increase in cutting speed leads to a significant increase in cutting power. In all probability, this is caused by the numerical influence of cutting speed, since the value of cutting speed is included as a multiplier in the dependence of cutting power determination.

The determination of the thermomechanical work effect of the end cutter on the machined material A_{cf} in the tertiary cutting zone was performed, as announced in Section 3, by simulating the stresses in the machined material and their corresponding strains. The adequacy of the developed numerical model of titanium alloy Ti-1023 milling for the studied range of cutting modes was checked by comparing the experimentally determined values of the resulting cutting force F_C and cutting power P_C with the corresponding simulated values, as was presented in Section 3.

The results of this comparison, exemplarily for cutting speeds from 60 m/min to 120 m/min, and for a radial depth of cut a_e = 1.5 mm and an axial depth of cut a_p = 5 mm, are shown in Figure 7. Together with the cutting force and cutting power values, the deviation values between the experimental and simulated values are presented in the figure. The specified deviations for a cutting speed V_C = 60 m/min and the entire range of cutter feed variation lie between about 10.3% and about 17.3%, as is presented in Figure 7a,d. The corresponding deviations between the experimental values of cutting force and cutting power for a cutting speed V_C = 90 m/min and the entire range of cutter feed range lie between about 11.9% and about 19.3%, as is presented in Figure 7b,e, and for a cutting speed V_C = 120 m/min and the entire range of cutter feed range, they lie between about 13.9% and about 21.5%, as is presented in Figure 7c,f. Thus, it can be assumed that the numerical model of titanium alloy milling is able to adequately simulate the characteristics of the machining process used later to match them with the mechanical characteristics of the machined subsurface layers of the workpiece.

The stresses and strains of the machined material in the tertiary cutting zone region, which are further used to calculate the A_{cf} thermomechanical impact of the end cutter on the machined material, were determined using five tracking points (see the diagram in Figure 5 and Section 3.2) as a result of the milling process simulation. The variations in effective strain ε_{cf} and effective stress σ_{cf}, determined at five specified tracking points, by simulation time and the dependence of effective stress σ_{cf} on effective strain ε_{cf}, exemplarily for a cutting speed V_C = 60 m/min and a radial depth of cut a_e = 1.5 mm, are presented in Figure 8.

The variation in effective strain ε_{cf} and effective stress σ_{cf} by simulation time for the considered five tracking points placed in the tertiary cutting zone region is presented in Figure 8a,b. According to the results of the change in ε_{cf} and σ_{cf} their relationship is determined and presented in Figure 8c. This relationship together with the material volume value V_m removed as a result of the milling, determined for the studied range of cutting modes, was used to calculate the cutting work A_{cf} in the tertiary cutting zone (see Equation (4), Section 3). In this case, the volume V_m was determined from the nominal chip shape, without taking chip compression into account. This assumption was made on the basis that the chip compression ratio of the used Ti-1023 titanium alloy for the studied range of cutting modes does not exceed 1.1 [84,86]. The effect of cutter feed and cutting speed on the cutting work A_{cf} is presented exemplarily for a cutting speed V_C = 60 m/min and a radial depth of cut a_e = 1.5 mm in Figure 9. The cutting work A_{cf} in the tertiary cutting zone increases almost linearly as the cutter feed increases from 0.06 mm/tooth to 0.12 mm/tooth—Figure 9a. This increase in cutting work seems logical because as the cutter feed rate increases, the amount of material removed per unit of time increases. An increase in cutting speed in the studied range causes a monotonic increase in cutting work A_{cf}—Figure 9b. This increase in A_{cf} is most likely also a consequence of the increase in the material removal volume with increasing cutting speed.

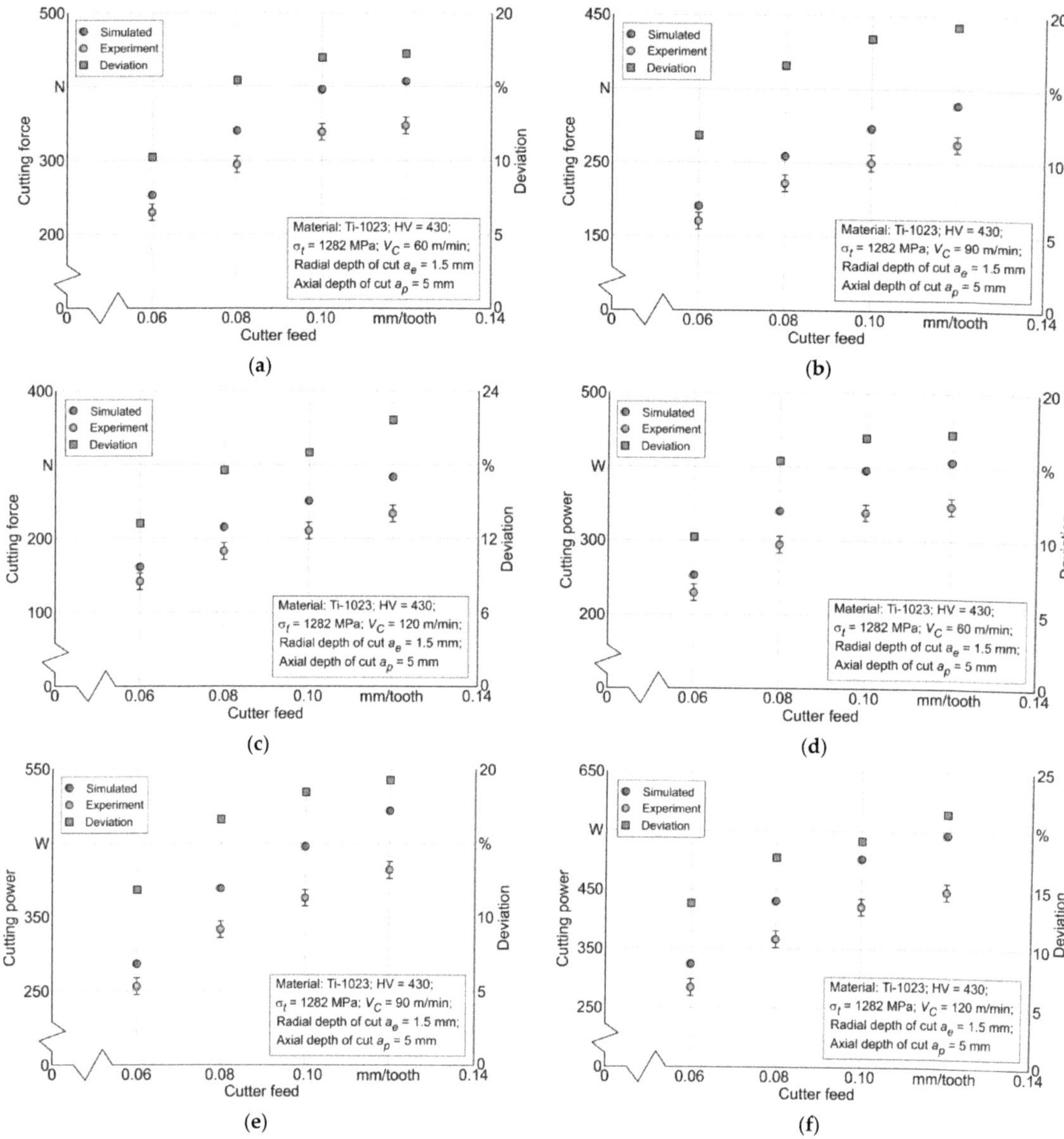

Figure 7. Comparison of experimental and simulated values of the resultant cutting force and cutting power with changing cutting modes: (**a**) cutting force dependence on cutter feed at cutting speed V_C = 60 m/min; (**b**) cutting force dependence on cutter feed at cutting speed V_C = 90 m/min; (**c**) cutting force dependence on cutter feed at cutting speed V_C = 120 m/min; (**d**) cutting power dependence on cutter feed at cutting speed V_C = 60 m/min; (**e**) cutting power dependence on cutter feed at cutting speed V_C = 90 m/min; and (**f**) cutting power dependence on cutter feed at cutting speed V_C = 120 m/min.

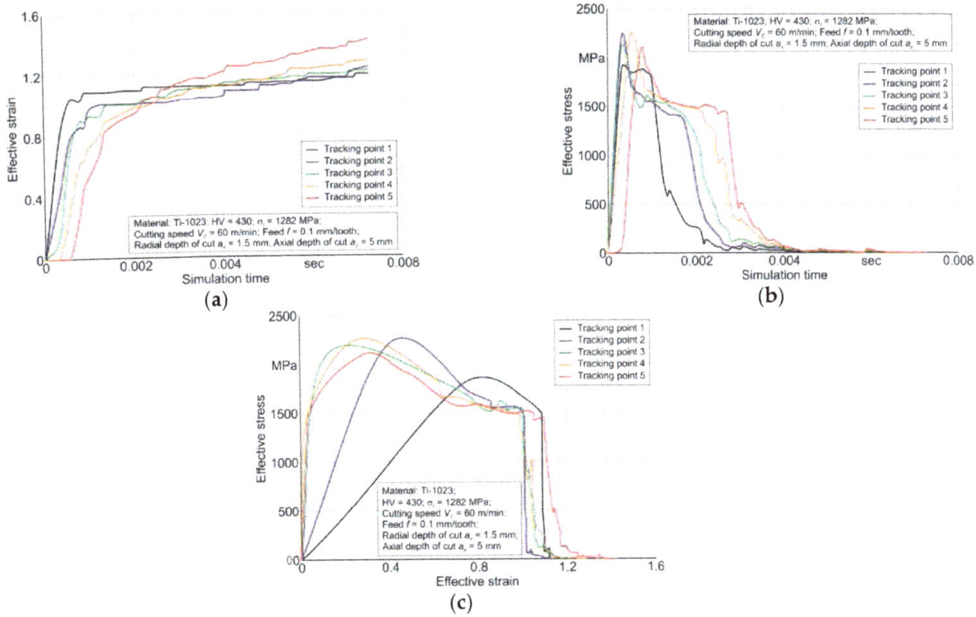

Figure 8. Change in effective stress and effective strain in the region of the tertiary cutting zone: (**a**) effective strain variation over the simulation time; (**b**) effective stress variation over the simulation time; (**c**) relationship between effective stress and effective strain for five tracking points.

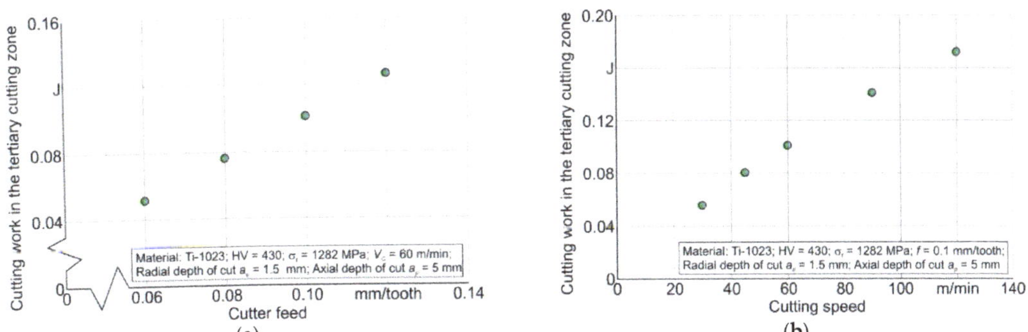

Figure 9. Effect of cutting modes on cutting work A_{cf} in the region of the tertiary cutting zone: (**a**) effect of cutter feed on cutting work; (**b**) effect of cutting speed on cutting work.

The values that predetermine the mechanical characteristics of the workpiece subsurface layers machined using end milling are the microhardness of these layers, the total indentation work, and the maximum indentation depth, as is presented in Figure 1 and Section 3. The first two values are determined by instrumented nanoindentation for a wide range of cutting conditions, as is presented in Figure 3 and Section 3.1. The results of the instrumented nanoindentation are exemplified by the radial depth of cut a_e = 1.5 mm in Figure 10. This figure demonstrates the effect of cutter feed and cutting speed on the microhardness of the milled subsurface layers and the total indenter penetration work. The microhardness of the machined subsurface layers monotonically increases with increasing cutter feed—Figure 10a. This effect of tool feed is logical because as the feed increases, the material removed volume per unit of time increases. The increase in the volume of removed material entails a corresponding increase in the strain degree of the machined subsurface layers of the workpiece, which leads in turn to the hardening of these layers and, naturally, to an increase in their microhardness [88,99,100]. At the same time, the microhardness of the subsurface layers decreases with increasing cutting speed—Figure 10a. In all probability, the reason for such an effect of cutting speed on microhardness is the increase in cutting temperature with increasing cutting speed, which entails the softening of subsurface layers [89,90,100]. The total indenter penetration work W_{IN} monotonically decreases with increasing cutter feed—Figure 10b. Such an effect of feed is quite understandable, since the consequence of increasing the cutter feed is (as already shown above) a corresponding increase in the volume of material removed per unit of time, entailing an increase in the strain degree of the machined material layers, and hence an increase in the hardening of these layers [89,90,92]. In turn, an increase in the hardening of the subsurface layers leads to a lower degree of indenter penetration into the studied material and, as a consequence, to a lower indenter penetration work W_{IN}. In this case, the consequence of increasing cutting speed is a monotonic increase in the indenter penetration work—Figure 10b. This increase is due to the softening of the machined subsurface layers of the workpiece caused by the increase in cutting temperature with increasing cutting speed [99,101]. The similar character of changes in the studied mechanical characteristics of subsurface layers is also observed at other values of the radial depth of cut a_e.

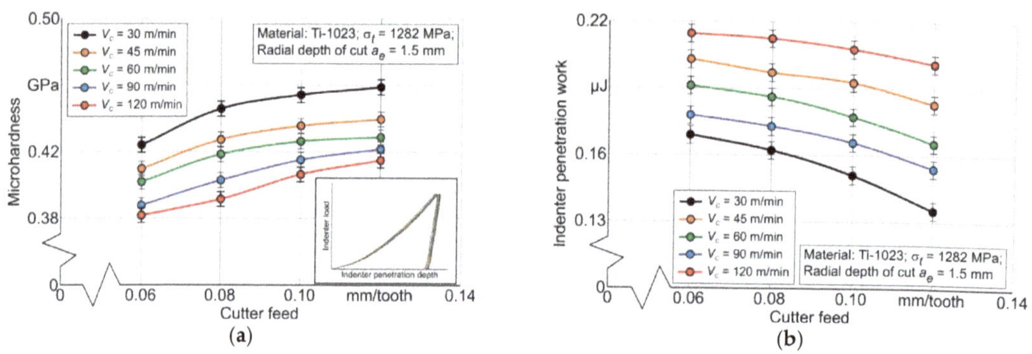

Figure 10. Effect of cutter feed and cutting speed on microhardness of machined subsurface layers of the workpiece and total indenter penetration work: (**a**) microhardness dependence on cutting modes; (**b**) indenter penetration work dependence on cutting modes.

The maximum indenter penetration depth was determined by the sclerometry of the specimen's milled surface, as is presented in Figure 4 and Section 3.1. The results of the sclerometry analysis are shown exemplarily for the radial depth of cut a_e = 1.5 mm in Figure 11.

The maximum depth of indenter penetration h_{max} into the milled surface as a result of sclerometry monotonically decreases with the cutter feed increase. This is due to the increased hardening of the workpiece subsurface layers as a result of machining with increasing cutter feed due to the corresponding increase in the volume of material removed per unit of time [99,100]. As a result of this increase in the hardening of the subsurface layers, the indenter penetration depth at constant load decreases [17]. At the same time, the maximum depth of indenter penetration increases with increasing cutting speed—Figure 11. This effect of cutting speed is a consequence of softening at the workpiece milled subsurface layers [88,89,99], caused by the increase in cutting temperature as a result of increasing cutting speed [89,101].

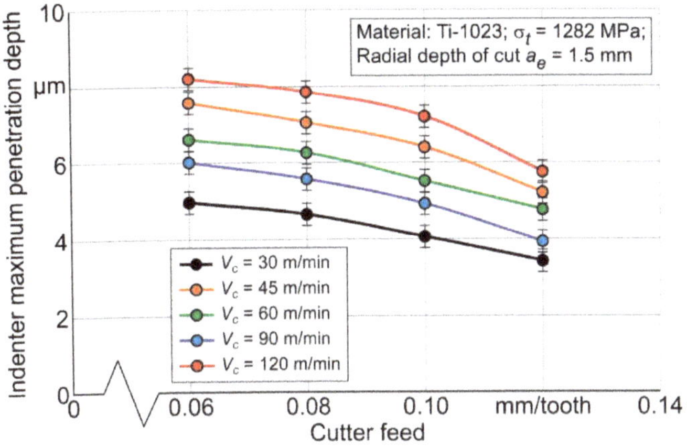

Figure 11. Indenter penetration depth dependence on cutter feed and cutting speed.

The evaluation of the mechanical characteristics of the milled subsurface layers, performed through instrumented nanoindentation and sclerometry with a significant change in cutting modes (cutter feed, cutting speed, and radial depth of cut), indicates the predominant effect of the milling process characteristics on the studied mechanical characteristics. This indicates the existence of a close correlation between the characteristics of the milling process and the mechanical characteristics of the machined subsurface layers. The presence of such a relationship makes it possible to compile the characteristics of the milling process with the studied mechanical characteristics according to the above-formulated postulates, as is presented in Section 3, Equations (1) and (2). To generalize these comparisons, the milling process characteristics should serve their energy values. In this case, such characteristics are cutting power and the work of thermomechanical interaction between the cutter and the workpiece in the tertiary cutting zone.

The coincidence results of both mechanical characteristics of the machined subsurface layers, indenter penetration work and microhardness, with cutting power P_C and cutting work A_{cf} in the tertiary zone are presented exemplarily for a cutting speed V_C = 60 m/min and a radial depth of cut a_e = 1.5 in Figure 12.

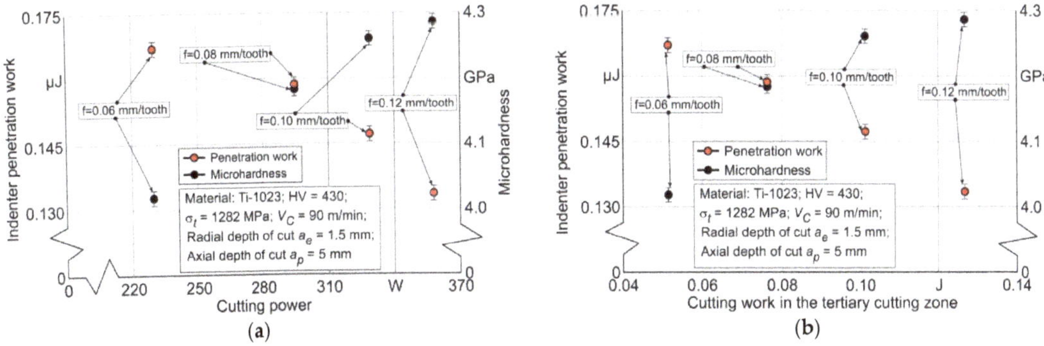

Figure 12. The coincidence of the indenter penetration work and microhardness with the cutting power and cutting work in the tertiary cutting zone: (**a**) depending on the cutting power; (**b**) depending on the cutting work in the tertiary cutting zone.

With increasing cutting power, the indenter penetration work W_{IN} monotonically decreases, and similarly, the microhardness of the machined subsurface layers of the workpiece increases with increasing cutter feed—Figure 12a. A similar effect on indenter penetration work and microhardness is due to the cutting work in the tertiary cutting zone region A_{cf}—Figure 12b. This influence effect can be explained by the fact that the increase in the interaction energy between the cutter and the machined material as a result of the increase in cutting power and cutting work in the tertiary cutting zone contributes to the hardening of the machined subsurface layers. The increased hardening of these layers inhibits indenter penetration, resulting in the observed increase in microhardness and the decrease in indenter penetration work.

When comparing the maximum indenter penetration depth h_{max} with the energy characteristics of the cutting process, with cutting power P_C and cutting work A_{cf} in the tertiary cutting zone, a monotonic decrease in h_{max} value is observed—Figure 13. This decrease occurs in conjunction with an increase in the indicated energy characteristics caused by an increase in cutter feed. The decrease in h_{max} with increasing cutting power P_C and cutting work A_{cf} in the tertiary cutting zone is also explained by the hardening of the milled subsurface layers of the workpiece. The increase in the hardening degree of these layers causes a decrease in the possibility of indenter penetration into the workpiece milled surface.

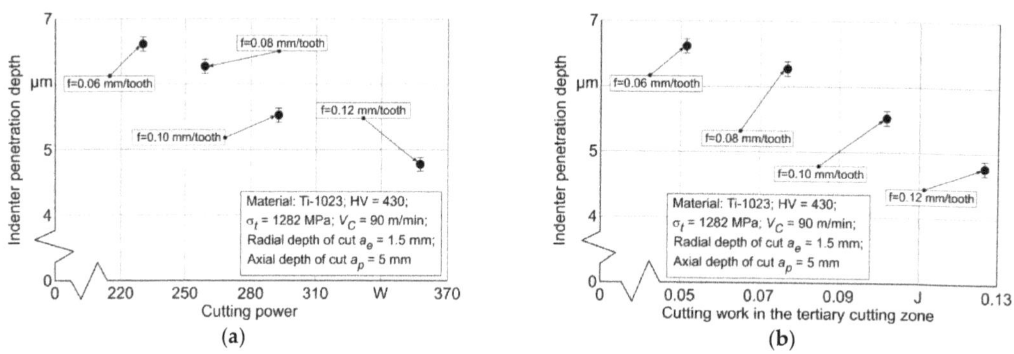

Figure 13. The coincidence of indenter penetration depth during sclerometry with the cutting power and cutting work in the tertiary cutting zone: (**a**) depending on the cutting power; (**b**) depending on the cutting work in the tertiary cutting zone.

The analysis of the coincidence of the mechanical characteristics from the milled subsurface layers with the characteristics of the cutting process makes it possible to confirm the above-formulated postulates about the significant influence of the considered energy characteristics of the end milling process on the studied integral mechanical characteristics. This creates the possibility of a purposeful selection of the contact conditions between the tool and the machined material through the assignment of cutting modes that ensure the necessary mechanical characteristics of the workpiece subsurface layers. The achievement of the specified mechanical characteristics of the machined by the milling subsurface layers may in turn enable the required service properties of machine parts.

5. Conclusions

The research performed here is devoted to the formation patterns of subsurface layers mechanical characteristics of titanium alloy Ti10V2Fe3Al (Ti-1023) workpieces as a result of their milling by an end milling cutter. The mentioned patterns are considered based on an analysis of the coincidence of the measured mechanical characteristics with the milling process characteristics. To generalize the studied patterns, integral characteristics were used as mechanical characteristics: indenter penetration work in the machined surface, the microhardness of these surfaces, and the maximum depth of indenter penetration. The measurements of the mentioned integral characteristics were carried out using nondestructive testing methods, namely, the instrumented nanoindentation and sclerometry of the studied subsurface layers. As generalized characteristics of the milling process, cutting power and the thermomechanical interaction work of the cutter with the machined material in the region of the tertiary cutting zone were used. To determine the cutting power of the milling process, the measured value of the resultant cutting force was used, and the thermomechanical interaction work of the cutter with the machined material was calculated using a numerical model of the milling process.

The following patterns of changes in the milling process kinetic characteristics of titanium alloy Ti-1023 and mechanical characteristics of the machined subsurface layers depending on the machining modes were established by experimental studies:

➢ The resulting cutting force monotonically increases with increasing cutter feed and decreases monotonically with increasing cutting speed;
➢ The microhardness of the workpiece subsurface layers monotonically increases with increasing cutter feed and decreases with increasing cutting speed;
➢ The indenter penetration work as a result of the instrumented nanoindentation of the workpiece subsurface layers decreases with increasing cutter feed and increases with increasing cutting speed;
➢ The indenter maximum penetration depth as a result of the sclerometry of the workpiece subsurface layers decreases with increasing cutter feed and increases with increasing cutting speed.

The machining modes' influence on the generalized characteristics of the cutting process is characterized by the following patterns:

➢ The cutting power increases both with increasing cutting speed and with increasing cutter feed;
➢ The cutting work in the tertiary cutting zone increases both with increasing cutter feed and with increasing cutting speed.

The regularities established as a result of experimental and simulation studies of the end tool milling process enable the possibility of identifying the coincidence of the machining process energy characteristics with the integral mechanical characteristics of the milled subsurface layers. The use of cutting process energy characteristics for this purpose provides a numerical characterization of machining technology as a way of generating mechanical characteristics of subsurface layers. This, in turn, ensures the possibility of analyzing the impact of the machining technology regardless of the technology used.

The analysis of the coincidence of the integral mechanical characteristics of the workpiece milled subsurface layers with the cutting process energy characteristics provides the possibility of a purposeful selection of the milling process conditions with the end tool of titanium alloy Ti 1023, conditioned by the appropriate choice of cutting modes. In turn, the possibility of such a choice enables the achievement of the required service properties of manufactured parts with mechanical characteristics of subsurface layers generated by the milling process.

The research direction presented in the paper is planned to be further developed by studying the coincidence of the characteristics of other machining processes with the mechanical characteristics of the machined subsurface layers. In addition, it is planned to expand the type of machined materials.

Author Contributions: Conceptualization, M.S.; methodology, M.S.; software, M.S.; validation, M.S., L.H. and O.M.; formal analysis, M.S.; investigation, M.S., L.H., and O.M.; resources, M.S. and O.M.; data curation, M.S. and L.H.; writing—original draft preparation, M.S.; writing—review and editing, M.S.; visualization, M.S. and L.H.; project administration, M.S and O.M.; funding acquisition, M.S. All authors have read and agreed to the published version of the manuscript.

Funding: This study was funded by the German Research Foundation (DFG) in the project HE-1656/153-1 "Development of a Concept for Determining the Mechanical Properties of the Cutting Material in Machining".

Institutional Review Board Statement: Not applicable.

Informed Consent Statement: Not applicable.

Data Availability Statement: Data are contained within the article.

Acknowledgments: The authors would like to thank the German Research Foundation (DFG) for their support, which is highly appreciated.

Conflicts of Interest: The authors declare no conflicts of interest. The funders had no role in the design of the study; in the collection, analyses, or interpretation of data; in the writing of the manuscript; or in the decision to publish the results.

References

1. Sagapuram, D.; Udupa, A.; Viswanathan, K.; Mann, J.B.; M'saoubi, R.; Sugihara, T.; Chandrasekar, S. On the Cutting of Metals: A Mechanics Viewpoint. *ASME J. Manuf. Sci. Eng.* **2020**, *142*, 110808. [CrossRef]
2. Guo, Y.; Saldana, C.; Compton, W.D.; Chandrasekar, S. Controlling deformation and microstructure on machined surfaces. *Acta Mater.* **2011**, *59*, 4538–4547. [CrossRef]
3. Bag, R.; Panda, A.; Sahoo, A.K.; Kumar, R. A Perspective Review on Surface Integrity and Its Machining Behavior of AISI 4340 Hardened Alloy Steel. *Mater. Today Proc.* **2019**, *18*, 3532–3538. [CrossRef]
4. Soori, M.; Arezoo, B. A Review in Machining-Induced Residual Stress. *J. New Technol. Mater.* **2022**, *12*, 64–83. Available online: https://hal.science/hal-03679993 (accessed on 11 February 2024).
5. Saptaji, K.; Afiqah, S.; Ramdan, R. A Review on Measurement Methods for Machining Induced Residual Stress. *Indonesian J. Comput. Eng. Des. (IJoCED)* **2019**, *1*, 106–120. [CrossRef]
6. Pan, Z.; Feng, Y.; Liang, S. Material microstructure affected machining: A review. *Manuf. Rev.* **2017**, *4*, 5. [CrossRef]
7. Luo, L.; Pang, J.; Song, Y.; Liu, S.; Yin, G.; Peng, H.; Pu, C.; Lin, Y.; Li, J.; Shi, X. Microstructure Evolution Mechanism of AISI 1045 Steel under High Speed Deformation. *Arch. Metall. Mater.* **2023**, *68*, 1525–1531. [CrossRef]
8. M'Saoubi, R.; Outeiro, J.C.; Chandrasekaran, H.; Dillon, O.W., Jr.; Jawahir, I.S. A review of surface integrity in machining and its impact on functional performance and life of machined products. *Int. J. Sustain. Manuf.* **2008**, *1*, 203–236. [CrossRef]
9. Malakizadi, A.; Bertolini, R.; Ducobu, F.; Kilic, Z.; Magnanini, M.C.; Shokrani, A. Recent advances in modelling and simulation of surface integrity in machining—A review. *Procedia CIRP* **2022**, *115*, 232–240. [CrossRef]
10. Davim, J.P. *Machining of Complex Sculptured Surfaces*; Springer: London, UK, 2012; 258p. [CrossRef]
11. Babichev, D.; Storchak, M. Synthesis of cylindrical gears with optimum rolling fatigue strength. *Prod. Eng. Res. Dev.* **2015**, *9*, 87–97. [CrossRef]
12. Ulutan, D.; Ozel, T. Machining induced surface integrity in titanium and nickel alloys: A review. *Int. J. Mach. Tools Manuf.* **2011**, *51*, 250–280. [CrossRef]
13. Kanaev, A.T.; Ramazanova, Z.M.; Biizhanov, S.K. Study of plasma-hardened wheel steel using nanoindentation. Industrial laboratory. *Diagn. Mater.* **2020**, *86*, 56–60. (In Russian) [CrossRef]
14. Oila, A.; Bull, S.J. Nanoindentation testing of gear steels. *Int. J. Mater. Res.* **2003**, *94*, 793–797. [CrossRef]
15. Randall, N.X. The current state-of-the-art in scratch testing of coated systems. *Surf. Coat. Technol.* **2019**, *380*, 125092. [CrossRef]
16. Li, J.; Beres, W. Scratch Test for Coating/Substrate Systems—A Literature Review. *Can. Metall. Q.* **2007**, *46*, 155–173. [CrossRef]
17. Storchak, M. Mechanical Characteristics Generation in the Workpiece Subsurface Layers through Cutting. *Crystals* **2023**, *13*, 761. [CrossRef]
18. Sinha, M.K.; Pal, A.; Kishore, K.; Singh, A.; Archana; Sansanwal, H.; Sharma, P. Applications of sustainable techniques in machinability improvement of superalloys: A comprehensive review. *Int. J. Interact. Des. Manuf.* **2023**, *17*, 473–498. [CrossRef]
19. Dai, X.; Zhuang, K.; Pu, D.; Zhang, W.; Ding, H. An Investigation of the Work Hardening Behavior in Interrupted Cutting Inconel 718 under Cryogenic Conditions. *Materials* **2020**, *13*, 2202. [CrossRef]
20. Ren, X.; Liu, Z. Influence of cutting parameters on work hardening behavior of surface layer during turning superalloy Inconel 718. *Int. J. Adv. Manuf. Technol.* **2016**, *86*, 2319–2327. [CrossRef]
21. Xu, D.; Ding, L.; Liu, Y.; Zhou, J.; Liao, Z. Investigation of the Influence of Tool Rake Angles on Machining of Inconel 718. *J. Manuf. Mater. Process.* **2021**, *5*, 100. [CrossRef]
22. Lu, X.; Jia, Z.; Wang, H.; Feng, Y.; Liang, S.Y. The effect of cutting parameters on micro-hardness and the prediction of Vickers hardness based on a response surface methodology for micro-milling Inconel 718. *Measurement* **2019**, *140*, 56–62. [CrossRef]
23. Xavior, A.; Manohar, M.; Madhukar, P.M.; Jeyapandiarajan, P. Experimental investigation of work hardening, residual stress and microstructure during machining Inconel 718. *J. Mech. Sci. Technol.* **2017**, *31*, 4789–4794. [CrossRef]
24. Suárez, A.; Veiga, F.; Polvorosa, R.; Artaza, T.; Holmberg, J.; de Lacalle, L.L.; Wretland, A. Surface integrity and fatigue of non-conventional machined Alloy 718. *J. Manuf. Process.* **2019**, *48*, 44–50. [CrossRef]
25. Hou, G.; Li, A. Effect of Surface Micro-Hardness Change in Multistep Machining on Friction and Wear Characteristics of Titanium Alloy. *Appl. Sci.* **2021**, *11*, 7471. [CrossRef]
26. Mathoho, I.; Akinlabi, E.T.; Mybiayi, M.P.; Mbohwa, C. Effect of milling parameters on microhardness and microstructure during dry and flood milling of Ti-6Al-4V. *IOP Conf. Ser. Mater. Sci. Eng.* **2018**, *423*, 012161. [CrossRef]

27. Monka, P.P.; Monkova, K.; Vasina, M.; Kubisova, M.; Korol, M.; Sekerakova, A. Effect of Machining Conditions on Temperature and Vickers Microhardness of Chips during Planing. *Metals* **2022**, *12*, 1605. [CrossRef]
28. Wang, X. Intelligent Prediction of Surface Micro-hardness after Milling Based on Smooth Support Vector Regression. In Proceedings of the International Symposium on Knowledge Acquisition and Modeling, Wuhan, China, 21–22 December 2008; pp. 728–731. [CrossRef]
29. Dos Santos, C.E.; Carneiro, J.R.G.; da Silva, G.C.; Brito, P.P.; dos Santos, B.; Campos, T.R. Residual stress and surface microhardness post-milling in 2205 duplex steel. *Int. J. Adv. Manuf. Technol.* **2021**, *113*, 3445–3455. [CrossRef]
30. Rangasamy, N.; Rakurty, C.S.; Balaji, A.K. A Multiscale Study on Machining Induced Surface Integrity in Ti-6Al-4V Alloy. *Procedia CIRP* **2022**, *108*, 787–792. [CrossRef]
31. Wang, Z.-Y.; Ren, J.-X.; Zhou, J.-H.; Cai, J. Correlation analysis of microstructure evolution on microhardness and residual stress for cutting Ti-6Al-4V titanium alloy. *Proc. Inst. Mech. Eng. Part B J. Eng. Manuf.* **2023**, *237*, 885–898. [CrossRef]
32. Mendas, M.; Benayoun, S.; Miloud, M.H.; Zidane, I. Microhardness model based on geometrically necessary dislocations for heterogeneous material. *J. Mater. Res. Technol.* **2021**, *15*, 2792–2801. [CrossRef]
33. Ameri, A.A.H.; Elewa, N.N.; Ashraf, M.; Escobedo-Diaz, J.P. General methodology to estimate the dislocation density from microhardness measurements. *Mater. Charact.* **2017**, *131*, 324–330. [CrossRef]
34. Alijani, A.; Amini, R.; Ghaffari, M.; Alizadeh, M.; Okyay, A.K. Effect of milling time on the structure, micro-hardness, and thermal behavior of amorphous/nanocrystalline TiNiCu shape memory alloys developed by mechanical alloying. *Mater. Des.* **2014**, *55*, 373–380. [CrossRef]
35. Chen, Z.; Huang, C.; Li, B.; Jiang, G.; Tang, Z.; Niu, J.; Liu, H. Experimental study on surface integrity of Inconel 690 milled by coated carbide inserts. *Int. J. Adv. Manuf. Technol.* **2022**, *121*, 3025–3042. [CrossRef]
36. Haddag, B.; Yameogo, D.; Nouari, M.; Makich, H. Multi-Physics Analysis of Machining Ti_6Al_4V Alloy: Experimental Characterization and a New Material Behavior Modeling. *Metals* **2022**, *12*, 581. [CrossRef]
37. da Silva, R.H.L.; Schoop, J.; Hassui, A.; Jawahir, I.S. Inconel 625 sustainable milling surface integrity and the dependence on alloy processing route. *Int. J. Adv. Manuf. Technol.* **2024**, *130*, 4493–4512. [CrossRef]
38. Rajguru, R.R.; Vasudevan, H. A study of micro hardness in the machining of Inconel 625 using TiAlSiN coated tools under dry cutting conditions. *Adv. Mater. Process. Technol.* **2022**, *8*, 120–130. [CrossRef]
39. Harun, S.; Burhanuddin, Y.; Ibrahim, G.A. The Effect of Cutting Parameters on Surface Roughness and Morphology of Ti_6Al_4V ELI Titanium Alloy during Turning with Actively Driven Rotary Tools. *J. Manuf. Mater. Process.* **2022**, *6*, 105. [CrossRef]
40. Pharr, G.M. Recent advances in small-scale mechanical property measurement by nanoindentation. *Curr. Opin. Solid State Mater. Sci.* **2015**, *19*, 315–316. [CrossRef]
41. Fischer-Cripps, A.C. Critical review of analysis and interpretation of nanoindentation test data. *Surf. Coat. Technol.* **2006**, *200*, 4153–4165. [CrossRef]
42. Li, X.; Bhushan, B. A review of nanoindentation continuous stiffness measurement technique and its applications. *Mater. Charact.* **2002**, *48*, 11–36. [CrossRef]
43. Wredenberg, F.; Larsson, P.-L. Scratch testing of metals and polymers: Experiments and numerics. *Wear* **2009**, *266*, 76–83. [CrossRef]
44. Kolawole, O.; Ispas, I. Evaluation of geomechanical properties via scratch tests: Where are we and where do we go from here? *SN Appl. Sci.* **2020**, *2*, 1633. [CrossRef]
45. Atkins, A.G.; Tabor, D. Plastic Indentation in Metals with Cones. *J. Mech. Phys. Solids* **1965**, *13*, 149–164. [CrossRef]
46. Doerner, M.; Nix, W. A method for interpreting the data from depth-sensing indentation instruments. *J. Mater. Res.* **1986**, *1*, 601–609. [CrossRef]
47. Oliver, W.C.; Pharr, G.M. An improved technique for determining hardness and elastic modulus using load and displacement sensing indentation experiments. *J. Mater. Res.* **1992**, *7*, 1564–1583. [CrossRef]
48. Pethica, J.B.; Hutchings, R.; Oliver, W.C. Hardness Measurement at Penetration Depths as Small as 20 nm. *Philos. Mag. A* **1983**, *48*, 593–606. [CrossRef]
49. Fischer-Cripps, A.C. *Nanoindentation*, 2nd ed.; Springer: Berlin/Heidelberg, Germany, 2011; 276p. [CrossRef]
50. Lin, C.K.; Berndt, C.C. Measurement and analysis of adhesion strength for thermally sprayed coatings. *J. Therm. Spray Technol.* **1994**, *3*, 75–104. [CrossRef]
51. Bull, S.J. Failure modes in scratch adhesion testing. *Surf. Coat. Technol.* **1991**, *50*, 25–32. [CrossRef]
52. Zivic, F.; Babic, M.; Adamovic, D.; Mitrovic, S.; Todorovic, P.; Favaro, G.; Pantić, M. Influence of the surface roughness on adhesion of chrome coatings on alloy tool steel x165crmov12. *J. Balk. Tribol. Assoc.* **2012**, *18*, 228–237.
53. Sousa, F.J.P.; Tridapalli, D.; Pereira, M.; Flesch, C.A.; Alarcon, O.E. Evaluation of measurement uncertainties for a scratching tester. *Measurement* **2006**, *39*, 594–604. [CrossRef]
54. Yildiz, F.; Asaran, A. Multi-pass scratch test behavior of modified layer formed during plasma nitriding. *Tribol. Int.* **2010**, *43*, 1472–1478. [CrossRef]
55. Storchak, M.; Zakiev, I.; Träris, L. Mechanical properties of subsurface layers in the machining of the titanium alloy $Ti_{10}V_2Fe_3Al$. *J. Mech. Sci. Technol.* **2018**, *32*, 315–322. [CrossRef]
56. Li, Z.; Herrmann, K.; Pohlenz, F. A comparative approach for calibration of the depth measuring system in a nanoindentation instrument. *Measurement* **2006**, *39*, 547–552. [CrossRef]

57. Peng, G.; Xu, F.; Chen, J.; Hu, Y.; Wang, H.; Zhang, T. A cost-effective voice coil motor-based portable micro-indentation device for in situ testing. *Measurement* **2020**, *165*, 108105. [CrossRef]
58. Ding, K.; Zhang, Y.; Birnbaum, A.J.; Michopoulos, J.G.; McDowell, D.L.; Zhu, T. Strain gradient plasticity modeling of nanoindentation of additively manufactured stainless steel. *Extrem. Mech. Lett.* **2021**, *49*, 101503. [CrossRef]
59. Fritz, R.; Kiener, D. Development and application of a heated in-situ SEM micro-testing device. *Measurement* **2017**, *110*, 356–366. [CrossRef]
60. Vargas, A.L.M.; Blando, E.; Hübler, R. Elasto—Plastic materials behavior evaluation according to different models applied in indentation hardness tests. *Measurement* **2019**, *139*, 134–139. [CrossRef]
61. Kang, J.J.; Becker, A.A.; Wen, W.; Sun, W. Extracting elastic-plastic properties from experimental loading-unloading indentation curves using different optimization techniques. *Int. J. Mech. Sci.* **2018**, *144*, 102–109. [CrossRef]
62. Guillonneau, G.; Kermouche, G.; Bec, S.; Loubet, J.-L. Determination of mechanical properties by nanoindentation independently of indentation depth measurement. *J. Mater. Res.* **2012**, *27*, 2551–2560. [CrossRef]
63. Harsono, E.; Swaddiwudhipong, S.; Liu, Z.S. The effect of friction on indentation test results. *Model. Simul. Mater. Sci. Eng.* **2008**, *16*, 065001. [CrossRef]
64. Wang, Y. Effects of indenter angle and friction on the mechanical properties of film materials. *Results Phys.* **2016**, *6*, 509–514. [CrossRef]
65. Sivaram, S.; Jayasinghe, J.A.S.C.; Bandara, C.S. Qualitative Study on Pile-up Effect on Hardness Test by Nano-Indentation. *Eng. J. Inst. Eng.* **2021**, *54*, 47–55. [CrossRef]
66. Farayibi, P.K.; Hankel, J.; Hassend, F.v.G.; Blüm, M.; Weber, S.; Röttger, A. Tribological characteristics of sintered martensitic stainless steels by nano-scratch and nanoindentation tests. *Wear* **2023**, *512–513*, 204547. [CrossRef]
67. Tsybenko, H.; Farzam, F.; Dehm, G.; Brinckmann, S. Scratch hardness at a small scale: Experimental methods and correlation to nanoindentation hardness. *Tribol. Int.* **2021**, *163*, 107168. [CrossRef]
68. England, J.; Uddin, M.J.; Ramirez-Cedillo, E.; Karunarathne, D.; Nasrazadani, S.; Golden, T.D.; Siller, H.R. Nanoindentation Hardness and Corrosion Studies of Additively Manufactured 316 L Stainless Steel. *J. Mater. Eng. Perform.* **2022**, *31*, 6795–6805. [CrossRef]
69. Moon, J.; Kim, S.; Jang, J.; Lee, J.; Lee, C. Orowan strengthening effect on the nanoindentation hardness of the ferrite matrix in microalloyed steels. *Mater. Sci. Eng. A* **2008**, *487*, 552–557. [CrossRef]
70. Li, C.; Zhao, H.; Sun, L.; Yu, X. In situ nanoindentation method for characterizing tensile properties of AISI 1045 steel based on mesomechanical analysis. *Adv. Mech. Eng.* **2019**, *11*, 1687814019862919. [CrossRef]
71. Paul, V.; Ameyama, K.; Ota-Kawabata, M.; Ohmura, T. Evaluation of Deformation and Fracture Behavior in 304 L Austenitic Steel Harmonic Structures through Nanoindentation. *Steel Res. Int.* **2023**, *94*, 2200354. [CrossRef]
72. Yang, L.; Sun, K.; Peng, W.; Li, X.; Zhang, L. Effects of Grain Boundary Angles on Initial Deformation of 304 Austenitic Stainless Steel under Nanoindentation: A Molecular Dynamics Simulation. *Crystals* **2022**, *12*, 58. [CrossRef]
73. Zhou, G.; Guo, J.; Zhao, J.; Tang, Q.; Hu, Z. Nanoindentation Properties of 18CrNiMo7-6 Steel after Carburizing and Quenching Determined by Continuous Stiffness Measurement Method. *Metals* **2020**, *10*, 125. [CrossRef]
74. Dean, J.; Aldrich-Smith, G.; Clyne, T.W. Use of nanoindentation to measure residual stresses in surface layers. *Acta Mater.* **2011**, *59*, 2749–2761. [CrossRef]
75. Wang, H.; Zhu, L.; Xu, B. *Residual Stresses and Nanoindentation Testing of Films and Coatings*; Springer Nature Singapore Pte Ltd.: Singapore; Science Press: Beijing, China, 2018; 207p, ISBN 978-981-10-7840-8. [CrossRef]
76. Zhang, W.; Wang, X.; Hu, Y.; Wang, S. Predictive modelling of microstructure changes, micro-hardness and residual stress in machining of 304 austenitic stainless steel. *Int. J. Mach. Tools Manuf.* **2018**, *130–131*, 36–48. [CrossRef]
77. Aurich, J.C.; Steffes, M. Single Grain Scratch Tests to Determine Elastic and Plastic Material Behavior in Grinding. *Adv. Mater. Res.* **2011**, *325*, 48–53. [CrossRef]
78. Fan, P.; Katiyar, N.K.; Zhou, X.; Goel, S. Uniaxial pulling and nano-scratching of a newly synthesised high entropy alloy. *APL Mater.* **2022**, *10*, 111118. [CrossRef]
79. Pratap, A.; Divse, V.; Goel, S.; Joshi, S.S. Understanding the surface generation mechanism during micro-scratching of Ti-6Al-4V. *J. Manuf. Process.* **2022**, *82*, 543–558. [CrossRef]
80. Liu, H.; Xu, X.; Zhang, J.; Liu, Z.; He, Y.; Zhao, W.; Liu, Z. The state of the art for numerical simulations of the effect of the microstructure and its evolution in the metal-cutting processes. *Int. J. Mach. Tools Manuf.* **2022**, *177*, 103890. [CrossRef]
81. Bezyazychnyy, V.F.; Prokofev, M.A.; Vinogradova, N.V. Research of the Influence of Technological Machining Conditions on the Accumulation of Latent Energy Deformation in the surface Parts. *Bull. PNIPU Aerosp. Eng.* **2015**, *43*, 131–144. [CrossRef]
82. Yamamoto, M.; Tanaka, M.; Furukimi, O. Hardness–Deformation Energy Relationship in Metals and Alloys: A Comparative Evaluation Based on Nanoindentation Testing and Thermodynamic Consideration. *Materials* **2021**, *14*, 7217. [CrossRef] [PubMed]
83. Wang, Q.; Liu, Z.; Wang, B.; Song, Q.; Wan, Y. Evolutions of grain size and micro-hardness during chip formation and machined surface generation for Ti-6Al-4V in high-speed machining. *Int. J. Adv. Manuf. Technol.* **2016**, *82*, 1725–1736. [CrossRef]
84. Storchak, M.; Stehle, T.; Möhring, H.-C. Numerical Modeling of Titanium Alloy $Ti_{10}V_2Fe_3Al$ Milling Process. *J. Manuf. Mater. Process.* **2023**, *7*, 132. [CrossRef]
85. Material Property Data. Available online: http://www.matweb.com (accessed on 19 April 2019).

86. Storchak, M.; Jiang, L.; Xu, Y.; Li, X. Finite element modeling for the cutting process of the titanium alloy $Ti_{10}V_2Fe_3Al$. *Prod. Eng. Res. Dev.* **2016**, *10*, 509–517. [CrossRef]
87. Storchak, M.; Stehle, T.; Möhring, H.-C. Determination of thermal material properties for the numerical simulation of cutting processes. *Int. J. Adv. Manuf.* **2021**, *118*, 1941–1956. [CrossRef]
88. Zorev, N.N. *Metal Cutting Mechanics*; Pergamon Press GmbH: Frankfurt am Main, Germany, 1966; 526p, ISBN 978-0080107233.
89. Oxley, P.L.B. Development and Application of a Predictive Machining Theory. *Mach. Sci. Technol.* **1998**, *2*, 165–189. [CrossRef]
90. Kushner, V.; Storchak, M. Modelling the Material Resistance to Cutting. *Int. J. Mech. Sci.* **2017**, *126*, 44–54. [CrossRef]
91. Heisel, U.; Kushner, V.; Storchak, M. Effect of machining conditions on specific tangential forces. *Prod. Eng.* **2012**, *6*, 621–629. [CrossRef]
92. Tsekhanov, J.; Storchak, M. Development of analytical model for orthogonal cutting. *Prod. Eng. Res. Dev.* **2015**, *9*, 247–255. [CrossRef]
93. Johnson, G.R.; Cook, W.H. A constitutive model and data for metals subjected to large strains, high strain and high temperatures. In Proceedings of the 7th International Symposium on Ballistics, The Hague, The Netherlands, 19–21 April 1983; pp. 541–547.
94. Heisel, U.; Krivoruchko, D.V.; Zaloha, W.A.; Storchak, M.; Stehle, T. Thermomechanical material models in the modeling of cutting processes. *ZWF Z. Fuer Wirtsch. Fabr.* **2009**, *104*, 482–491. [CrossRef]
95. Heisel, U.; Krivoruchko, D.V.; Zaloha, W.A.; Storchak, M.; Stehle, T. Thermomechanical exchange effects in machining. *ZWF Z. Fuer Wirtsch. Fabr.* **2009**, *104*, 263–272. [CrossRef]
96. Storchak, M.; Möhring, H.-C.; Stehle, T. Improving the friction model for the simulation of cutting processes. *Tribol. Int.* **2022**, *167*, 107376. [CrossRef]
97. Heisel, U.; Krivoruchko, D.V.; Zaloha, W.A.; Storchak, M.; Stehle, T. Breakage models for the modeling of cutting processes. *ZWF Z. Fuer Wirtsch. Fabr.* **2009**, *104*, 330–339. [CrossRef]
98. Cockroft, M.G.; Latham, D.J. Ductility and workability of metals. *J. Inst. Met.* **1968**, *96*, 33–39. [CrossRef]
99. Kushner, V.; Storchak, M. Determining mechanical characteristics of material resistance to deformation in machining. *Prod. Eng. Res. Dev.* **2014**, *8*, 679–688. [CrossRef]
100. Kushner, V.; Storchak, M. Determination of Material Resistance Characteristics in Cutting. *Procedia CIRP* **2017**, *58*, 293–298. [CrossRef]
101. Storchak, M.; Kushner, V.; Möhring, H.-C.; Stehle, T. Refinement of temperature determination in cutting zones. *J. Mech. Sci. Technol.* **2021**, *35*, 3659–3673. [CrossRef]

materials

MDPI

Article

Advanced Thermal Imaging Processing and Deep Learning Integration for Enhanced Defect Detection in Carbon Fiber-Reinforced Polymer Laminates

Renan Garcia Rosa [1], Bruno Pereira Barella [1], Iago Garcia Vargas [1], José Ricardo Tarpani [2], Hans-Georg Herrmann [3,4] and Henrique Fernandes [1,5,*]

[1] Faculty of Computing, Federal University of Uberlandia, Uberlandia 38408-100, Brazil; renan.garcia@ufu.br (R.G.R.); brunobarella@ufu.br (B.P.B.); iagogarcia@ufu.br (I.G.V.)
[2] Department of Materials, Sao Carlos School of Engineering, University of Sao Paulo, Sao Carlos 13566-590, Brazil; jrpan@sc.usp.br
[3] Fraunhofer IZFP Institute for Non-Destructive Testing, Campus E3 1, 66123 Saarbrucken, Germany; hans-georg.herrmann@izfp.fraunhofer.de
[4] Chair for Lightweight Systems, Saarland University, Campus E3 1, 66123 Saarbrucken, Germany
[5] IVHM Centre, Faculty of Engineering and Applied Sciences, Cranfield University, Cranfield MK43 0AL, UK
[*] Correspondence: h.fernandes@cranfield.ac.uk

Abstract: Carbon fiber-reinforced polymer (CFRP) laminates are widely used in aerospace, automotive, and infrastructure industries due to their high strength-to-weight ratio. However, defect detection in CFRP remains challenging, particularly in low signal-to-noise ratio (SNR) conditions. Conventional segmentation methods often struggle with noise interference and signal variations, leading to reduced detection accuracy. In this study, we evaluate the impact of thermal image preprocessing on improving defect segmentation in CFRP laminates inspected via pulsed thermography. Polynomial approximations and first- and second-order derivatives were applied to refine thermographic signals, enhancing defect visibility and SNR. The U-Net architecture was used to assess segmentation performance on datasets with and without preprocessing. The results demonstrated that preprocessing significantly improved defect detection, achieving an Intersection over Union (IoU) of 95% and an F1-Score of 99%, outperforming approaches without preprocessing. These findings emphasize the importance of preprocessing in enhancing segmentation accuracy and reliability, highlighting its potential for advancing non-destructive testing techniques across various industries.

Keywords: pulsed thermography; carbon fiber-reinforced polymer; thermal image preprocessing; non-destructive testing (NDT); deep learning; polynomial approximation

1. Introduction

Carbon fiber-reinforced polymers (CFRPs) have revolutionized industries where lightweight and high-strength materials are essential, such as aerospace, automotive, and renewable energy sectors [1]. Their exceptional combination of mechanical and physical properties has made them indispensable in applications demanding high-performance solutions. However, the structural integrity of CFRPs can be compromised during manufacturing or in-service use due to defects such as delaminations, cracks, and voids, posing significant safety and reliability risks [1]. The formation and progression of these defects are influenced by various factors, including manufacturing inconsistencies and prolonged exposure to harsh environmental conditions. High temperatures, corrosive agents, and mechanical stresses (both dynamic and static) can degrade the resin matrix and weaken the fiber–resin interface, leading to structural failures over time [2]. In addition, sustained loading combined with temperature cycling has been shown to significantly weaken CFRP bonds, accelerating failure. Exposure to saline environments further exacerbates degradation through galvanic corrosion, particularly when CFRP is in direct contact with steel. Moisture ingress can plasticize adhesives and lower bond strength, while fatigue loading can cause progressive debonding at CFRP–steel interfaces. Studies have indicated that CFRP systems subjected to long-term environmental exposure can lose up to 60% of their original bond strength [3]. Understanding these degradation mechanisms is crucial for improving both defect detection and preventive maintenance strategies.

Non-destructive testing (NDT) techniques play a pivotal role in detecting internal defects without compromising the functionality of the tested components. Several NDT methods have been employed to inspect CFRP laminates, each with distinct advantages and limitations. Ultrasonic testing (UT) is widely used due to its sensitivity to subsurface defects. Yet, it faces challenges when inspecting highly attenuative composite materials [4]. X-ray computed tomography (XCT) provides detailed internal imaging but is time-consuming and costly, limiting its practical application in large-scale inspections [5]. Eddy current testing (ECT) is effective for conductive composites but has limited applicability to non-conductive CFRPs. In this context, pulsed thermography (PT) has emerged as a preferred method due to its ability to provide rapid and comprehensive assessments of composite materials' structural health [5]. PT operates by applying a heat pulse to the material's surface and analyzing the thermal response, allowing defect detection based on variations in heat diffusion properties [6]. It offers a non-invasive and efficient solution for defect detection. Despite its advantages, raw thermographic data often exhibit substantial noise and thermal variations, complicating accurate defect identification [7]. Consequently, there is a growing need for advanced preprocessing methods to enhance the quality and interpretability of thermographic data.

A recent study demonstrated the effectiveness of combining self-organizing maps (SOMs) with bio-inspired parameter optimization through bee colony optimization (BCO) techniques in improving thermographic image analysis [8]. However, these methods often require extensive computational resources and can lack generalizability across different CFRP structures and defect types, limiting their broader adoption [9]. This methodology highlights the potential of advanced algorithms in improving non-destructive testing workflows and supports the exploration of similar innovations in the field.

Thermographic Signal Reconstruction (TSR) has been proposed as a robust preprocessing technique to improve the signal-to-noise ratio (SNR) in thermographic images [7]. By employing polynomial fitting and derivative analysis, TSR refines the thermal signals, enabling clearer visualization of defect features. Nonetheless, one of the primary challenges in thermographic image processing lies in balancing noise reduction while preserving critical defect-related thermal signatures. Over-smoothing can obscure defect boundaries, whereas inadequate filtering may fail to suppress noise effectively [10]. With the rise of deep learning methods, particularly convolutional neural networks (CNNs) such as U-Net, the ability to preprocess data effectively has become even more critical [11,12].

Deep learning models like U-Net have shown significant promise in the segmentation of thermographic images, achieving high accuracy in identifying complex patterns and defects [13]. However, the performance of these networks heavily depends on the quality of input data. Despite advancements in PT and deep learning, studies explicitly quantifying the impact of preprocessing methods like TSR on U-Net performance remain scarce. While prior research has successfully applied deep learning techniques to defect detection, the direct influence of preprocessing on segmentation accuracy and model robustness has not been fully addressed. A systematic investigation of how preprocessing steps enhance deep learning-based defect segmentation is still lacking in the literature. Addressing this gap is crucial to advancing the integration of NDT with artificial intelligence for composite materials.

This work aims to systematically investigate the influence of TSR preprocessing on the performance of the U-Net architecture in detecting and segmenting defects in CFRPs. By comparing the segmentation results of U-Net models trained on raw and TSR-preprocessed thermographic data, we seek to demonstrate how preprocessing can significantly enhance defect detection capabilities. Metrics such as F1-Score and Intersection over Union (IoU) will be employed to evaluate the effectiveness of the proposed approach. In comparison to existing studies, this research uniquely quantifies the impact of TSR preprocessing on deep learning-driven defect segmentation, providing a structured analysis of its benefits and trade-offs. By bridging the gap between thermographic preprocessing and deep learning-based defect detection, this study offers new insights into optimizing defect identification pipelines for high-performance composite materials.

2. Materials and Methods

This study focuses on a unidirectional carbon fiber-reinforced polymer (CFRP) laminate composed of carbon/PEEK (polyether ether ketone) APC-2/AS4, a material known for its high strength-to-weight ratio, making it widely used in aerospace structural applications. CFRP structures are susceptible to defects such as delaminations and inclusions, which can compromise their mechanical performance and pose risks of operational failure. To simulate these defects, polyimide film inserts with a thickness (Kapton® from 3M, St. Paul, MN, USA) were embedded in the laminate during manufacturing. These inserts had dimensions of 4 × 4 mm, 3 × 3 mm, and 2 × 2 mm and were placed at specific depths to replicate internal anomalies, as shown in Figure 1.

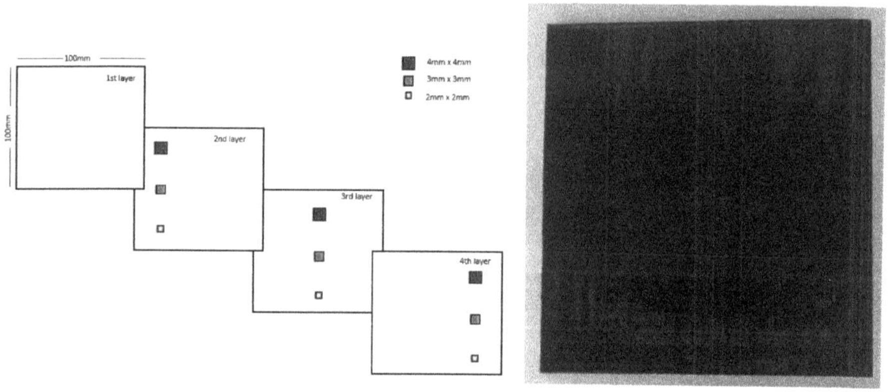

(**a**) (**b**)

Figure 1. (**a**) Positions and sizes of defects in the laminate layers. (**b**) Inspected sample.

The laminate sample used in this experiment consisted of a flat CFRP panel with nine artificial defects. The laminate was manufactured using APC-2/AS4 (© Syensqo, Brussels, Belgium), which contains a fiber volume fraction of 61% and a $[0_2/90_2]_6$ stacking sequence. The defect positions within the laminate are illustrated in Figure 1, with nominal depths of 0.13 mm (D1), 0.26 mm (D2), and 0.39 mm (D3). The thermal properties of APC-2/AS4 are presented in Table 1 while mechanical properties are presented in Table 2.

Table 1. Thermal properties of APC-2/AS4 [14].

Property	Value	Unit
k: Thermal conductivity (Longitudinal)	5.65	W/mK
k: Thermal conductivity (Transverse)	0.35	W/mK
C_p: Specific heat [(a)]	1310	J/kgK
ρ: Density	1584	kg/m³

[(a)] at constant pressure.

Table 2. Mechanical properties of APC-2/AS4 [15].

Property	VAlue	Unit
Tensile Properties		
Longitudinal Tensile Modulus	127.6	GPa
Transverse Tensile Modulus	10.3	GPa
Longitudinal Tensile Strength	2132	MPa
Transverse Tensile Strength	95.2	MPa
Shear Properties		
In-Plane Shear Strength	82	MPa
Poisson's Ratio		
Longitudinal	0.32	
Transverse	0.022	
Thermal Property		
Glass Transition Temperature	143	°C

The approach employed in this study involves a four-phase process, outlined in Figure 2: capturing thermographic images via pulsed thermography, enhancing data quality through Thermographic Signal Reconstruction (TSR), annotating defects manually (process detailed in Section 2.5), and utilizing neural networks for defect segmentation. This methodology facilitates a thorough investigation of how TSR preprocessing affects the performance of deep learning models.

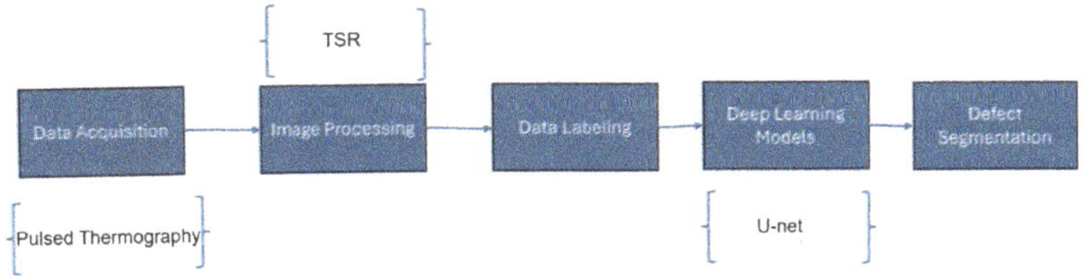

Figure 2. Defect segmentation methodology.

2.1. *Pulsed Infrared Thermography*

Pulsed thermography is an active infrared non-destructive testing (NDT) technique that utilizes thermodynamic principles and infrared imaging to assess material properties and identify internal defects. This method involves applying a short-duration heat pulse to the material's surface, resulting in a localized temperature rise. The heat propagates through the material, and internal discontinuities, such as delaminations and inclusions, alter the thermal diffusion, causing localized variations in the thermal response. The material's thermal behavior is recorded over time using an infrared camera, capturing the emitted radiation. By analyzing the cooling process, defects can be identified based on temperature contrast, as these anomalies create variations in the thermal response [16,17]. Figure 3 illustrates the setup used in a typical pulsed thermography inspection.

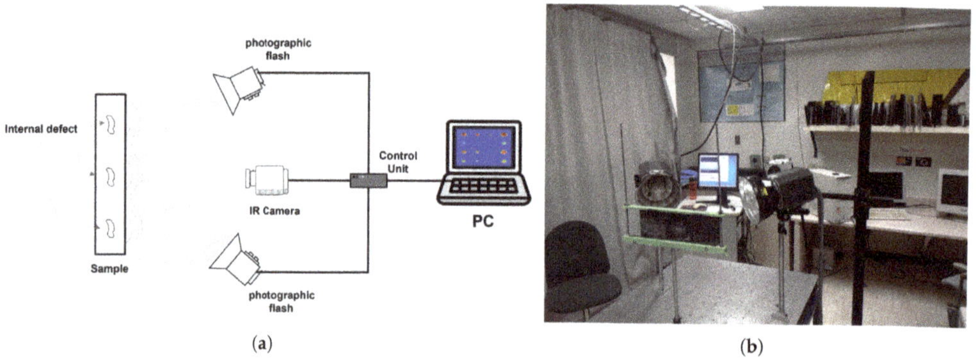

Figure 3. (**a**) Schematic of the experimental setup for pulsed thermography inspection. (**b**) Photograph of the work's experimental setup.

Pulsed thermography has been used for years, and its main advantages and disadvantages are as follows [10]:

Advantages

- Fast surface inspection: While traditional ultrasound inspection can take several minutes (even hours), a thermographic inspection takes only a few seconds.
- Ease of installation: In many cases, no special preparation is required for the inspection. Only an infrared camera and a heat source are sufficient. Additionally, no contact with the surface is necessary.
- Safety: Unlike X-ray inspections, no harmful waves are used. However, stimulation using strong heat sources (such as photographic flashes) requires eye protection, and mechanical wave-generated heat requires ear protection.
- Inspection requires access to only one surface: Often, the inspected piece is installed and in use, making it impossible to move. Infrared thermography allows for on-site inspection with access to just one surface.
- Easy numerical thermal modeling: A numerical model of a thermographic experiment typically involves only heat transfer in solids, which can be easily solved using the finite element method and simulation software.
- Easy interpretation of thermograms: Unlike ultrasound inspection, where results are often 1D waveforms, infrared thermography provides 2D images, making it easier to distinguish defective and non-defective areas.
- Wide range of applications: Infrared thermography is used in various fields, including
  - Monitoring and diagnostics of electrical components, thermal comfort, buildings, and artwork.
  - Process control, such as sealing line inspection of Tetra Pak® packaging (Meyrin, Switzerland), automotive brake system efficiency, and heat loss in electronic modules.
  - Detection of discontinuities, metal corrosion, cracks, and impact damage.
  - Material characterization, including thermal properties, moisture content, and fiber orientation.

Disadvantages

- Different emissivity of inspected materials: Low-emissivity materials reflect a lot of ambient thermal radiation, which can interfere with inspections. When possible, surface painting with spray can adjust emissivity for better results.
- Thermal losses: Heat loss due to conduction and radiation may lead to misinterpretation of results.

- Equipment cost: Infrared cameras and thermal stimulation units used in active thermography are more expensive than some other non-destructive testing techniques (e.g., visual inspection and basic ultrasound equipment). However, costs are competitive when compared to advanced technologies like phased arrays (ultrasound and eddy currents) and X-ray systems.
- Limited to detecting defects that alter thermal properties: Only defects that cause measurable thermal property changes can be detected.
- Reduced inspection depth: Infrared thermography is limited to a certain depth below the material surface. However, defects a few centimeters beneath the surface can be detected using low-frequency excitation in modulated thermography.
- Difficulty in achieving uniform heating: Achieving uniform heating, especially with photographic flashes, can be challenging.
- Transient nature of inspections: The transient thermal contrast requires infrared cameras capable of capturing sequential images.
- Need for a clear line of sight: The inspected object must be visible to the infrared camera without obstructions; otherwise, the inspection cannot be performed.

The following equation can mathematically describe the thermal behavior during pulsed thermography:

$$y(z,t) = T_0 + \frac{Q}{e\sqrt{\pi t}} \exp\left(-\frac{z^2}{4\alpha t}\right), \tag{1}$$

where α is the thermal diffusivity of the material, k is the thermal conductivity, c_p is the heat capacity, $e = \sqrt{k\rho c_p}$ is the thermal effusivity, z represents the depth of the defect, and t is time.

At the material's surface ($z = 0$), the equation simplifies to

$$y(t) = T_0 + \frac{Q}{e\sqrt{\pi t}}, \tag{2}$$

where T_0 is the initial temperature and Q is the applied heat flux [18].

The parameter e describes the material's ability to exchange thermal energy with its surroundings. In a sound region with homogeneous material properties, thermal diffusivity and conductivity are consistent. However, defective regions exhibit distinct thermal behaviors due to different parameters. This temperature variation is what allows PT to successfully differentiate between sound and defective areas, serving as the foundation for non-destructive evaluation [16].

Additionally, the thermal effusivity e can be calculated as

$$e = \sqrt{k\rho c_p}, \tag{3}$$

where ρ represents the material's density. This parameter plays a crucial role in heat transfer analysis and is fundamental in interpreting thermographic data for defect detection [17].

For the experimental setup, a high-power optical heating system was used to generate the thermal pulse. The heat source was positioned at a controlled distance from the CFRP surface to ensure uniform heat distribution. An infrared camera with a resolution of 640 × 512 pixels and a frame rate of 55 Hz was used to record the images. Calibration procedures were performed before each test to maintain consistency in the experimental conditions.

Pulsed thermography has been widely applied in various industries, including aerospace, automotive, and composite materials inspection. It is particularly effective for detecting internal defects in carbon fiber-reinforced polymers (CFRPs) and other advanced materials. Recent advancements in signal processing techniques, such as independent component analysis, have further enhanced its capability to detect and quantify defects with improved accuracy and reliability [11,18].

2.2. Thermographic Signal Reconstruction—TSR

Thermographic Signal Reconstruction (TSR) is a widely adopted algorithm in the field of infrared thermography. As described by [7], TSR focuses on reducing both the spatial and temporal resolution of thermographic data sequences, significantly streamlining the volume of information to be processed. This approach mitigates common challenges in pulsed thermography by transitioning the data into a logarithmic domain.

TSR is based on Fourier's law of heat conduction, which describes the thermal diffusion process in a material. To facilitate data analysis, the classical one-dimensional Fourier equation is transformed into a logarithmic domain, which enables the representation of temperature decay in a linearized form. By applying this transformation, the temperature variation over time can be expressed as follows:

$$\ln(\Delta T) = \ln\left(\frac{Q}{e}\right) - \frac{1}{2}\ln(\pi t) \tag{4}$$

This logarithmic transformation of Fourier's one-dimensional solution is adjusted to fit a time series using a polynomial function of degree n. This polynomial approximation enables the reconstruction of temperature decay curves for each pixel, allowing the identification of defective regions that deviate from the expected thermal response. The resulting thermographic sequence is then converted into images representing the $n+1$ polynomial coefficients, facilitating the creation of synthetic thermograms.

$$\ln(\Delta T) = c_0 + c_1 \ln(t) + c_2 \ln^2(t) + \cdots + c_n \ln^n(t) \tag{5}$$

In Equation (5), ΔT denotes the temperature change over time t for each pixel (i,j). This polynomial-based representation transforms the thermographic data into a series of coefficients, from $c_0(i,j)$ to $c_n(i,j)$, which are used to generate images reflecting the temperature variation (i,j,t). These coefficients also allow for the computation of derivatives, accounting for temporal noise and changes.

The TSR method offers notable advantages, including noise reduction, analytical flexibility, and efficient data compression. It also enables the interpolation of temperature values between data acquisition intervals. The first derivative of the polynomial reveals the cooling rate, while the second derivative highlights variations in this rate, providing deeper insights into the thermal behavior of materials [19].

Several alternative preprocessing techniques have been explored in the literature to enhance defect detection in infrared thermography. Principal Component Thermography (PCT) [18] and Independent Component Analysis (ICA) [11] are commonly used to extract key features from thermal sequences by decomposing the data into orthogonal or statistically independent components. While these methods effectively reduce noise and emphasize defect-related patterns, they often require parameter tuning and may suffer from information loss during dimensionality reduction.

In contrast, TSR preserves the original thermal decay behavior by utilizing polynomial approximations, allowing direct interpretation of defect-related variations in thermal signals. The polynomial-based approach enhances defect contrast while maintaining spatial integrity, which is particularly advantageous for integration with deep learning models such as U-Net. Additionally, unlike PCT and ICA, TSR does not rely on statistical assumptions about data distribution, making it more adaptable to different materials and imaging conditions.

However, TSR is not without its challenges. The effectiveness of the polynomial reconstruction depends on the selection of parameters, such as polynomial degree and derivative order, which may require optimization for different defect types. Furthermore, environmental factors, such as temperature fluctuations and reflections, can introduce uncertainties in real-world applications. To mitigate these issues, future research could explore adaptive TSR models that dynamically adjust reconstruction parameters based on material properties and experimental conditions.

Despite these challenges, the results presented in this study demonstrate that TSR preprocessing significantly enhances defect segmentation accuracy. By refining the thermal signal before deep learning-based segmentation, the proposed method improves the robustness and reliability of defect detection in CFRP materials, reinforcing its potential for advancing non-destructive testing techniques [7].

2.3. Deep Learning-Based Methods Used in Thermography for NDT

Recent advancements in deep learning have significantly enhanced the classification and segmentation of pulsed thermography (PT) data. For instance, Mask-RCNN has been employed to analyze synthetic PT datasets, effectively detecting abnormal regions in composite materials [20]. Similarly, U-Net models have been applied to segment defect regions in curved CFRP samples inspected via PT, demonstrating strong performance in handling complex geometries [20]. Faster-RCNN architectures, incorporating Inception V2 and Inception ResNet V2, have been utilized to identify defects in composite materials through thermographic images. Comparisons of average precision reveal that the Inception V2-based model achieves superior accuracy compared to the Inception ResNet V2 model [20].

Further research has explored hydrogen-based deep neural networks, which integrate temporal and spatial features to detect defects in composites and coatings [21]. Another study applied machine learning classification techniques to detect impact damage in composite samples using PT, achieving classification accuracies ranging from 78.7% to 93.5% [22]. Recurrent and feed-forward neural networks have been investigated for identifying defects in non-planar CFRP components, with LSTM networks outperforming feed-forward networks in handling temporal dependencies [23]. In addition, neural networks trained on raw thermographic data were compared to those trained on TSR-preprocessed data, with the latter significantly improving segmentation accuracy [24].

Generative adversarial networks (GANs) have also shown promise in processing thermographic data. A GAN-based approach for thermal image enhancement has been developed to improve defect visibility in CFRP components [25]. The IRT-GAN model, trained on six datasets of simulated thermographic data, leverages TSR coefficients as input to produce accurate and segmented images, further demonstrating the potential of GANs in automated defect detection [26].

Expanding on these innovations, a novel approach has been developed by combining DeepLabv3 and BiLSTM models for segmenting thermographic images of CFRP composites. This integration, applied for the first time in infrared imaging, has demonstrated substantial improvements in defect detection accuracy. Experimental comparisons revealed that the DeepLabv3-BiLSTM combination achieved an F1-Score of 0.96 and an IoU of 0.83, surpassing other methods such as U-Net. These findings emphasize the importance of incorporating both temporal and spatial features to analyze complex thermal patterns, thereby advancing the inspection of composite materials through pulsed thermography [27].

Despite significant advancements, the literature persists in a critical gap regarding the integration of TSR as a core preprocessing technique in thermal analysis workflows. This study seeks to address this gap by exploring the impact of TSR on improving the accuracy and reliability of deep neural networks for defect detection in NDT. Additionally, it introduces novel approaches for inspecting high-performance materials like CFRP, advancing the capabilities of existing methodologies.

2.4. U-Net

The U-Net neural network has emerged as a cornerstone in image segmentation, extensively utilized in medical imaging and industrial inspection applications [28,29]. Initially proposed by Ronneberger et al. [29], the architecture has gained recognition for its ability to deliver high accuracy even with limited training data. Its dual capability to capture global context and preserve fine-grained localization details has made it indispensable in various domains [28].

The architecture of U-Net is defined by its symmetric encoder–decoder structure, as depicted in Figure 4. The encoder, often referred to as the contracting path, progressively reduces the spatial resolution of input images while extracting hierarchical features through convolutional and max-pooling layers [29]. This step enables the network to model complex high-level abstractions, which are essential for identifying significant patterns within the data [28].

In parallel, the decoder, or expansive path, reverses this process by gradually restoring the spatial resolution through upsampling and convolutional layers. Crucially, skip connections between corresponding layers in the encoder and decoder allow the network to combine precise localization information from earlier stages with the contextual features learned during downsampling [28,29]. This integration enhances segmentation accuracy by ensuring the alignment of detailed and contextual features.

Beyond its initial application in biomedical imaging, U-Net has been adapted to various industrial tasks, including the detection of defects in materials inspected through infrared thermography [28]. Its resilience to data scarcity and effectiveness in handling diverse segmentation challenges underscore its versatility and enduring importance in modern image analysis pipelines [29].

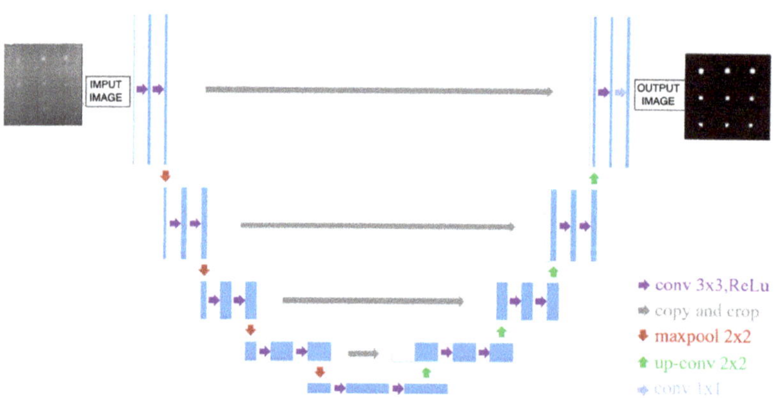

Figure 4. The architecture of the U-Net.

The implementation of the model was performed using the PyTorch 2.6 library, which enabled efficient processing of large-scale image datasets through GPU acceleration [30]. The model architecture is built from sequential blocks, each consisting of two 2D convolutional layers followed by batch normalization and ReLU activation functions, forming the core computational operations.

The encoder, or contraction path, reduces the spatial resolution of the input features while increasing the depth of representation. This path consists of four levels, each including a double convolutional block and a max-pooling operation. The number of filters increases progressively, starting from 64 and doubling at each level, reaching 512 at the deepest layer. Conversely, the decoder, or expansion path, restores spatial resolution using bilinear upsampling. Each upsampling step is followed by the concatenation of the corresponding feature maps from the encoder and additional convolutional blocks, refining the representations and ensuring accurate segmentation.

The final output is produced by a single convolutional layer with one filter, mapping the refined features into a binary segmentation mask. This output highlights the regions of interest within the input image, enabling precise defect detection.

Optimization of the model was achieved using the Adam optimizer, with an initial learning rate of 10^{-5}. To enhance the training process, the learning rate was dynamically adjusted using the ReduceLROnPlateau scheduler, which reduces the learning rate when the validation loss stagnates, inspired by adaptive learning rate strategies proposed by [31]. The Binary Cross Entropy with Logits Loss (BCEWithLogitsLoss) function was used to compute the loss. The training was conducted for 60 epochs in a GPU-accelerated environment, ensuring efficient computation and faster convergence.

2.5. Training, Validation, and Test Data

The dataset collected in the experimental setup illustrated in Figure 3 comprises 1053 preprocessed thermographic images using the TSR technique. These images were systematically divided into three distinct subsets: training, validation, and testing. A total of 737 images (70%) were allocated for training, ensuring the model could learn effectively from a diverse set of examples. Another 158 images (15%) were set aside for validation, enabling the tuning of hyperparameters and monitoring performance during training. Finally, the remaining 158 images (15%) were designated for testing, ensuring an unbiased evaluation of the model's segmentation capabilities. This stratified division guarantees that each subset maintains a balanced distribution of defects, contributing to a robust and generalizable model.

To facilitate model training, all collected images were meticulously annotated by an expert. The annotation process was carried out using the VGG Image Annotator (VIA), a web-based tool developed by the Visual Geometry Group at the University of Oxford [32]. The VIA tool was chosen for its user-friendly interface, flexibility, and capability to handle diverse annotation tasks efficiently.

The annotation workflow consisted of four main stages to ensure precision and consistency in defect labeling:

- Image Processing and Standardization: Before annotation, all thermal images were uploaded to the VIA tool and pre-processed to maintain uniform resolution and format across the dataset. This step was crucial to ensure consistency in defect representation.
- Manual Region Marking: An expert manually outlined the Regions of Interest (ROIs) that corresponded to defect locations within the CFRP material. These delineations served as the basis for accurate defect segmentation.
- Defect Classification: Each annotated ROI was assigned a specific label to categorize the defect type. This structured labeling approach helped distinguish different types of defects clearly.
- Validation and Quality Assurance: To ensure annotation reliability, a cross-validation process was conducted involving multiple annotators. Any inconsistencies were resolved through discussions, and necessary adjustments were made to maintain a high standard of accuracy.

The involvement of multiple annotators and the previous knowledge of the positions of the polyimide film tapes prior to molding ensured that all defects were correctly labeled.

Once the annotation process was finalized, the labeled regions were compiled into a ground truth dataset, which served as a reference for training and validating the segmentation models. The creation of precise ground truth masks was essential for aligning model predictions with actual defect structures, thereby improving the reliability of the defect detection system. Figure 5 illustrates a visual comparison of the original thermal image, the annotated regions, and the corresponding segmentation mask, showcasing the transformation of raw data into a structured dataset optimized for deep learning applications.

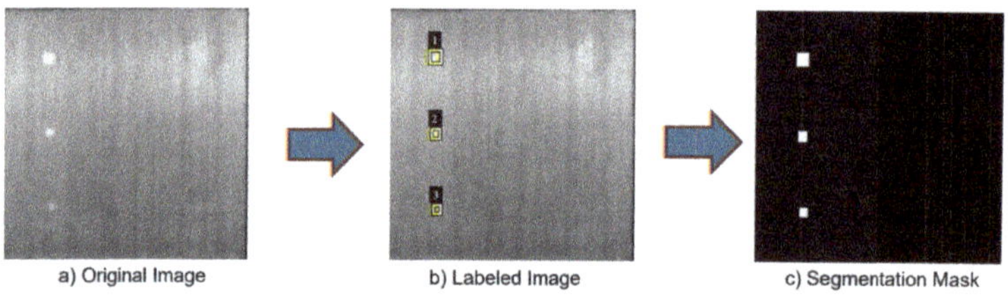

Figure 5. Visual comparison of the manually annotated image and the generated segmentation mask.

2.6. Evaluation Metrics

To evaluate the efficiency of the data processing approach in defect detection, both qualitative and quantitative analyses were conducted. The qualitative analysis involved comparing the processed images step-by-step to observe visual improvements. Quantitatively, metrics such as the signal-to-noise ratio (SNR) were used to compare the original images with those enhanced by the TSR technique. Additionally, segmentation performance was assessed using metrics like the F1-Score and Intersection over Union (IoU).

The SNR measures the contrast between defective regions and surrounding intact areas, offering a dynamic range representation. For each defect, a region within the defect and a noise region representing the defect-free area are selected. In this study, SNR values are calculated using Equation (6), following the methodology described by [33].

$$SNR = \frac{S}{N} = 20 \cdot \log_{10}\left(\frac{\text{abs}(S_{\text{area(mean)}} - N_{\text{area(mean)}})}{\sigma}\right) [\text{dB}] \quad (6)$$

In Equation (6), $N_{\text{area(mean)}}$ represents the average pixel intensity within the defective region, $S_{\text{area(mean)}}$ is the average intensity in the defect-free area, and σ is the standard deviation of pixel intensities in the defect-free region.

The F1-Score evaluates model accuracy by balancing precision and recall, making it particularly useful for imbalanced datasets. It is computed using Equation (7), as the harmonic mean of precision and recall [34].

$$F1 = 2 \cdot \frac{\text{Precision} \cdot \text{Recall}}{\text{Precision} + \text{Recall}} \quad (7)$$

IoU is another widely used metric for segmentation tasks, quantifying the overlap between the predicted regions and the actual ground truth. It is calculated using Equation (8) by dividing the intersection area by the union area.

$$IoU = \frac{\text{Intersection Area}}{\text{Union Area}} \quad (8)$$

Higher *IoU* values, closer to 1, indicate greater accuracy in identifying defects, while values near 0 suggest poor performance. In this study, *IoU* was adapted to calculate overlaps using defective and non-defective pixels instead of bounding boxes, ensuring the metric's relevance to thermal image segmentation.

3. Results

The application of the TSR technique demonstrates a substantial improvement in the visualization of defects within CFRP samples, as shown in Figure 6. The comparison focuses on three distinct defects, labeled D1, D2, and D3, located at varying depths within the material. In Figure 6a, which presents the raw image, D1, the shallowest defect, is moderately visible, while D2 and D3, corresponding to intermediate and deeper defects, respectively, become increasingly difficult to detect due to reduced contrast and the effects of thermal diffusion.

(a) (b)

Figure 6. Comparison between images with and without processing. (**a**) Top section of raw data, (**b**) data processed with TSR.

In contrast, Figure 6b, representing the TSR-processed image, reveals a significant enhancement in defect visibility. D1 appears with sharper definition, and both D2 and D3 exhibit noticeably improved contrast, making them easier to identify. The deeper defect, D3, in particular, benefits substantially from TSR processing, as its thermal signature is amplified, compensating for the challenges associated with its depth. These results highlight TSR's capability to enhance thermal contrast, improving defect detectability across various depths and enabling more reliable inspection and analysis of composite materials.

The signal-to-noise ratio (SNR) was determined following the methodology described in Equation (6). As shown in Figure 7, defective regions (highlighted in red) and intact regions (highlighted in green) were selected for the SNR calculation. This process was performed on a representative thermal image and subsequently repeated across all frames in the thermal sequence.

The results of the SNR calculations for each identified defect are summarized in Table 3, providing a quantitative assessment based on the defined regions in the image. This approach ensures consistency in evaluating the contrast between defective and defect-free areas, facilitating a detailed analysis of the thermal sequence.

Figure 7. Defect areas (red) and intact areas (green).

Table 3. Calculation of SNR in decibels (dB).

Defects	SNR Original (dB)	SNR TSR (dB)
D1	14.03	18.41
D2	13.84	15.72
D3	2.98	17.80

The results presented in Figure 7 and Table 3 illustrate the impact of the TSR technique on improving the Signal-to-Noise Ratio (SNR) for defect detection in CFRP samples. In the raw images, the SNR values for defects D1, D2, and D3 were 14.03 dB, 13.84 dB, and 2.98 dB, respectively, indicating that deeper defects (e.g., D3) are harder to detect due to their weaker thermal signal. After applying the TSR processing, significant improvements in SNR were observed, with D1, D2, and D3 reaching values of 18.41 dB, 15.72 dB, and 17.80 dB, respectively.

This demonstrates the effectiveness of TSR in enhancing the contrast and clarity of defects, particularly for deeper anomalies like D3, which showed a remarkable increase in detectability. The improved SNR values reflect the enhanced ability to distinguish defective regions from intact areas, confirming TSR's capability to process thermal signals effectively and provide more reliable insights for defect identification in materials with varying depths. These results underscore the importance of incorporating advanced preprocessing techniques in thermographic analyses to address challenges related to signal attenuation and noise.

To evaluate the segmentation performance, the processed data were utilized as input to the U-Net architecture, aiming to delineate the defective regions within the thermal images. The U-Net model underwent two distinct training and validation phases to enable a comparative analysis of its performance. In the initial phase, the network was trained using raw thermal images acquired directly from the experimental setup, without any preprocessing. In the subsequent phase, the same model was trained and validated using thermal images enhanced through the TSR technique.

Figure 8 showcases the results of this comparison, displaying three sets of segmentation outputs for all nine defects: (a) ground truth segmentation manually annotated by an expert; (b) the predicted segmentation produced by the U-Net model trained with raw thermal data; and (c) the segmentation results obtained when the model was trained with TSR-preprocessed images. In these images, red pixels indicate regions where the model failed to detect defects (false negatives), while blue pixels represent areas where the model incorrectly classified non-defective regions as defective (false positives). This comparison highlights the influence of preprocessing on the network's ability to accurately segment and identify defective areas across varying depths within the material.

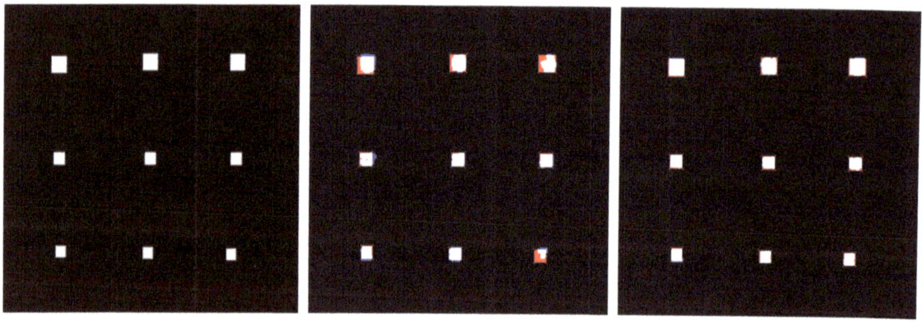

(**a**) Manual segmentation. (**b**) Unprocessed images. (**c**) TSR-processed images.

Figure 8. Comparison of segmentation results. Red pixels indicate false negatives (regions where defects were present but not detected), while blue pixels indicate false positives (regions where defects were incorrectly detected). (**a**) Ground truth segmentation, (**b**) segmentation from unprocessed images, (**c**) segmentation from TSR-processed images.

The results presented in Figure 8 provide a direct visual comparison of segmentation performance under three conditions: manual annotation, model predictions using raw thermal images, and predictions with TSR-processed images. The manual segmentation (a) serves as the ideal reference, outlining the precise defect locations. When comparing (b) and (c), a higher density of red and blue pixels is observed in the segmentation from raw thermal data, indicating greater difficulty in distinguishing defective from non-defective regions. This effect is particularly pronounced in deeper defects (rightmost column), where low thermal contrast leads to misclassification. However, the segmentation results from TSR-processed images (c) show a notable reduction in false positives and false negatives, demonstrating that preprocessing significantly improves model performance. These results reinforce the effectiveness of TSR in enhancing defect segmentation accuracy.

To assess the segmentation performance quantitatively, thermal image sequences were processed by the models, as shown in Figure 9. Each frame in the sequence was segmented, and evaluation metrics, including F1-Score and IoU, were calculated for every segmented frame. These metrics provide a frame-by-frame assessment of how well the models identify defective regions. Subsequently, the average values of these metrics across the entire sequence were computed to summarize the overall performance of the models.

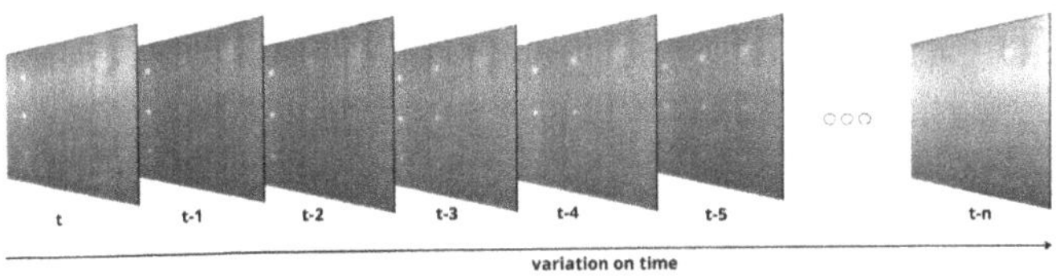

(**a**) Thermal sequence over time.

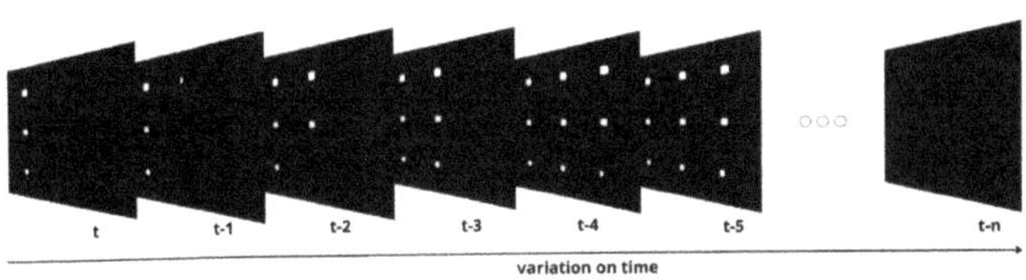

(**b**) Predicted mask over time.

Figure 9. Comparison of original and predicted images over time [27].

Figure 9 illustrates the original thermal image sequence and the corresponding segmented results. Panel (a) depicts the original thermal frames over time, capturing the evolution of heat distribution across the material. Panel (b) shows the predicted segmentation outputs, highlighting the regions identified as defects. This comparison enables a comprehensive evaluation of the model's ability to accurately segment defects at different depths and time frames.

The quantitative results are summarized in Table 4, which compares the U-Net model trained on TSR-processed data with models previously reported in the literature, including U-Net and DeepLabv3+BILSTM from [27]. This comparative analysis underscores the impact of TSR preprocessing on model accuracy and demonstrates the advantages of advanced architectures in capturing temporal and spatial features effectively.

Table 4. Comparison of segmentation metrics for different models.

Model	Preprocessing	F1-Score	IoU
U-Net *	Without TSR	0.9142	0.6141
DeepLabv3+BILSTM *	Without TSR	0.9629	0.8312
U-Net	With TSR	0.9903	0.9516

* Adapted from [27].

The results presented in Table 4 highlight the impact of preprocessing and model architecture on segmentation performance. The U-Net model, when trained without TSR preprocessing, achieved an F1-Score of 0.9142 and an IoU of 0.6141, demonstrating its baseline capability in segmenting defects. However, the DeepLabv3+BILSTM model, even without TSR preprocessing, outperformed the U-Net with an F1-Score of 0.9629 and an IoU of 0.8312, likely due to its ability to incorporate temporal and spatial features. When TSR preprocessing was applied to the U-Net model, there was a significant improvement, with an F1-Score of 0.9903 and an IoU of 0.9516.

This demonstrates the critical role of TSR in enhancing thermal image quality and segmentation accuracy. Notably, the U-Net model with TSR preprocessing surpassed the DeepLabv3+BILSTM model without TSR, emphasizing the importance of preprocessing techniques in improving segmentation performance. Despite these benefits, TSR preprocessing introduces an additional computational cost compared to raw image processing. The polynomial fitting and derivative calculations required for TSR increase both memory usage and processing time. However, the impact remains manageable for most applications, as the preprocessing stage is performed offline before model inference. For real-time scenarios, optimization strategies such as parallel processing and hardware acceleration could be explored to minimize computational overhead. These findings underscore the value of combining advanced preprocessing methods like TSR with neural networks to achieve superior results in defect detection and segmentation tasks.

4. Conclusions

The findings of this study have direct implications for the detection of CFRP defects in industrial applications. By leveraging TSR preprocessing, the accuracy and reliability of defect segmentation in pulsed thermography can be significantly enhanced. The ability to amplify defect visibility, particularly in low signal-to-noise ratio (SNR) conditions, suggests that this methodology could be effectively deployed in aerospace, automotive, and infrastructure industries where CFRP integrity is critical. Improved segmentation accuracy translates into more reliable non-destructive testing (NDT) procedures, reducing the likelihood of undetected structural damage and enabling predictive maintenance strategies.

This study demonstrated the critical role of thermographic signal reconstruction (TSR) using polynomial approximations in enhancing the defect detection and segmentation of CFRP laminates through pulsed thermography. By preprocessing thermal images with TSR, significant improvements were observed across all evaluation metrics, highlighting its ability to enhance the signal-to-noise ratio (SNR) and improve the overall quality of input data for neural networks. The most notable SNR improvements were observed in defect D4, which increased from 9.33 dB to 22.36 dB, and defect D9, which improved from −1.48 dB to 9.02 dB, illustrating the technique's effectiveness in amplifying defect visibility and contrast.

The U-Net model trained with TSR-processed images achieved exceptional results, with an IoU of 95.16% and an F1-Score of 99.03%, far outperforming the unprocessed model, which achieved an IoU of 61.41% and an F1-Score of 91.42%. These findings underscore the importance of preprocessing techniques in non-destructive testing workflows, enabling more precise and reliable segmentation of defects, particularly for deeper anomalies that are traditionally challenging to detect. Furthermore, the TSR-preprocessed U-Net model achieved comparable, and in some cases superior, performance to more complex architectures like DeepLabv3+BILSTM.

Despite these advancements, some limitations should be acknowledged. TSR preprocessing introduces an additional computational cost due to the polynomial fitting and derivative calculations, which increase processing time and memory usage. While this is manageable for offline analysis, further optimizations would be necessary for real-time applications. Additionally, the selection of TSR parameters, such as polynomial degree, has a direct impact on segmentation performance and may require tuning for different material properties and defect characteristics. Another limitation is that this study was conducted under controlled experimental conditions. To ensure broader applicability, future work should validate the approach across a wider range of CFRP structures, defect types, and real-world environmental conditions.

Future research could explore the integration of TSR with complementary preprocessing strategies, such as wavelet transforms or feature extraction techniques, to further enhance defect segmentation. Additionally, the application of Transformer-based models and hybrid deep learning approaches could improve the accuracy and generalization capability of segmentation networks. Another promising avenue is the optimization of computational efficiency, either through parallel processing techniques or the development of lightweight neural architectures, making TSR preprocessing viable for real-time inspections. Finally, expanding this study to include validation in industrial settings, such as automated CFRP defect detection in aerospace maintenance, could bridge the gap between research and practical deployment.

This research highlights the value of integrating advanced preprocessing methods with neural networks, providing a robust framework for defect detection in high-performance materials such as CFRP laminates. The combination of TSR with deep learning-based segmentation models significantly improves defect detection accuracy and robustness, demonstrating its potential for practical applications in industrial non-destructive testing. Further developments in this field could lead to more efficient and scalable defect detection systems, contributing to safer and more reliable composite material inspections.

Author Contributions: Conceptualization, R.G.R., B.P.B. and I.G.V.; methodology, R.G.R. and H.F.; software, R.G.R.; validation, B.P.B. and I.G.V.; investigation, J.R.T., H.-G.H. and H.F.; resources, H.F.; data curation, I.G.V.; writing—original draft preparation, R.G.R.; writing—review and editing, B.P.B., I.G.V., J.R.T., H.-G.H. and H.F.; supervision, H.F.; project administration, H.F.; funding acquisition, H.F. All authors have read and agreed to the published version of the manuscript.

Funding: This study was financed in part by the *Coordenação de Aperfeiçoamento de Pessoal de Nível Superior*—Brazil (CAPES)—Finance Code 001. H.F. gratefully acknowledges the financial support of CNPq (Grants #312530/2023-4 and #407140/2021-2).

Institutional Review Board Statement: Not applicable.

Informed Consent Statement: Not applicable.

Data Availability Statement: The dataset in this study is available at: Garcia Vargas, Iago; Fernandes, Henrique (2025), "Thermal Inspection Dataset for Defect Segmentation in CFRP Laminates", Mendeley Data, V1, doi:10.17632/jrsb4b9yy5.1.

Conflicts of Interest: The authors declare no conflicts of interest.

References

1. Teng, H.; Li, S.; Cao, Z.; Li, C.; Ko, T.J. Carbon fiber composites for large-scale wind turbine blades: Applicability study and comprehensive evaluation in China. *J. Mar. Sci. Eng.* **2023**, *11*, 624. [CrossRef]
2. Golewski, P.; Sadowski, T.; Kneć, M.; Budka M. The effect of thermal aging degradation of CFRP composite on its mechanical properties using destructive and non-destructive methods and the DIC system. *Polym. Test.* **2023**, *118*, 107902. [CrossRef]
3. Borrie, D.; Al-Saadi, S.; Zhao, X.-L.; Raman, R.K.S.; Bai, Y. Bonded CFRP/Steel Systems, Remedies of Bond Degradation and Behaviour of CFRP Repaired Steel: An Overview. *Polymers* **2021**, *13*, 1533. [CrossRef]
4. Torbali, M.E.; Zolotas, A.; Avdelidis, N.P.; Alhammad, M.; Ibarra-Castanedo, C.; Maldague, X.P. A Complementary Fusion-Based Multimodal Non-Destructive Testing and Evaluation Using Phased-Array Ultrasonic and Pulsed Thermography on a Composite Structure. *Materials* **2024**, *17*, 3435. [CrossRef] [PubMed]
5. Pedrayes, O.D.; Lema, D.G.; Usamentiaga, R.; Venegas, P.; García, D.F. Semantic segmentation for non-destructive testing with step-heating thermography for composite laminates. *Measurement* **2022**, *200*, 111653. [CrossRef]

6. Zhang, H.; Fernandes, H.; Hassler, U.; Ibarra-Castanedo, C.; Genest, M.; Robitaille, F.; Joncas, S.; Maldague, X. Comparative study of microlaser excitation thermography and microultrasonic excitation thermography on submillimeter porosity in carbon fiber reinforced polymer composites. *Opt. Eng.* **2016**, *56*, 041304. [CrossRef]
7. Shepard, S.M.; Beemer, M.F. Advances in thermographic signal reconstruction. In Proceedings of the Thermosense: Thermal Infrared Applications XXXVII, SPIE, Baltimore, MD, USA, 20–24 April 2015; Volume 94850R, pp. 204–210. [CrossRef]
8. Pereira Barella, B.; Garcia Rosa, R.; Maira Barbosa de Oliveira, G.; Fernandes, H. Enhancing fault characterisation in composites using infrared thermography: A bee colony optimisation approach with self-organising maps. *Quant. Infrared Thermogr. J.* **2024**, 1–19, *ahead-of-print*. [CrossRef]
9. Saragadam, V.; Dave, A.; Veeraraghavan, A.; Baraniuk, R.G. Thermal Image Processing via Physics-Inspired Deep Networks. In Proceedings of the IEEE/CVF International Conference on Computer Vision Workshops (ICCVW), Montreal, BC, Canada, 11–17 October 2021; pp. 3998–4007. [CrossRef]
10. Maldague, X. *Theory and Practice of Infrared Technology for Nondestructive Testing*; Wiley-Interscience: Hoboken, NJ, USA, 2001.
11. Wei, Z.; Fernandes, H.; Herrmann, H.-G.; Tarpani, J.R.; Osman, A. A deep learning method for the impact damage segmentation of curve-shaped CFRP specimens inspected by infrared thermography. *Sensors* **2021**, *21*, 395. [CrossRef]
12. Zhang, Z.; Hu, J.; Wang, R.; Chen, X.; Yang, D.; Vavilov, V.P.; Duan, Y.; Zhang, H. Automatic segmentation of microporous defects in composite film materials based on the improved attention U-Net module. *Quant. Infrared Thermogr. J.* **2024**, 1–16, *ahead-of-print*. [CrossRef]
13. Zhou, X.; Li, Y.; Wang, P.; Chen, M.; Zhang, T. Attention U-Net for segmentation of defects in CFRP thermographic images with high accuracy. *Nondestruct. Test. Eval.* **2024**, *39*, 238–257. [CrossRef]
14. Ageorges, C.; Ye, L.; Mai, Y.-W. ; Hou, M. Characteristics of resistance welding of lap shear coupons. Part I: Heat transfer. *Compos. Part Appl. Sci. Manuf.* **1998**, *29*, 899–909. [CrossRef]
15. Cogswell, F. *Thermoplastic Aromatic Polymer Composites*; Woodhead: Cambridge, UK, 1992.
16. Chung, Y.; Lee, S.; Kim, W. Latest advances in common signal processing of pulsed thermography for enhanced detectability: A review. *Appl. Sci.* **2021**, *11*, 12168. [CrossRef]
17. Fleuret, J.R.; Ebrahimi, S.; Ibarra-Castanedo, C.; Maldague, X.P. Independent component analysis applied on pulsed thermographic data for carbon fiber reinforced plastic inspection: A comparative study. *Appl. Sci.* **2021**, *11*, 4377. [CrossRef]
18. Ibarra-Castanedo, C.; Genest, M.; Servais, P.; Maldague, X.P.; Bendada, A. Qualitative and quantitative assessment of aerospace structures by pulsed thermography. *Non-Destr. Test. Eval.* **2007**, *22*, 199–215. [CrossRef]
19. Fernandes, H.; Zhang, H.; Figueiredo, A.; Ibarra-Castanedo, C.; Guimaraes, G.; Maldague, X. Carbon fiber composites inspection and defect characterization using active infrared thermography: Numerical simulations and experimental results. *Appl. Opt.* **2016**, *55*, D46–D53. [CrossRef]
20. Bang, H.T.; Park, S.; Jeon, H. Defect identification in composite materials via thermography and deep learning techniques. *Compos. Struct.* **2020**, *246*, 112405. [CrossRef]
21. Luo, Q.; Gao, B.; Woo, W.L.; Yang, Y. Temporal and spatial deep learning network for infrared thermal defect detection. *NDT E Int.* **2019**, *108*, 102164. [CrossRef]
22. Alhammad, M.; Avdelidis, N.; Ibarra-Castanedo, C.; Torbali, M.; Genest, M.; Zhang, H.; Zolotas, A.; Maldague, X. Automated Impact Damage Detection Technique for Composites Based on Thermographic Image Processing and Machine Learning Classification. *Sensors* **2022**, *22*, 9031. [CrossRef]
23. Tao, Y.; Hu, C.; Zhang, H.; Osman, A.; Ibarra-Castanedo, C.; Fang, Q.; Sfarra, S.; Dai, X.; Maldague, X.; Duan, Y. Automated Defect Detection in Non-planar Objects Using Deep Learning Algorithms. *J. Nondestruct. Eval.* **2022**, *41*, 14. [CrossRef]
24. Duan, Y.; Liu, S.; Hu, C.; Hu, J.; Zhang, H.; Yan, Y.; Tao, N.; Zhang, C.; Maldague, X.; Fang, Q.; et al. Automated defect classification in infrared thermography based on a neural network. *NDT E Int.* **2019**, *107*, 102147. [CrossRef]
25. Liu, K.; Li, Y.; Yang, J.; Liu, Y.; Yao, Y. Generative principal component thermography for enhanced defect detection and analysis. *IEEE Trans. Instrum. Meas.* **2020**, *69*, 8261–8269. [CrossRef]
26. Cheng, L.; Tong, Z.; Xie, S.; Kersemans, M. IRT-GAN: A generative adversarial network with a multi-headed fusion strategy for automated defect detection in composites using infrared thermography. *Compos. Struct.* **2022**, *290*, 115543. [CrossRef]
27. Garcia Vargas, I.; Fernandes, H. Spatial and temporal deep learning algorithms for defect segmentation in infrared thermographic imaging of carbon fiber-reinforced polymers. *Non-Destr. Test. Eval.* **2025**, 1–21, *ahead-of-print*. [CrossRef]
28. He, Y.; Deng, B.; Wang, H.; Cheng, L.; Zhou, K.; Cai, S.; Ciampa, F. Infrared machine vision and infrared thermography with deep learning: A review. *Infrared Phys. Technol.* **2021**, *116*, 103754. [CrossRef]
29. Ronneberger, O.; Fischer, P.; Brox, T. U-Net: Convolutional networks for biomedical image segmentation. In Proceedings of the Medical Image Computing and Computer-Assisted Intervention–MICCAI 2015: 18th International Conference, Munich, Germany, 5–9 October 2015; Springer: Cham, Switzerland, 2015; Volume 9351, pp. 234–241. [CrossRef]

30. Ravikumar, A.; Sriraman, H. Acceleration of image processing and computer vision algorithms. In *Handbook of Research on Computer Vision and Image Processing in the Deep Learning Era*; Srinivasan, A., Ed.; IGI Global: Hershey, PA, USA, 2023; pp. 1–18. [CrossRef]
31. Smith, L.N. Cyclical Learning Rates for Training Neural Networks. In Proceedings of the 2017 IEEE Winter Conference on Applications of Computer Vision (WACV), Santa Rosa, CA, USA, 24–31 March 2017; pp. 464–472. [CrossRef]
32. Dutta, A.; Gupta, A.; Zissermann, A. VGG Image Annotator (VIA). 2016. Available online: https://www.robots.ox.ac.uk/vgg/software/via/ (accessed on 19 March 2025).
33. Hidalgo-Gato García, R.; Andrés Álvarez, J.R.; López Higuera, J.M.; Madruga Saavedra, F.J. Quantification by signal-to-noise ratio of active infrared thermography data processing techniques. *Opt. Photonics J.* **2013**, *3*, 7–11. [CrossRef]
34. Obi, J.C. A comparative study of several classification metrics and their performances on data. *World J. Adv. Eng. Technol. Sci.* **2023**, *8*, 308–314. [CrossRef]

materials

MDPI

Article

A Statistical Assessment of Drilling Effects on Glass Fiber-Reinforced Polymeric Composites

Ana Martins [1,2], Alda Carvalho [1,3,4], Ivo M. F. Bragança [1,5], Inês C. J. Barbosa [1,5], Joaquim Infante Barbosa [1,5] and Maria A. R. Loja [1,5,*]

1 Centro de Investigação em Modelação e Otimização de Sistemas Multifuncionais (CIMOSM), Instituto Superior de Engenharia de Lisboa (ISEL/IPL), Av. Conselheiro Emídio Navarro 1, 1959-007 Lisboa, Portugal; ana.martins@isel.pt (A.M.); alda.carvalho@uab.pt (A.C.); ivo.braganca@isel.pt (I.M.F.B.); ines.barbosa@isel.pt (I.C.J.B.); joaquim.barbosa@isel.pt (J.I.B.)
2 Research Centre for Mathematics and Applications (CIMA), Instituto Superior de Engenharia de Lisboa (ISEL/IPL), Av. Conselheiro Emídio Navarro 1, 1959-007 Lisboa, Portugal
3 DCeT, Universidade Aberta, Palácio Ceia, Rua da Escola Politécnica, n.° 147, 1269-001 Lisboa, Portugal
4 CEMAPRE/ISEG Research, ULisboa, Rua Quelhas, 6, 1200-781 Lisboa, Portugal
5 Instituto de Engenharia Mecânica (IDMEC), IST-Instituto Superior Técnico, Universidade de Lisboa, Av. Rovisco Pais, 1, 1049-001 Lisboa, Portugal
* Correspondence: amelia.loja@isel.pt

Abstract: Fiber-reinforced composites are extensively used in many components and structures in various industry sectors, and the need to connect and assemble such types of components may require drilling operations. Although drilling is a common machining process; when dealing with fiber-reinforced composite materials, additional and specific problems may arise that can com-promise mechanical integrity. So, the main goal of this work is to assess how various input variables impact two main outcomes in the drilling process: the exit-adjusted delamination factor and the maximum temperature on the bottom surface where the drilling tool exits. The input variables include the type of drilling tools used, the operating speeds, and the thickness of the plates being drilled. By using Analysis of Variance (ANOVA), the analysis aims to identify which factors significantly influence damage and exit temperature. The results demonstrate that the influence of tools and drilling parameters is critical, and those selections impact the quality of the hole and the extent of the induced damage to the surrounding area. In concrete, considering the initially selected set of tools, the BZT03 tool does not lead to high-quality holes when drilling medium- and high-thickness plates. In contrast, the Dagger tool shows potential to reduce exit hole damage while also lowering temperature.

Keywords: composite materials; drilling; delamination characterization; thermographic characterization; statistical assessment; ANOVA

1. Introduction

As the use of composite materials continues to grow, the need to connect parts made of these materials increases the interest in understanding the drilling damage that occurs when preparing those parts. Because of the heterogeneous characteristics of composite materials, the damage that occurs during drilling, such as delamination and tearing, as well as thermomechanical changes due to the drilling effects on the composite, deteriorate the quality of the composite [1].

Several studies regarding the drilling of Fiber-Reinforced Polymer (FRP) composites can be found, among them, reviews on drilling Carbon Fiber-Reinforced Polymers (CFRP). An extensive review of drilling on CFRP presents relevant information found on drilling mechanisms, thermomechanical responses, drilling-induced damage, and the effects of various process conditions. High cutting speeds and low feed rates were found to improve the hole quality of CFRPs. Also, this review indicates that developing suitable tool geometries/materials, and the optimization of cutting parameters effectively decreases the drilling damage of cut CFRP holes [2]. The type of matrix material highly influences the thrust force and torque in conventional dry drilling of CFRP composites, while higher cutting speeds result in lesser torques developed during drilling [3].

To study the influence of machining parameters on the delamination damage of Glass Fiber-Reinforced Polymers (GFRP) during drilling, several drilling processes were explored and it was found that delamination is most influenced by feed rate, tool material, and cutting speed in conventional machining, while vibration assisted drilling and ultrasonic assisted drilling are more appropriate for drilling of GFRP [4].

As delamination is important damage resulting from drilling FRPs, many authors have investigated its mechanism and contributing factors [5–12]. In a review focused on delamination quantification and measurement techniques, several delamination measurement methods along with their advantages and drawbacks are compared and discussed [8]. The more accurate measurement techniques are X-ray radiography and computerized tomography, but are used less because of the high initial cost, the need for a secured area for inspection, and a high sample preparation time. Although with a lower accuracy, microscopy is the most generally used method due to its simplicity. Higher dimensionality of the delamination factor improves its accuracy but also increases computation time. The calculation of the delamination factor differs depending on the method used and many permutations and combinations are possible to define a more accurate delamination factor. Different models are used to predict drilling delamination, using different drilling parameters, such as cutting speed or cutting sequences [6,9].

The effect of tool material and geometry on the damage induced during drilling glass or carbon FRP composites is another interesting topic addressed by some authors [3,12–15]. Although extensive studies on the effect of cutting parameters and tool geometry on the quality of the hole have been made, the shearing of fiber-reinforced polymers needs a better understanding. Also, specific tools with special geometry need to be developed to achieve better performances [15]. In another article comparing different drilling tools on GFRPs, delamination was observed in the form of matrix debonding, uncut fibers, and fiber pull-out [7]. Results showed that the solid carbide tools had the best drilling performance for a low feed rate and a high speed, and high laminate thickness. Different drill types were also studied in drilling GFRP pipes and several tests were performed at a constant speed and different feed rates [16]. Thrust forces were measured and hole exit surface damage and borehole surface damage were examined with a digital microscope and scanning electron microscope, after the drilling operations. Damage is very much influenced by the tool geometry and feed rate, observing increased delamination for a conventional twist drill at lower feed rates, in comparison with a brad and spur drill, and a brad center drill. The latter generated less damage. New drilling tools are also presented in different works to improve hole quality, diminish delamination, and decrease overall damage on FRPs [14,17–20]. Hole quality is an important indicator of good drilling on FRPs, as it is related to less damage and smaller temperature-affected areas. Several studies seek to evaluate this indicator through different techniques such as studying bore quality factors' influence on the progress of tool wear and the thrust force [21], the effect of speed, feed rate, and drill angle on hole quality [22], and the application of artificial neuron networks to assess the effects of drilling parameters on drilling temperature and hole quality [23].

New methods for predicting damage in FRPs have been devised. Several models were developed to predict the delamination factor [6,24,25]. The use of Deep Neural Networks resulted in very small errors in the predicted delamination factor and circularity, demonstrating that this technique is very suitable for predicting drilling damage on FRPs [6]. Other more simple modelling techniques have been proposed to predict delamination with great accuracy, such as AutoCAD Image Processing [24]. Other models focus on predicting the temperature distribution during the drilling process. In one study, a model to predict the temperature distribution during the drilling process of unidirectional CFRP is used to explore the influence of feed rate and spindle speed, and the results show that increasing the feed rate decreases the drilling temperature and the opposite occurs when increasing the spindle speed [26].

The variation of temperature during drilling of FRPs is an important parameter for the quality of the hole and, consequently, it is related to damage. Thus, many studies focus on thermal damage, cutting heat accumulation and tool wear extent during the drilling of FRP structures. Several strategies are used to diminish the wear of tool bits, such as variable feed rates and reducing the number of holes performed [27]. The surface integrity is also used as an indicator of the quality of the hole to assess the performance of special drills made for drilling CFRPs with less tool wear [28]. Many techniques are used to obtain information on temperature, such as infrared cameras [5,29,30] and thermocouples [26,31]. When considering the temperature-dependent material properties of CFRP laminates to study the problem of contact at the drill margin–borehole surface interface during dry drilling [31], it is known that the increase in temperature during dry drilling reduces the elastic modulus of the CFRP and causes thermal expansion of the drill. This causes significant contact length at the drilling margin and borehole surface interface, which, in turn, increases damage. The temperature increase also increases the thrust force and torque, indicating that low feed rates is disadvantageous to dry drilling because of the temperature rise due to inefficient material cutting.

The use of non-destructive testing methods for estimating damage in composite materials has been an established practice in research for years. However, several studies have noted the increase in research on non-destructive testing techniques, driven by the growing use of composites in industries like aerospace, automotive, and energy sectors, where ensuring the integrity of structures is critical [32]. Furthermore, advancements in computing power, sensor technology, and imaging techniques have made these methods more accessible and reliable, promoting broader applications [33]. Many of these methods, including ultrasonic testing, infrared thermography, shearography, and acoustic emission, have evolved, offering greater precision, faster data processing, and automated systems for inspecting composite structures [32–38].

In another complementary perspective, exploratory data analysis, although consisting of an important research area by itself, plays a very important role when linked to specific engineering problems, as it may provide important insights into the experimental and/or modelling approaches observations, identifying the components, parameters, that most contribute to explain the observed experimental or simulated responses. To illustrate this, one may refer to the work developed by some co-authors, in complementary areas. Carvalho et al. [39] investigated the variability in the static and dynamic response of fiber-reinforced composites, considering multivariable linear regression models, to characterize the contribution of each modelling parameter to the explanation of those variabilities. Focusing on another type of composite materials, Rosa et al. [40] analyzed how material and geometrical uncertainty may modify the foreseen deterministic response of a structure built from dual-phase functionally graded materials. In that work, the authors proposed the constitution of statistic models to allow their use as alternative prediction models for such structures under similar operating conditions. More recently Carvalho et al. [41] investigated the influence that the uncertainty associated with carbon nanotubes' material and geometrical characteristics may have in the static behavior of functionally graded plates, where this gradient is dictated by the weight volume fraction of these nanoparticles. The study considered the constitution of multiple regression models which allow concluding on the influence of the characteristic parameters and also can be used as alternative prediction tools within the domain of the study.

In the present work, drilling parameters, such as the spindle speed, the feed rate, the GFRPs laminates' thicknesses, and tool bits, are considered and their influence is assessed on a set of usual delamination factors that are often used as damage metrics, and on the thermally influenced area.

Regarding the authors' knowledge, this work has an innovative character, considering the joint glass fiber composites' drilling experimental work and the detailed statistical study performed upon the results achieved. This statistical approach enables an extensive analysis and corresponding conclusions regarding the engineering problem this work addresses, namely considering the influence of tool selection, material thicknesses on final part drill quality, as well as the importance of drilling parameters in minimizing undesirable effects.

The remainder of this manuscript presents the following structure: the second section presents the fundamental theoretical aspects required for the study's development, followed by the third section where the case studies and the results achieved are presented and discussed and more specific conclusions are drawn. The last section refers to the conclusions of the present work.

2. Materials and Methods

2.1. Composites Plates and Experimental Setup Characterization

In the present work, four sets of glass fiber polymeric laminated plates were produced via a wet layup method: a set of plane laminates consisting of 10 layers of short glass fibers; and three sets of plane laminates made of fiberglass textile and epoxy resin, with 10, 20, and 30 layers, respectively. All sets had a quadrilateral configuration with a 250 mm edge. After the curing process, the plates were characterized and identified, as presented in Table 1.

Table 1. Plate characterization.

Reinforcement	Number of Layers	Thickness (mm)	Glass Fiber
Fabric (long fiber)	10 10 10 20 30	2.88 3.00 2.95 5.54 8.36	
Mat (short fiber)	10	6.37	

Glass fiber was herein considered either in the form of fabric, produced with interwoven long fibers, or in the form of mat, which consists of short fibers randomly overlapped, nonwoven. These two types of fiber dispositions represent the most common alternatives used in glass fiber composite manufacturing.

Different plates were prepared as indicated in Table 1. The wet layup technique was used, with Sicomin's SR1500 resin mixed with hardener SD2505 in a 3:1 ratio. The various layers were impregnated with the resin and cured at room temperature under compression between metal plates.

Figure 1 presents the tools used in the drilling experiments.

As the material properties of GFRC are highly dependent on the fabrication method, and on the uncertainty of associated sources, thermographic analysis was used to find correlations between the material defects and the drilling outcome.

To perform the drilling operations, a specific support was developed to be adjusted to the computerized numerical control machine (CNC, Cincinnati Arrow 500 VMC, Cincinnati, OH, USA), and two thermographic cameras (ThermaVue CSI 735R INFRARED (Southampton, UK) and TROTEC ec060 (Heinsberg, Germany)) were used to record the operation from the upper and lower laminate surfaces' perspectives.

The camera recording the lower surface of the laminate was aligned with the tool axis, while the upper surface camera had a slight deviation to this axis due to being put near the tool. The experimental apparatus is shown in Figure 2.

	Tool	Number of Teeth (z)	Diameter (mm)
(a)	BZT01	2	6
(b)	BZT03	10	
(c)	*Dagger*	2	
(d)	Helical	2	

Figure 1. Cutting tools: (**a**) BZT01 Tool (Manufacturer BZT, Leopoldshöhe, Germany, Ref. 751080060F); (**b**) BZT03 Tool (Manufacturer BZT, Ref. 751070060F); (**c**) Dagger Tool (Manufacturer GANDTRACK, Oldham, UK, Ref. GT-50-6.0 63089); (**d**) Helicoidal Tool (Manufacturer SECO, Fagersta, Sweden, Ref. SD205A-6.0-32-6R1-C2).

Figure 2. Experimental apparatus: (**a**) drilling tool; (**b**) lower thermographic camera; (**c**) plate; (**d**) upper thermographic camera.

After conducting preliminary experimental tests for the thicker laminate using the BZT03 tool, the temperature range was verified to be within [23–250 °C].

2.2. Delamination Quantification Metrics

To perform the statistical assessment of the results, we considered the various affected areas, including those related to delamination induced by drilling and those associated with thermally affected regions in the laminate. The quantification of delaminated areas has been explored by various authors, who have employed different methods to assess the extent of the damage. One usual indicator is the damage factor Fd [42,43]; however, in this work, the adjusted delamination factor, F_{daj}, that considers the contribution of the fissure size and the damage area [44] is more extensively used:

$$F_{daj} = F_d + \frac{A_d}{(A_{max} - A_0)}\left({F_d}^2 - F_d\right) \tag{1}$$

where A_d is the delaminated area, A_{max} and A_0 are, respectively, the areas corresponding to the maximum damage diameter, D_{max}, and to the hole diameter, D_0. An alternative assessment of the delaminated area was proposed by Ahn et al. [45], which considers the existing proportionality between the image areas and the corresponding number of pixels:

$$F_{da} = 1 + \frac{N_p(A_d)}{N_p(A_0)} \tag{2}$$

where $N_p(A_d)$ is the number of pixels of the delaminated area, and $N_p(A_0)$ is the number of pixels corresponding to the hole area. The maximum delamination diameter D_{max} and the delaminated area A_d are obtained as:

$$D_{max} = D_0\frac{N_p(D_{max})}{N_p(D_0)} \tag{3}$$

where $N_p(D_{max})$ is the number of pixels of the maximum diameter area, and D_0 is the hole diameter. This last calculation of the damaged area, F_{da}, has proven to be more reliable than the damage factor, F_d, as the latter is not sensitive to the in-plane shape of the delamination profile. The studies conducted take these different measurements into account to characterize the results of the experimental drilling tests.

To identify the delaminated and thermally affected regions resulting from drilling operations, scans were performed on both the upper and lower surfaces of the laminate. The entire drilling process was also recorded using thermographic cameras. The captured digital data were subsequently processed with the open-source image processing software Fiji (ImageJ 1.53q, public domain (Bethesda, MD, USA) [46], the images were examined, and the extent of the damage was quantified.

After calibrating the initial image dimensions, Figure 3a, the photograph was converted to binary grayscale to isolate the damaged areas, Figure 3b. The FFT bandpass filter corrected shadows and smoothed the image, and the threshold tool was applied using the iterative procedure known as the IsoData algorithm [47], as seen in Figure 3c. The resulting binary image was used to determine the maximum damage diameter through a line that passes through the hole's center, Figure 3d.

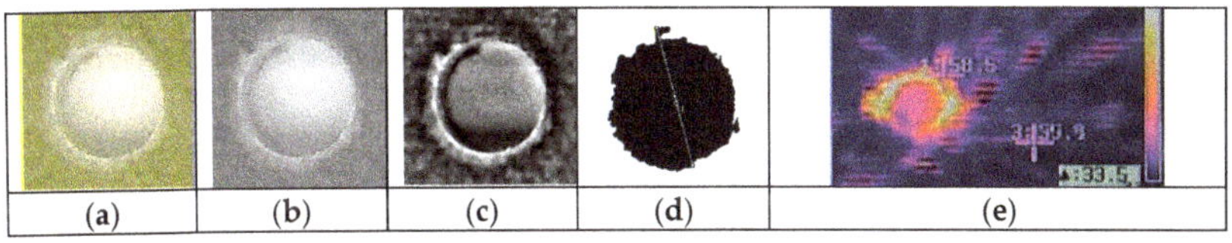

Figure 3. Procedure to evaluate the delaminated and thermally area affected by drilling: (**a**) original photograph; (**b**) binary image; (**c**) image after FFT bandpass filtering; (**d**) diameter measured after thresholding. (**e**) Thermographic image from lower camera.

In addition to the discrete temperature data provided by the thermographic cameras at specific points (see crosses in Figure 3e), a complementary methodology based on the RGB color system was employed to characterize the maximum temperature reached during the drilling process. The color scale of the images was also taken into account. This procedure ensures that the maximum temperature in the selected image is measured, as it may be somewhat distant from the points marked by the crosses. However, the presence of uncut material, fibers, expelled material, as well as the determination of the frame considered to reach the maximum value, can influence the determined value.

2.3. Statistical Analysis Methodology

One way to compare a variable across different categories is through a graphical representation using multiple boxplots [48]. This graphical tool shows the distributions side by side, allowing for a clear comparison of their medians, ranges, and variability.

An important graphical tool for visualizing a multivariate dataset is a composition of several plots, referred to here as a matrix plot. This plot allows for the analysis of both the individual distributions and the different correlations between variables. In this type of plot, the individual distribution of the variables under analysis is shown along the diagonal of the matrix. An empirical density line is added, providing a suggestion of the appropriate probability model for each variable. This type of analysis is important for assessing the assumptions often required for inferential purposes. Additionally, matrix plots allow for the analysis of correlations between pairs of variables. Below the diagonal, scatter plots are displayed, while above it, the values of Pearson's linear correlation coefficient are shown [48].

Analysis of Variance (ANOVA) is a statistical method used to determine if there are statistically significant differences between the means of three or more independent groups [48]. ANOVA tests the null hypothesis that all group means are equal,

$$H_0 : \mu_1 = \mu_2 = \cdots = \mu_p \qquad vs \qquad H_1 : \exists\, \mu_i \neq \mu_j$$

and if this hypothesis is rejected, it indicates that at least one group differs from the others. However, ANOVA does not specify which groups differ from each other, so post-hoc tests, such as Tukey's Honestly Significant Difference (HSD) test, are often used to perform pairwise comparisons between group means. Tukey's HSD test controls for Type I errors by adjusting the significance level in multiple comparisons, ensuring a more reliable identification of specific differences between group means. It is possible to visualize these differences through the graphical representation of confidence intervals for the difference between means [48].

ANOVA relies on several key assumptions to ensure the validity of its results [48]. It assumes that the data in each group follow a normal distribution. This assumption can be checked using normality tests such as Shapiro–Wilk [49]. Also, it assumes the independence of observations, meaning that the data points in each group are independent of each other, which is often guaranteed by proper experimental design. All ANOVA results presented in this article meet these assumptions with a significance level of 1%.

When the assumptions of a parametric test are violated, a non-parametric approach is applied, such as the Kruskal–Wallis test [50]. This method is used to compare three or more independent groups and to assess whether there are statistically significant differences in their medians.

3. Results and Discussion

As an introductory note in this section and with no prejudice to more detailed descriptions that will be provided in each sub-section, it is considered adequate to make an overall brief presentation of the data analysis flow which was supported by intermediate conclusions.

Hence, due to some experimental limitations, and issues encountered during the experimental tests, the statistical analysis was conducted sequentially.

In the first phase, the aim was to determine if there were differences among the four tools used in terms of damage indicators or temperature within the considered range of values (spindle speed of 5300 rpm and feed rate of 640 mm/min). This initial case allowed to identify high correlations in the output variable values at the entrance and exit. These values remained similar throughout the tests, which led us to focus on the exit-adjusted delamination factor, the variable Fdaj_out, for damage analysis. Furthermore, based on the results obtained, it was decided to continue the study using only the BZT03 and the Dagger tools, as they represent two very different performances in terms of damage.

In the second phase, given the distinct nature of the two tools, a study of both tools was conducted in parallel. The aim of this study was to understand the influence of using low, medium, or high ranges in a combination of speed and feed rate on the damage factor. In the third phase, different thickness values were then analyzed.

In the end, a discussion was elaborated and provided for each of the two tools (BZT03 and Dagger) regarding the relationship between damage and temperature across the different categorical dimensions used (feed rate and thickness).

3.1. Variables Characteristics

The input variables considered in the present work essentially comprise the drilling tools used, the operating speeds, and the thickness of the plates to be drilled.

The drilling of the samples, considering the different variables, was performed and the effect on the samples was registered, as shown in Figure 4 for one of the samples. Despite being by far the drill that makes the best holes and causes significantly less damage, the Dagger tool sometimes leaves some fibers uncut at the exit.

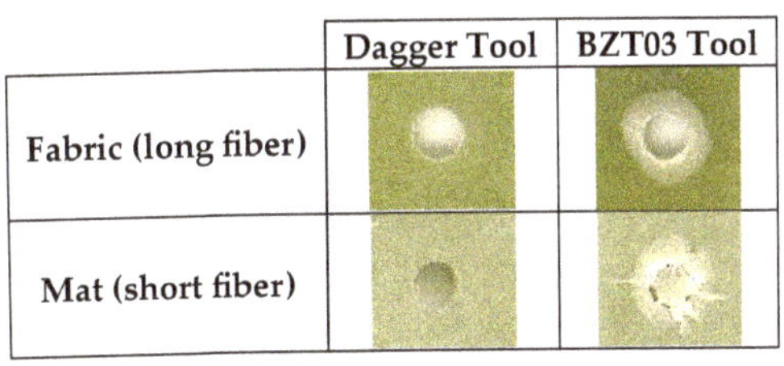

Figure 4. Holes' exits performed by Dagger and BZT03 in plates reinforced with small or long fibers (spindle speed 5300 rpm; feed rate 640 mm/min).

After conducting some exploratory tests and observing minimal variation in the effect of drilling parameters on damage of different plates with different fiberglass textile, it was decided to evaluate only the holes made in the long-fiber plates. This choice helped reduce the variation in measured exit damage, often caused by factors like a single fiber being pulled out rather than delamination between layers.

The output variables are coefficients calculated according to damage characterization indicators, and the maximum temperature at the bottom surface where the drilling tool exits the plate. The input variables and data associated as well as the output variables and corresponding acronyms are presented in Table 2.

Table 2. Description of input and output variables.

Input Variables	Data
Tool	BZT01, BZT03, Dagger, Seco
Spindle speed (rpm)	1300, 2650, 5300
Feed rate (mm/min)	78, 156, 234, 318, 640
Thickness (mm)	2.88, 2.95, 3, 5.54, 8.36
Output variables	**Acronym**
Exit delamination factor	Fd_out
Exit delaminated area factor	Fda_out
Exit-adjusted delamination factor	Fdaj_out
Exit delaminated area	Area_out
Exit maximum temperature	Tmax_inf

Considering the characteristics of the input variables, they were all considered categorical variables. From the experimental tests developed, 76 (see Appendix A) were considered valid, being those summarized in Table 3.

Table 3. Experiment characteristics for output variables assessment using the tools BZT01, BZT03, Seco, and Dagger.

Tool	Thickness (mm)	Spindle Speed (rpm)	Feed Rate (mm/min)	Number of Tests
BZT01	2.88 (low)	5300	640	5
Seco	2.88 (low)	5300	640	5
BZT03	2.88 (low)	5300	640	10
	5.54 (medium)	5300	640	5
	8.36 (high)	5300	640	3
	3 (low)	2650	318	4
	3 (low)	1300	156	5
	3 (low)	1300	78	5
	3 (low)	1300	234	5
Dagger	2.88 (low)	5300	640	9
	2.95 (low)	2650	318	5
	2.95 (low)	1300	156	5
	2.95 (low)	1300	78	5
	2.95 (low)	1300	234	5

Concerning the information in Table 3, it is relevant to note that the column corresponding to the thickness variable, besides containing the values themselves, also contains the indication of a qualitative classification {low, medium, high}. This is because, for a set of situations where the thickness assumes the values of 2.88 mm, 2.95 mm, and 3 mm, this was not intentional and corresponds to uncertainties in the manufacturing process. So, this set of thicknesses is considered to pertain to the same class of thinner plates.

Cutting tools with the same diameter but different characteristics were selected to evaluate the influence of the tools' geometry. Some of the selected tools resemble milling cutters, while others are more similar to traditional drill bits. This selection aimed to analyze a wider range of tool types for drilling holes in glass fiber epoxy plates.

3.2. Case 1: Assessing the Influence of Input Variables on the Exit-Adjusted Delamination Factor for an Initial Set of Drilling Tools

In the first stage of this work, we began by considering five experimental tests for each of the four tools under study. All the drilling tools were tested under the same conditions (spindle speed of 5300 rpm and feed rate of 640 mm/min) on plates with a uniform thickness of 2.88 mm.

In this preliminary case, it is also important to say that the output variables at the tool entrance on the top surface of the plate were also initially registered, but they were shown to be less significant when compared to the results obtained at the exit. So, only the exit output variables will be considered.

The correlation coefficients between the different variables can be observed in Figure 5, where the matrix plots corresponding to the entrance and exit output variables are presented. The red line corresponds to a smooth curve called Locally Estimated Scatterplot Smoothing (LOESS). This method is useful for visualization and is constructed by fitting multiple quadratic (or possibly linear) regression lines as a moving window passes along the x-axis. The black dots are the data and the red one is the corresponding mass center.

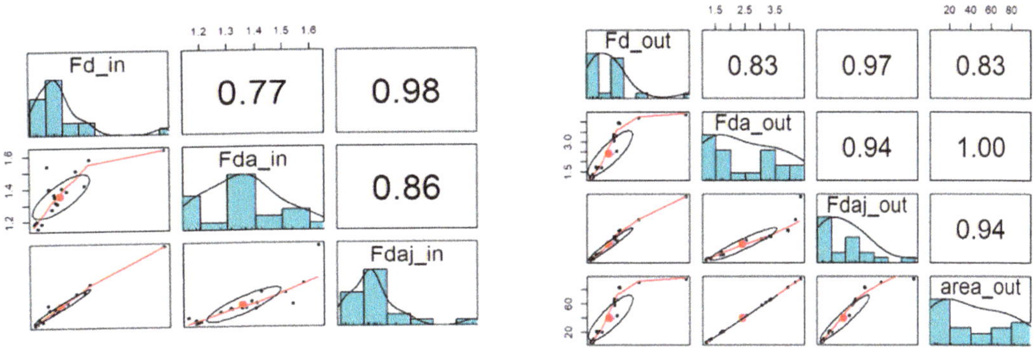

(**a**) (**b**)

Figure 5. Matrix plots for the output variables; (**a**) entrance, and (**b**) exit.

Considering the results in the matrix plot of the exit output variables, it is visible that the variables are strongly correlated, namely the exit delamination factor (Fd_out) and the exit-adjusted delamination factor (Fdaj_out), the exit-delamination area factor (Fda_out), and the exit-delaminated area (area_out). This is an expected result considering how these metrics are obtained. Because of this, we have decided to keep only one of these metrics to characterize the damage, being this metric, the Fdaj_out variable. So, from this point on, only the exit output variables will be considered, more specifically the exit-adjusted delamination factor.

3.2.1. Influence of a Tool on the Exit-Adjusted Delamination Factor at the Tool's Exit

Following the previous analysis, it was important to understand which tool would be more adequate for the drilling operation. If one builds the boxplots of the exit-adjusted delamination factor for each tool, one obtains the representations in Figure 6.

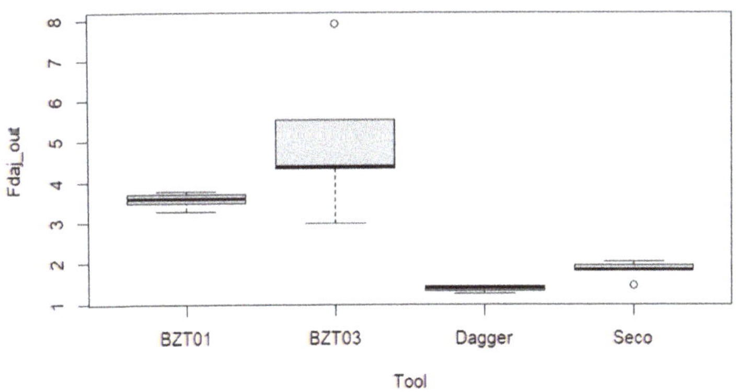

Figure 6. Multiple boxplots for the exit-adjusted delamination factor as a function of the tools BZT01, BZT03, Dagger, and Seco.

From Figure 6, we conclude that there is a significant difference among the Fdaj_out variable mean values associated with each tool. It is possible to verify that the Dagger and Seco tools enable drilling with a lower exit damage value for the plates. In contrast, BZT tools exhibit not only a higher exit-adjusted delamination factor but also higher uncertainty. So, Dagger tools offer a more consistent drilling performance with greater uniformity and quality, providing superior control over the process.

Table 4 presents the ANOVA results from the comparison of the Fadj_out mean values for the four tools, allowing for the conclusion that there is a significant difference among them. The first column lists the different sources of variability (Tool, Error, Total), indicating where the variation in the data originates. The second column shows the degrees of freedom (df), representing the number of independent observations that can vary for each source of variation. The third column contains the Sum of Squares (SS), which measures the total variability attributed to each source and quantifies how much variation is explained by each factor. The fourth column presents the Mean Square (MS), calculated by dividing the Sum of Squares by the degrees of freedom; this gives the average variation for each source of variability (MS = SS/df). The fifth column provides the F-Statistic, the ratio of the mean square between groups to the mean square within groups, which tests whether the variability between groups is greater than that within groups, indicating a significant effect. Finally, the last column displays the p-value, which represents the probability of obtaining a result at least as extreme as the observed one, assuming that the null hypothesis (the hypothesis that there is no difference) is true. A low p-value suggests that the differences between groups are statistically significant. To aid in interpreting the results, a code is added to the *p*-value to indicate whether the statistical significance is below 0.001 or at 0.01.

Table 4. ANOVA results from the comparison of the Fadj_out mean values for the four tools: BZT 01, BZT 03, Dagger, and Seco.

	Degrees of Freedom	Sum of Squares	Mean Squares	F	*p*-Value [1]
Tool	3	42.87	14.29	16.38	<0.001 **
Residuals	16	13.96	0.87		

[1] Significance codes: <0.001 '**' 0.01 '*'.

Given that the equality of the mean values has been rejected, it is appropriate to examine the differences between each pair. So, a multiple comparisons test (Tukey test) was performed. The results obtained are presented in Figure 7.

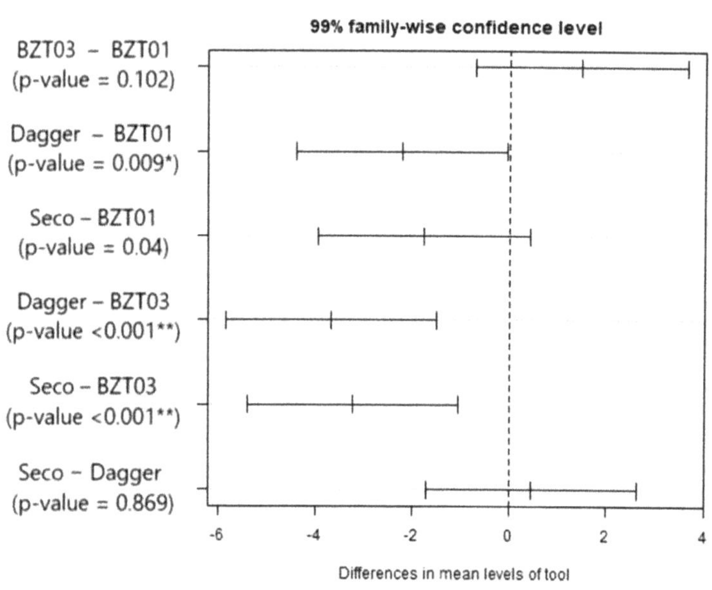

Figure 7. Results from the multiple comparisons test, between pairs of tools: confidence intervals for mean differences. Significance codes: <0.001 '**' 0.01 '*'.

Considering the results in Figure 7, it is possible to conclude that the BZT01 and BZT03 tools do not differ significantly, the same applying to the Dagger and Seco pair and the Seco and BZT01 pair. Supported by these results, we proceeded with further tests considering only the BZT03 and Dagger tools, which is in agreement with what was already observed in Figure 6.

3.2.2. Influence of Spindle Speed and Feed Rate on the Exit-Adjusted Delamination Factor at the Tool's Exit

The experimental data were increased to evaluate the influence of the spindle speed and the feed rate on the exit-adjusted delamination factor at the tool's exit, as shown in Table 5.

Table 5. Experiment characteristics for exit-adjusted delamination factor assessment using BZT03 and Dagger tools.

Tool	Thickness (mm)	Spindle Speed (rpm)	Feed Rate (mm/min)	Number of Tests
BZT 03	2.88	5300	640	10
	3	2650	318	4
	3	1300	156	5
	3	1300	78	5
	3	1300	234	5
Dagger	2.88	5300	640	9
	2.95	2650	318	5
	2.95	1300	156	5
	2.95	1300	78	5
	2.95	1300	234	5

The analysis of the boxplots in Figure 8, relating the exit-adjusted delamination factor at the tool's exit with the spindle speed for each one of the tools, raises doubts regarding the hypothesis of equality of the mean value of the exit-adjusted delamination factor in the three ranges of the spindle speed considered (low, medium, and high).

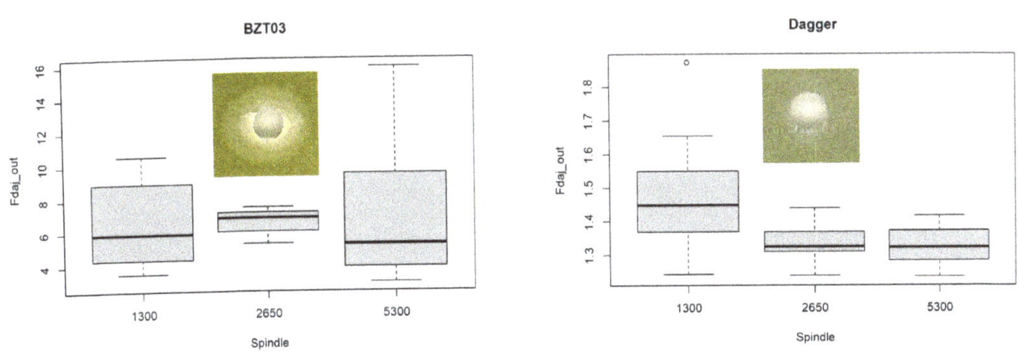

(**a**) (**b**)

Figure 8. Multiple boxplots for the exit-adjusted delamination factor at the tool's exit as a function of the spindle speed for the (**a**) BZT03, and (**b**) Dagger tools.

To confirm, an ANOVA was performed and the p-values obtained from the ANOVA conducted for the BZT03 and Dagger tools (Tables 6 and 7) were respectively 0.99 and 0.02. Considering a significant level of 1%, there was no significant difference in the mean value of the exit-adjusted delamination factor in the three ranges of the spindle speed considered (low, medium, and high).

Table 6. Regarding the BZT 03 tool, ANOVA results from the comparison of the Fadj_out mean values for the three ranges of the spindle speed.

BZT03	Degrees of Freedom	Sum of Squares	Mean Squares	F	*p*-Value [1]
Spindle	2	0.10	0.07	0.01	0.99
Residuals	34	327.30	9.63		

[1] Significance codes: <0.001 '**' 0.01 '*'.

Table 7. Regarding the Dagger tool, ANOVA results from the comparison of the Fadj_out mean values for the three ranges of the spindle speed.

Dagger	Degrees of Freedom	Sum of Squares	Mean Squares	F	*p*-Value [1]
Spindle	2	0.15	0.07	4.27	0.02
Residuals	26	0.45	0.02		

[1] Significance codes: <0.001 '**' 0.01 '*'.

A similar approach was conducted regarding the feed rate. The corresponding boxplots are presented in Figure 9.

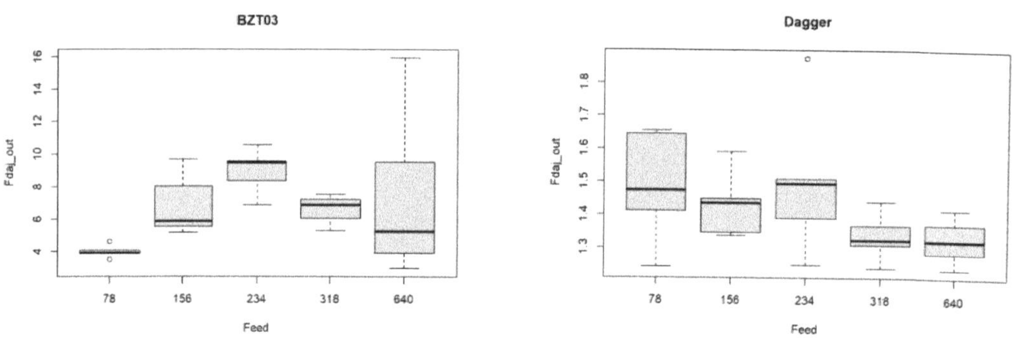

(**a**) (**b**)

Figure 9. Multiple boxplots for the exit-adjusted delamination factor at the tool's exit as a function of the feed rate for the (**a**) BZT03, and (**b**) Dagger tools.

Although there are differences in Figure 9, they are not significant. The ANOVA results (Tables 8 and 9) indicate that when considering a significant level of 1%, there are no significant differences in Fdaj_out, for all the feed rate values considering both the Dagger and the BZT 03 tools.

Table 8. Regarding the BZT 03 tool, ANOVA results from the comparison of the Fadj_out mean values for the five ranges of the feed rate.

BZT03	Degrees of Freedom	Sum of Squares	Mean Squares	F	*p*-Value [1]
Feed rate	4	62.35	15.59	1.88	0.14
Residuals	32	265.11	8.285		

[1] Significance codes: <0.001 '**' 0.01 '*'.

Table 9. Regarding the Dagger tool, ANOVA results from the comparison of the Fadj_out mean values for the five ranges of the feed rate.

Dagger	Degrees of Freedom	Sum of Squares	Mean Squares	F	p-Value [1]
Feed rate	4	0.16	0.04	2.22	0.10
Residuals	24	0.44	0.02		

[1] Significance codes: <0.001 '**' 0.01 '*'.

3.3. Case 2: Assessing the Maximum Temperature at the Tool's Exit as a Function of the Spindle Speed and the Feed Rate for the BZT03 and Dagger Tools

In the sequel of the previous case study, where two tools (BZT03 and Dagger) were selected considering their differentiated characteristics regarding the delamination assessment, the present case study proceeds with those same tools to assess how the input parameters may influence the maximum temperature achieved at the tool's exit location, in the bottom surface of the plates.

The experiment characteristics used to assess the influence of the spindle speed and the feed rate on the maximum temperature at the tool's exit were already presented in Table 3.

From this experimental data obtained, we have built the boxplots in Figure 10, where the exit-adjusted delamination factor at the tool's exit is related to the three spindle speed values tested for each one of the tools. Those boxplots in Figure 10a suggest that for the BZT03 tool, some differences in the mean value of the maximum temperature at the tool's exit, for the three spindle speed values. However, the ANOVA confirms that they are not significant (p-value = 0.03) (Table 10).

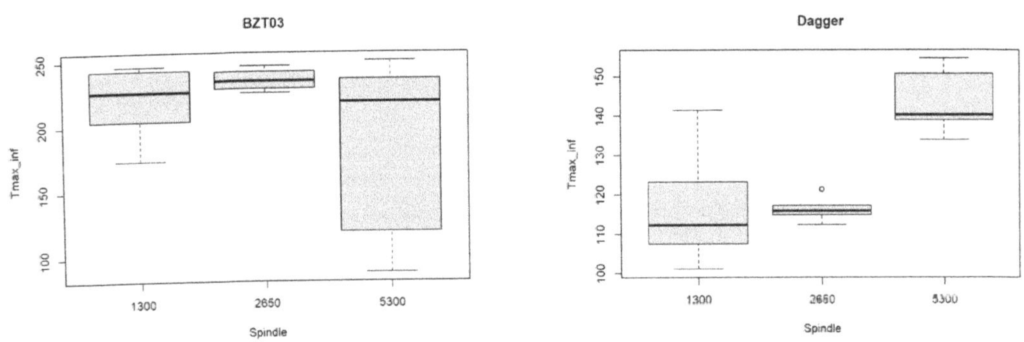

(a) (b)

Figure 10. Multiple boxplots for the maximum temperature at the tool's exit as a function of the spindle speed for the (**a**) BZT03 and (**b**) Dagger tools.

Table 10. Regarding the BZT03 tool, ANOVA results from the comparison of the Tmax_inf mean values for the three ranges of the spindle speed.

BZT03	Degrees of Freedom	Sum of Squares	Mean Squares	F	p-Value [1]
Spindle	2	21,371	10,685	4.11	0.03
Residuals	33	85,827	2601		

[1] Significance codes: <0.001 '**' 0.01 '*'.

However, Figure 10b and Table 11 show that for the Dagger tool, there exists a significant difference among the three spindle speed values (ANOVA p-value < 0.001). The temperature for the higher speed significantly differs from the mean temperature for the other two spindle speed values.

Table 11. Regarding the Dagger tool, ANOVA results from the comparison of the Tmax_inf mean values for the three ranges of the spindle speed.

Dagger	Degrees of Freedom	Sum of Squares	Mean Squares	F	*p*-Value [1]
Spindle	2	4347	2173.6	21.57	<0.001 **
Residuals	26	2620	100.8		

[1] Significance codes: <0.001 '**' 0.01 '*'.

Figure 11 depicts the multiple comparison tests between the pairs of spindle speed values considered in the experiments. It can be concluded that a significant difference exists for the cases: 5300–1300 rpm and 5300–2650 rpm. It can be inferred that using the maximum spindle rotation speed in this study will lead to significant differences when compared to intermediate or lower speeds, making it more relevant to evaluate the extreme values.

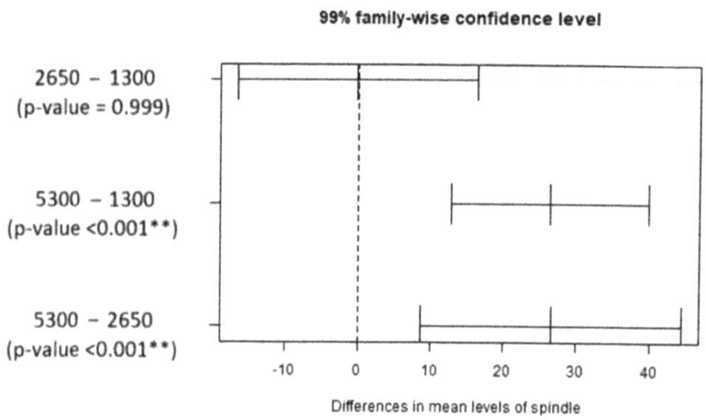

Figure 11. Results from the multiple comparisons test for pairs of spindle speed: confidence intervals for mean differences. Significance codes: <0.001 '**' 0.01 '*'.

The boxplots for the maximum temperature at the tool's exit at the bottom of the plates, as a function of the feed rate, are presented in Figure 12.

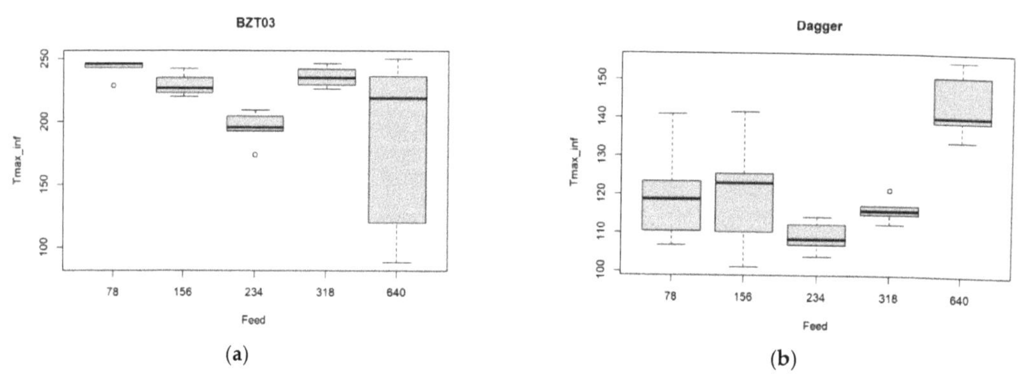

Figure 12. Multiple boxplots for the maximum temperature at the tool's exit as a function of the feed rate for the (**a**) BZT03 and (**b**) Dagger tools.

Regarding the BZT 03 tool boxplots presented in Figure 12a, the null hypotheses of the Shapiro normality test were rejected, so the non-parametric Kruskal–Wallis test was used.

The mean temperature value is significantly different for the five feed rate values (p-value < 0.001). The multiple comparison tests allow us to conclude that there exists only a significant difference for the cases 78–234 mm/min.

For the Dagger tool results presented in Figure 12b one observes that the mean temperature value is significantly different for the five feed rate values considered (ANOVA p-value < 0.001, see Table 12).

Table 12. Regarding the Dagger tool, ANOVA results from the comparison of the Tmax_inf mean values for the five ranges of the feed rate.

Dagger	Degrees of Freedom	Sum of Squares	Mean Squares	F	p-Value [1]
Spindle	4	4755	1188.7	12.89	<0.001 **
Residuals	24	2213	92.2		

[1] Significance codes: <0.001 '**' 0.01 '*'.

The multiple comparison tests (Figure 13) show that a significant difference exists for the cases: 78–640 mm/min, 156–640 mm/min, 234–640 mm/min, and 318–640 mm/min. Concerning Shapiro normality test, it was verified.

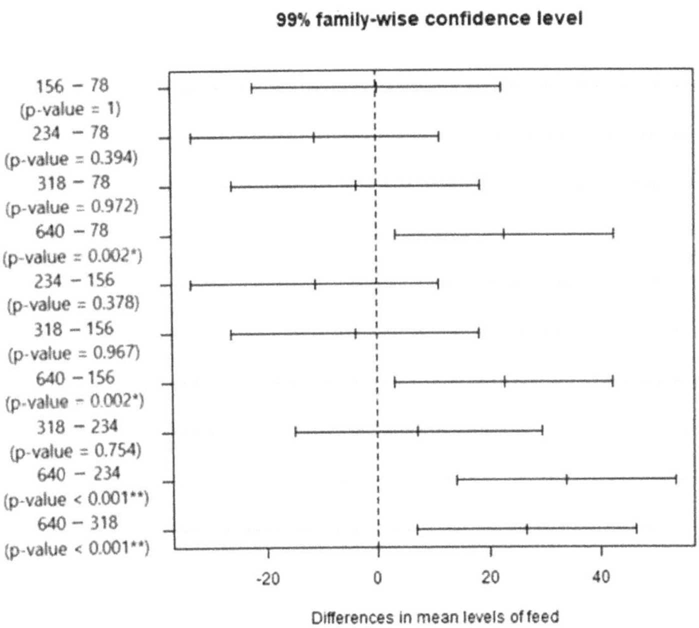

Figure 13. Results from the multiple comparisons test between pairs of feed rate values: confidence intervals for mean differences. Significance codes: <0.001 '**' 0.01 '*'.

3.4. Case 3: Influence of Plates' Thickness on the Exit-Adjusted Delamination Factor and the Maximum Temperature at the Tool's Exit

To analyze the influence of the plates' thickness on the exit-adjusted delamination factor and on the maximum temperature at the tool's exit neighborhood, a few more experiments were conducted using only the BZT 03 tool. Table 13 summarizes these experimental characteristics.

Table 13. Set of experimental tests with the BZT03 tool for the maximum temperature at the tool's exit.

Thickness (mm)	Spindle Speed (rpm)	Feed Rate (mm/min)	Number of Tests
2.88	5300	640	10
5.54	5300	640	5
8.36	5300	640	3
3	2650	318	4
3	1300	156	5
3	1300	78	5
3	1300	234	5

The boxplots corresponding to the results obtained from these experimental tests are presented in Figure 14.

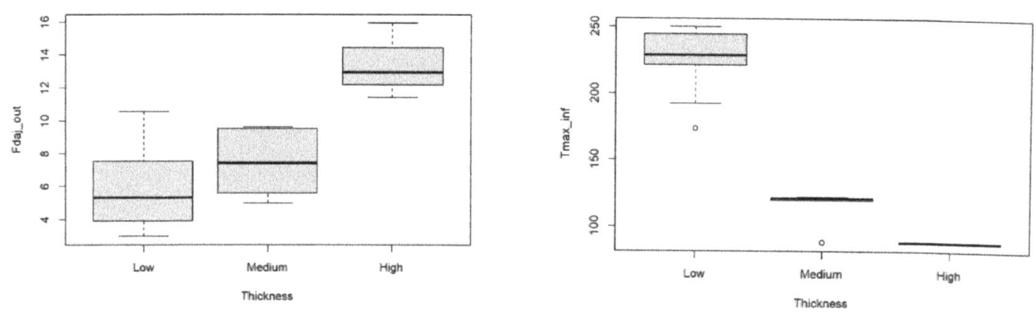

(a) (b)

Figure 14. Multiple boxplots for the (**a**) exit-adjusted delamination factor, and for the (**b**) maximum temperature at the BZT3 tool's exit as a function of three classes of thicknesses—low, medium, and high.

The results from the ANOVA (Table 14) and multiple comparison tests (Figure 15) indicate that the mean of Fdaj_out is significantly different for the high and medium thickness as for the high and low thickness. An inverse relation is observed between the damage factor and the maximum temperature as the plate thickness varies. The damage values increase with greater thicknesses, even though the temperature on the lower side of the plate decreases. This may be caused by the tool's geometry, which does not promote good chip removal, potentially leading to poorer cutting performance in the final stage. This may result in greater compressive forces and contribute to greater delamination of the composite.

However, regarding the maximum temperature at the tool's exit, the null hypothesis of the Shapiro normality test was rejected, so we resorted to the non-parametric Kruskal–Wallis test. This latter test indicates that the mean of the output variable, Tmax_inf, is significantly different for the three thickness classes. The corresponding non-parametric multiple comparison tests indicate that the mean of Tmax_inf is significantly different for the low and medium thicknesses as well as for high and low thicknesses.

Table 14. ANOVA results from the comparison of the Fdaj_out mean values for the three ranges of thickness.

BZT03	Degrees of Freedom	Sum of Squares	Mean Squares	F	p-Value [1]
Thickness	2	159.7	79.83	16.17	<0.001**
Residuals	34	167.8	4.94		

[1] Significance codes: <0.001 '**' 0.01 '*'.

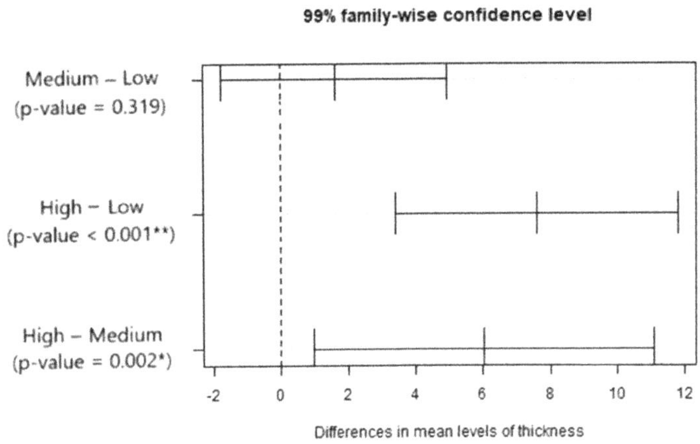

Figure 15. Results from the multiple comparisons test between pairs of mean levels of thicknesses: confidence intervals for mean differences. Significance codes: <0.001 '**' 0.01 '*'.

3.5. *Case 4: The Relations Among Temperature, Exit-Adjusted Delamination Factor, and Other Input Variables, for the BZT03 and Dagger Tools*

This final sub-section devoted to the presentation and discussion of the results is intended to analyze if and how the exit-adjusted delamination factor (Fdaj_out) and the maximum temperature (Tmax_inf) at the bottom surface are related.

To avoid redundancy, the presentation of results is limited to some sufficiently illustrative studies to demonstrate the behaviors' trends. The BZT03 and Dagger tools were used separately to analyze the correlation between the two output variables (Tmas_inf and Fdaj_out).

3.5.1. Influence of Plates' Thicknesses and Spindle Speed on the Exit-Adjusted Delamination Factor and Maximum Temperature at the Tool's Exit for the BZT03 Tool

According to the analysis performed in Section 3.3, it was possible to conclude that the plates' thickness classes (low, medium, and high) differently affect the exit-adjusted delamination factor and the maximum temperature at the tool's exit. In the present study, we aim to introduce an additional input variable into this analysis, namely the spindle speed. The relations among this wider set of variables are presented in Figure 16.

The bubble plot in Figure 16 illustrates very clearly that for medium- to high-thickness plates, we achieved the highest values for the exit-adjusted delamination factor, although for lower maximum temperatures. For the thickest plates ("high" class of thicknesses) the lowest temperatures were achieved although presenting the highest adjusted delamination factors.

Figure 16 also shows that when dealing with thinner plates in the so-called "low" class, although attaining the highest temperatures, the use of the highest spindle speed (5300 rpm) provides the lowest exit-adjusted delamination factor values. For lower values of spindle speed, namely for the lower one, the exit-adjusted delamination factor starts to increase despite presenting a greater dispersion in the delamination factor.

Since the thicker plates, medium and high, show exit-adjusted delamination factor values assessed as too high to be considered successfully executed, these variations will not be taken into account in the upcoming comparisons.

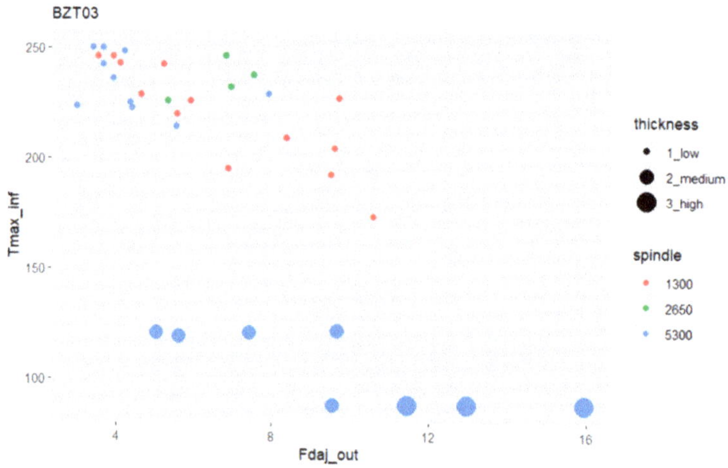

Figure 16. Bubble plot relating the maximum temperature with the exit-adjusted delamination factor, the thickness class, and spindle speed for the BZT03 tool.

3.5.2. Influence of Spindle Speed and Feed Rate on the Exit-Adjusted Delamination Factor and Maximum Temperature at the Tool's Exit, for the BZT03 Tool

Regarding the BZT03 tool, the Pearson's correlation coefficient between the variables (Tmax_inf) and (Fdaj_out) is −0.730, so there is an inverse correlation.

Figure 17 presents for the BZT03 tool a bubble plot with the relative influence of the input variables, spindle speed, and feed rate in the exit-adjusted delamination factor and the maximum temperature at the tool's exit in the plates' bottom surfaces.

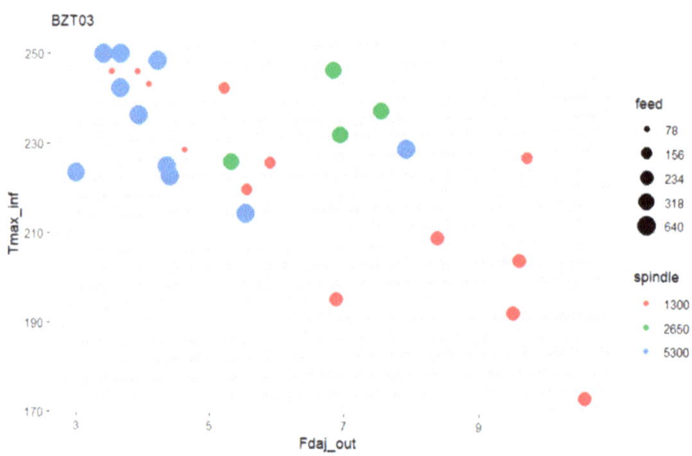

Figure 17. Bubble plot relating the maximum temperature with the exit-adjusted delamination factor and the feed rate and spindle speed for BZT03 tool.

As the bubble plot in Figure 17 illustrates, when the BZT03 tool operates at 5300 rpm with a feed rate of 640 mm/min, higher values of temperature at the tool's exit (near the maximum value for the camera) are attained, although with a lower exit-adjusted delamination factor. For the spindle speed of 2650 rpm the temperatures are similarly high although the damage presents in general a higher metric value.

The lower spindle speed (1300 rpm) when combined with the lowest feed rate (78 mm/min) yields very high-temperature values although the lowest damage values; however, when the feed rate is augmented until 234 mm/min, the temperature progressively diminishes and the exit-adjusted delamination factor increases, assuming values within the range of 6.5 to 11. Even considering only the tests that used the same feed per tooth with the BZT03 tool, it can be stated that as the drilling speed increases, the damage at the exit of the hole decreases, despite the temperature rising.

3.5.3. Influence of Spindle Speed and Feed Rate Speed on the Exit-Adjusted Delamination Factor and Maximum Temperature at the Tool's Exit for the Dagger Tool

After a more detailed study in the previous sub-sections regarding the BZT03 tool, this sub-section aims to focus on the Dagger tool, to identify the main influences and relations among the output variables and the input variables associated with the drilling operating parameters.

Regarding the Dagger tool, the Pearson's correlation coefficient between the maximum temperature and the exit-adjusted delamination factor at the tool's exit is -0.367, in this case also an inverse correlation, although less strong when compared with the results obtained for the BZT03 tool.

Figure 18 presents the bubble plot where the relations among these output and input variables can be observed.

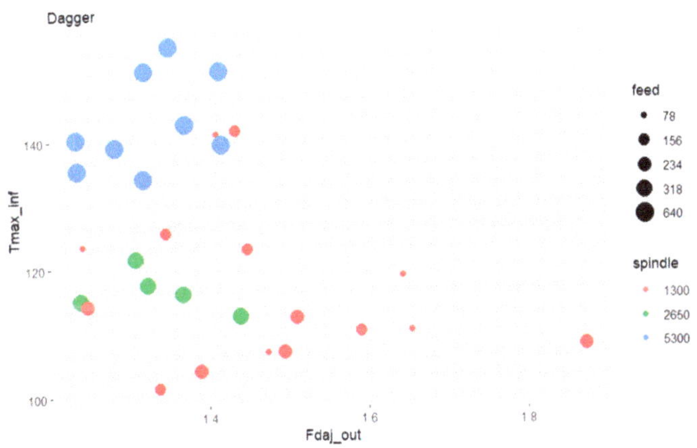

Figure 18. Bubble plot relating the maximum temperature with the exit-adjusted delamination factor, the feed rate and spindle speed for Dagger tool.

With the Dagger tool it is possible to understand that overall, regardless of the spindle speed used, the exit-adjusted delamination factor is significantly minor compared with those presented when the drilling tool is the BZT03. It is also visible that the highest values of maximum temperature appear for the higher spindle speed values, although with a minor magnitude when compared with the values attained with the BZT03 tool.

Variation in the machining parameters appears to have a limited influence on the damage at the tool exit, despite some changes in temperature. Thus, for these ranges and this tool, there does not seem to be a significant correlation between these two outputs.

4. Conclusions

In this study, an assessment of damage and temperature in holes drilled in glass fiber composite plates has been successfully conducted using various statistical tools. An experimental setup was established to evaluate these parameters, and the statistical tools proved essential for interpreting the numerous tests performed. This approach helped to identify differences between the various tests and excluded those that do not significantly vary.

As an overall conclusion, the results show that the tool selection and the composite thickness show a great impact on the output parameters analyzed.

The influence of tools and drilling parameters is critical, as these selections directly impact the quality of the holes and the extent of damage at the hole exit. The determination of operational parameters to minimize undesirable effects is strongly correlated with the type of drill selected, indicating that the optimal parameters for one tool may differ significantly from those for another.

The Tukey test revealed that the Dagger and Seco tools resulted in lower exit damage values for the plates, while the BZT tools displayed a higher exit-adjusted delamination factor and greater variability. In contrast, the Dagger tools provided more consistent drilling performance, with greater uniformity and control, leading to superior overall quality in the drilling process. Also, the results indicate no significant difference between the BZT01 and BZT03 tools, as well as between the Dagger and Seco tools, and similarly between the Seco and BZT01 tools.

The BZT03 tool does not allow the manufacture of good-quality holes when drilling medium and high thickness plates. An inverse relation is observed between the damage factor and maximum temperature as plate thickness changes. Damage increases with thicker plates, while temperatures on the lower side decrease. This may be caused by the tool's geometry, which does not promote good chip removal, potentially leading to poorer cutting performance in the final stage. This may result in greater compressive forces and contribute to greater delamination of the composite. When focusing exclusively on tests that used the same feed per tooth with the BZT03 tool, the findings suggest that increasing drilling speed reduces damage at the hole's exit, despite the associated rise in temperature.

When comparing the tools, the BZT03 tool exhibits a wider range of results in contrast to the Dagger. With the Dagger tool, the exit-adjusted delamination factor remains consistently lower than with the BZT03, regardless of spindle speed. Additionally, while maximum temperature increases at higher spindle speeds, these temperatures are notably lower than those observed when using the BZT03 tool. The Dagger tool demonstrates an ability to improve damage at the hole exit while simultaneously reducing temperature. However, establishing strong correlations between these two variables remains challenging.

Author Contributions: Conceptualization, A.M., A.C. and M.A.R.L.; Data curation, A.M., A.C., I.M.F.B. and I.C.J.B.; Formal analysis, A.M., A.C., I.M.F.B., I.C.J.B., J.I.B. and M.A.R.L.; Investigation, A.M., A.C., I.M.F.B., I.C.J.B., J.I.B. and M.A.R.L.; Methodology, A.M., A.C. and M.A.R.L.; Resources, I.M.F.B. and I.C.J.B.; Software, A.M. and A.C.; Supervision, A.M., A.C. and M.A.R.L.; Validation, I.M.F.B., I.C.J.B. and M.A.R.L.; Visualization, A.M., A.C., I.M.F.B. and I.C.J.B.; Writing—original draft, A.M., A.C., I.M.F.B., I.C.J.B., J.I.B. and M.A.R.L.; Writing—review & editing, A.M., A.C., J.I.B. and M.A.R.L. All authors have read and agreed to the published version of the manuscript.

Funding: This research received no external funding.

Institutional Review Board Statement: Not applicable.

Informed Consent Statement: Not applicable.

Data Availability Statement: The original contributions presented in the study are included in the article, further inquiries can be directed to the corresponding author.

Acknowledgments: The authors acknowledge the support of FCT, through IDMEC, under LAETA, project UIDB/50022/2020. The author A. Carvalho acknowledges the support of FCT, through CEMAPRE/ISEG Research project UIDB/05069/2020. The authors acknowledge Marisa Alves for her assistance.

Conflicts of Interest: The authors declare no conflicts of interest.

Appendix A

Tool	Thickness [mm]	Spindle Speed (rpm)	Feed Rate [mm/min]	Fdaj_out	Tmax_inf
BZT01	2.88 (low)	5300	640	3.595	250
				3.481	250
				3.699	250
				3.769	241.638
				3.277	233.044
Seco	2.88 (low)	5300	640	1.434	79.116
				1.802	69.268
				1.933	75.524
				1.818	87.314
				2.026	86.646
				1.434	79.116
BZT03	2.88 (low)	5300	640	3.647	242.387
				3.653	250
				3.395	250
				4.199	248.443
				3.921	236.274
				4.385	222.716
				5.529	214.491
				7.907	229.016
				2.988	223.528
				4.340	224.933
	5.54 (medium)	5300	640	5.592	119.338
				9.636	122
				9.544	88
				7.407	121
				5.008	121
	8.36 (high)	5300	640	11.436	87.982
				15.935	88
				12.956	88
	3 (low)	2650	318	6.935	232.101
				6.819	246.467
				7.527	237.575
				5.316	225.816
		1300	156	5.896	225.668
				8.028	-----
				9.689	227.145
				5.207	242.438
				5.556	219.759
			78	4.081	243.046
				4.623	228.579
				3.914	246
				3.522	246
				3.908	246
			234	6.882	195.256
				8.365	209.159
				9.493	192.476
				9.586	204.38
				10.567	173.591

Tool	Thickness [mm]	Spindle Speed (rpm)	Feed Rate [mm/min]	Fdaj_out	Tmax_inf
Dagger	2.88 (low)	5300	640	1.415	150.771
				1.319	133.857
				1.235	140.142
				1.416	139.302
				1.371	142.372
				1.415	150.771
				1.319	133.857
				1.235	140.142
				1.416	139.302
	2.95 (low)	2650	318	1.368	115.851
				1.308	121.286
				1.238	114.886
				1.323	117.232
				1.440	112.424
		1300	156	1.347	125.457
				1.338	101.128
				1.591	110.194
				1.449	122.96
				1.435	141.387
			78	1.655	110.338
				1.474	106.805
				1.644	118.796
				1.242	123.43
				1.411	140.866
			234	1.247	114.181
				1.494	106.892
				1.874	108.239
				1.389	103.857
				1.509	112.236

References

1. Fleischer, J.; Teti, R.; Lanza, G.; Mativenga, P.; Möhring, H.-C.; Caggiano, A. Composite materials parts manufacturing. *CIRP Ann.* **2018**, *67*, 603–626. [CrossRef]
2. Xu, J.; Geier, N.; Shen, J.; Krishnaraj, V.; Samsudeensadham, S. A review on CFRP drilling: Fundamental mechanisms, damage issues, and approaches toward high-quality drilling. *J. Mater. Res. Technol.* **2023**, *24*, 9677–9707. [CrossRef]
3. Merino-Pérez, J.L.; Royer, R.; Merson, E.; Lockwood, A.; Ayvar-Soberanis, S.; Marshall, M.B. Influence of workpiece constituents and cutting speed on the cutting forces developed in the conventional drilling of CFRP composites. *Compos. Struct.* **2016**, *140*, 621–629. [CrossRef]
4. Kavad, B.V.; Pandey, A.B.; Tadavi, M.V.; Jakharia, H.C. A Review Paper on Effects of Drilling on Glass Fiber Reinforced Plastic. *Procedia Technol.* **2014**, *14*, 457–464. [CrossRef]
5. Cao, S.; Li, Y.; Zhang, K.; Hou, G.; Luo, B.; Liu, S.; Long, T.; Liu, P. Investigation of CFRP damages induced by the interface high temperature and mixed tool wear mechanism in drilling of thin-walled CFRP/Ti stacks. *Compos. Struct.* **2023**, *323*, 117438. [CrossRef]
6. Panchagnula, K.K.; Jasti, N.V.K.; Panchagnula, J.S. Prediction of drilling induced delamination and circularity deviation in GFRP nanocomposites using deep neural network. *Mater. Today Proc.* **2022**, *62*, 7118–7123. [CrossRef]
7. Malik, K.; Ahmad, F.; Keong, W.T.; Gunister, E. The effects of drilling parameters on thrust force, temperature and hole quality of glass fiber reinforced polymer composites. *Polym. Polym. Compos.* **2022**, *30*, 096739112211311. [CrossRef]
8. Patel, P.; Chaudhary, V. Delamination evaluation in drilling of composite materials—A review. *Mater. Today Proc.* **2022**, *56*, 2690–2695. [CrossRef]
9. Xu, J.; Lin, T.; Li, L.; Ji, M.; Davim, J.P.; Geier, N.; Chen, M. Numerical study of interface damage formation mechanisms in machining CFRP/Ti6Al4V stacks under different cutting sequence strategies. *Compos. Struct.* **2022**, *285*, 115236. [CrossRef]
10. Mustari, A.; Chakma, P.; Sobhan, R.; Dhar, N. Investigation on roundness and taper of holes in drilling GFRP composites with variable weight percentages of glass fiber. *Mater. Today Proc.* **2021**, *38*, 2578–2583. [CrossRef]
11. An, Q.; Dang, J.; Li, J.; Wang, C.; Chen, M. Investigation on the cutting responses of CFRP/Ti stacks: With special emphasis on the effects of drilling sequences. *Compos. Struct.* **2020**, *253*, 112794. [CrossRef]

12. Xu, J.; Li, L.; Geier, N.; Davim, J.P.; Chen, M. Experimental study of drilling behaviors and damage issues for woven GFRP composites using special drills. *J. Mater. Res. Technol.* **2022**, *21*, 1256–1273. [CrossRef]
13. Jia, Z.-Y.; Zhang, C.; Wang, F.-J.; Fu, R.; Chen, C. An investigation of the effects of step drill geometry on drilling induced delamination and burr of Ti/CFRP stacks. *Compos. Struct.* **2020**, *235*, 111786. [CrossRef]
14. Hao, J.; Wang, F.; Zhao, M.; Bai, Y.; Jia, Z. Drill bit with clip-edges based on the force control model for reducing the CFRP damage. *J. Reinf. Plast. Compos.* **2020**, *40*, 206–219. [CrossRef]
15. Abrão, A.M.; Faria, P.E.; Rubio, J.C.C.; Reis, P.; Davim, J.P. Drilling of fiber reinforced plastics: A review. *J. Mech. Work. Technol.* **2007**, *186*, 1–7. [CrossRef]
16. Gemi, L.; Morkavuk, S.; Köklü, U.; Gemi, D.S. An experimental study on the effects of various drill types on drilling performance of GFRP composite pipes and damage formation. *Compos. Part B Eng.* **2019**, *172*, 186–194. [CrossRef]
17. Shu, L.; Li, S.; Fang, Z.; Kizaki, T.; Kimura, K.; Arai, G.; Arai, K.; Sugita, N. Study on dedicated drill bit design for carbon fiber reinforced polymer drilling with improved cutting mechanism. *Compos. Part A Appl. Sci. Manuf.* **2021**, *142*, 106259. [CrossRef]
18. Sugita, N.; Shu, L.; Kimura, K.; Arai, G.; Arai, K. Dedicated drill design for reduction in burr and delamination during the drilling of composite materials. *CIRP Ann.* **2019**, *68*, 89–92. [CrossRef]
19. Su, F.; Zheng, L.; Sun, F.; Wang, Z.; Deng, Z.; Qiu, X. Novel drill bit based on the step-control scheme for reducing the CFRP delamination. *J. Mech. Work. Technol.* **2018**, *262*, 157–167. [CrossRef]
20. Jia, Z.; Fu, R.; Niu, B.; Qian, B.; Bai, Y.; Wang, F. Novel drill structure for damage reduction in drilling CFRP composites. *Int. J. Mach. Tools Manuf.* **2016**, *110*, 55–65. [CrossRef]
21. Voß, R.; Henerichs, M.; Rupp, S.; Kuster, F.; Wegener, K. Evaluation of bore exit quality for fibre reinforced plastics including delamination and uncut fibres. *CIRP J. Manuf. Sci. Technol.* **2016**, *12*, 56–66. [CrossRef]
22. Seif, A.; Sadoun, A.; Fathy, A.; Megahed, A. Evaluation of hole quality in drilling process of GF/Aluminum wire mesh reinforced epoxy composites. *Alex. Eng. J.* **2024**, *94*, 257–273. [CrossRef]
23. Kolesnyk, V.; Peterka, J.; Alekseev, O.; Neshta, A.; Xu, J.; Lysenko, B.; Sahul, M.; Martinovič, J.; Hrbal, J. Application of ANN for Analysis of Hole Accuracy and Drilling Temperature When Drilling CFRP/Ti Alloy Stacks. *Materials* **2022**, *15*, 1940. [CrossRef] [PubMed]
24. Khashaba, U. A novel approach for characterization of delamination and burr areas in drilling FRP composites. *Compos. Struct.* **2022**, *290*, 115534. [CrossRef]
25. Joshi, S.; Rawat, K.; Balan, A.S.S. A novel approach to predict the delamination factor for dry and cryogenic drilling of CFRP. *J. Mech. Work. Technol.* **2018**, *262*, 521–531. [CrossRef]
26. Li, S.; Teng, H.; Dai, L.; Zhou, Y.; Li, C.; Li, P.; Ko, T.J. Comprehensive prediction model of drilling temperature of UD-CFRP laminates considering the combined action of main cutting edge and chisel edge. *Compos. Struct.* **2023**, *313*, 116899. [CrossRef]
27. Li, M.; Soo, S.L.; Aspinwall, D.K.; Pearson, D.; Leahy, W. Study on tool wear and workpiece surface integrity following drilling of CFRP laminates with variable feed rate strategy. *Procedia CIRP* **2018**, *71*, 407–412. [CrossRef]
28. Xu, J.; Lin, T.; Davim, J.P.; Chen, M.; El Mansori, M. Wear behavior of special tools in the drilling of CFRP composite laminates. *Wear* **2021**, *476*, 203738. [CrossRef]
29. Hou, G.; Zhang, K.; Fan, X.; Luo, B.; Cheng, H.; Yan, X.; Li, Y. Analysis of exit-ply temperature characteristics and their effects on occurrence of exit-ply damages during UD CFRP drilling. *Compos. Struct.* **2020**, *231*, 111456. [CrossRef]
30. Loja, M.A.R.; Alves, M.S.F.; Bragança, I.M.F.; Rosa, R.S.B.; Barbosa, I.C.J.; Barbosa, J.I. An assessment of thermally influenced and delamination-induced regions by composites drilling. *Compos. Struct.* **2018**, *202*, 413–423. [CrossRef]
31. Karpat, Y.; Karagüzel, U.; Bahtiyar, O. A thermo-mechanical model of drill margin-borehole surface interface contact conditions in dry drilling of thick CFRP laminates. *Int. J. Mach. Tools Manuf.* **2020**, *154*, 103565. [CrossRef]
32. Wang, B.; Zhong, S.; Lee, T.-L.; Fancey, K.S.; Mi, J. Non-destructive testing and evaluation of composite materials/structures: A state-of-the-art review. *Adv. Mech. Eng.* **2020**, *12*, 1687814020913761. [CrossRef]
33. Kumpati, R.; Skarka, W.; Ontipuli, S.K. Current Trends in Integration of Nondestructive Testing Methods for Engineered Materials Testing. *Sensors* **2021**, *21*, 6175. [CrossRef] [PubMed]
34. Ciecieląg, K.; Kęcik, K.; Skoczylas, A.; Matuszak, J.; Korzec, I.; Zaleski, R. Non-Destructive Detection of Real Defects in Polymer Composites by Ultrasonic Testing and Recurrence Analysis. *Materials* **2022**, *15*, 7335. [CrossRef] [PubMed]
35. Graziano, F.; Tortora, C.; Vespini, V.; Rippa, M.; Dentico, V.; Leone, F.; Gallo, N.; Stella, E.; Russo, P.; Coppola, S.; et al. Non-destructive techniques for quality control of composite materials. In Proceedings of the 2023 IEEE 10th International Workshop on Metrology for AeroSpace (MetroAeroSpace), Milan, Italy, 19–21 June 2023; pp. 529–533. [CrossRef]
36. Yang, H.; Yang, L.; Yang, Z.; Shan, Y.; Gu, H.; Ma, J.; Zeng, X.; Tian, T.; Ma, S.; Wu, Z. Ultrasonic detection methods for mechanical characterization and damage diagnosis of advanced composite materials: A review. *Compos. Struct.* **2023**, *324*, 117554. [CrossRef]
37. Wang, B.; He, P.; Kang, Y.; Jia, J.; Liu, X.; Li, N. Ultrasonic Testing of Carbon Fiber-Reinforced Polymer Composites. *J. Sens.* **2022**, *2022*, 5462237. [CrossRef]
38. Yurov, V.; Goncharenko, V.; Vasiliev, S.; Dmitriev, S.; Yurgenson, S. X-Ray Computed Tomography-Based Analysis of Impact Damage Propagation in Composite Materials. *Eurasian Phys. Tech. J.* **2019**, *16*, 31–35. [CrossRef]
39. Carvalho, A.; Silva, T.A.N.; Loja, M.A.R. Assessing Static and Dynamic Response Variability due to Parametric Uncertainty on Fibre-Reinforced Composites. *J. Compos. Sci.* **2018**, *2*, 6. [CrossRef]

40. Rosa, R.D.S.B.; Loja, M.A.R.; Carvalho, A.C.J.V.N. Toward Variability Characterization and Statistic Models' Constitution for the Prediction of Exponentially Graded Plates' Static Response. *J. Compos. Sci.* **2018**, *2*, 59. [CrossRef]
41. Carvalho, A.; Martins, A.; Mota, A.F.; Loja, M.A.R. Variability on Functionally Graded Plates' Deflection Due to Uncertainty on Carbon Nanotubes' Properties. *Math. Comput. Appl.* **2024**, *29*, 22. [CrossRef]
42. Chen, W.-C. Some experimental investigations in the drilling of carbon fiber-reinforced plastic (CFRP) composite laminates. *Int. J. Mach. Tools Manuf.* **1997**, *37*, 1097–1108. [CrossRef]
43. Durão, L.M.P.; Tavares, J.M.R.S.; Marques, A.T.; Freitas, M.; Magalhães, A.G. Caracterização de Danos de Maquinagem em Placas Compósitas. In *Actas do CMCE2004-Congresso de Métodos Computacionais em Engenharia, Proceedings of the CMCE2004—Congress of Computational Methods in Engineering (Incorporating the VIII National Congress of Applied and Computational Mechanics and the VI Congress of Numerical Methods in Engineering), National Laboratory of Civil Engineering, Lisbon, Portugal, 31 May–2 June 2004*; APMTAC: Lisbon, Portugal, 2004; pp. 294–304.
44. Davim, J.P.; Rubio, J.C.; Abrao, A. A novel approach based on digital image analysis to evaluate the delamination factor after drilling composite laminates. *Compos. Sci. Technol.* **2007**, *67*, 1939–1945. [CrossRef]
45. Ahn, D.-K.; Choi, J.-H.; Kweon, J.-H. Relationship between the drilling condition and the damage (delamination) zone of glass-fiber-reinforced plastic composites. *Adv. Compos. Mater.* **2014**, *24*, 297–305. [CrossRef]
46. Schindelin, J.; Arganda-Carreras, I.; Frise, E.; Kaynig, V.; Longair, M.; Pietzsch, T.; Preibisch, S.; Rueden, C.; Saalfeld, S.; Schmid, B.; et al. Fiji: An open-source platform for biological-image analysis. *Nat. Methods* **2012**, *9*, 676–682. [CrossRef] [PubMed]
47. Ridler, T.W.; Calvard, S. Picture Thresholding Using an Iterative Selection Method. *IEEE Trans. Syst. Man Cybern.* **1978**, *8*, 630–632. [CrossRef]
48. Montgomery, D.C. *Design and Analysis of Experiments*; John Wiley& Sons Inc.: New York, NY, USA, 1997.
49. Shapiro, S.S.; Wilk, M.B. An analysis of variance test for normality (complete samples). *Biometrika* **1965**, *52*, 591–611. [CrossRef]
50. Siegel, S.; Castellan, N.J., Jr. *Nonparametric Statistics for the Behavioral Sciences*, 2nd ed.; Mcgraw-Hill Book Company: New York, NY, USA, 1988.

materials

MDPI

Article

A Statistical Study on the Influence of Drilling Process in Delamination Observed in Composite Plates

Hugo R. C. Cerqueira [1], João E. Matos [1], José L. Esteves [2,3], Susana C. F. Fernandes [1,2] and Luis M. P. Durão [1,2,*]

1 ISEP, Polytechnic of Porto, rua Dr. António Bernardino de Almeida, 4249-015 Porto, Portugal; 1220546@isep.ipp.pt (H.R.C.C.); jem@isep.ipp.pt (J.E.M.); scf@isep.ipp.pt (S.C.F.F.)
2 INEGI Instituto de Ciência e Inovação em Engenharia Mecânica e Engenharia Industrial, 4200-465 Porto, Portugal; jesteves@fe.up.pt
3 FEUP Faculdade de Engenharia da Universidade do Porto, 4200-465 Porto, Portugal
* Correspondence: lmd@isep.ipp.pt

Abstract: Composite materials are increasingly being implemented in various solutions, ranging from conventional applications, like furniture, to more advanced ones, such as aerospace, based on their excellent properties, such as high mechanical strength and low weight. There are applications in which these materials are coupled to other parts. To achieve this connection, drilling processes are commonly used. Drilling causes irreversible damage to the material, which influences the mechanical strength of the plates. This study was conducted on 48 carbon/epoxy plates, each with two drilled holes, based on DOE (design of experiments) and the Taguchi method to design the experimental plan and to validate the results. Three control factors were considered for drilling: drill bit type, cutting speed, and feed rate, as it is expected that a low feed rate and a high cutting speed is the drilling configuration that inflicts the least damage. Subsequently, these specimens were subjected to enhanced radiography and an image analysis processing tool based on MatLab® to assess the data collected and compute damage results. At the end, in analyzing the results of the Taguchi method, it is possible to validate the assumptions on the influence of the drilling process in delamination extension.

Keywords: composite materials; drilling damage; damage assessment; non-destructive testing; Taguchi method

1. Introduction

In a progressively developed world, materials with enhanced properties are increasingly sought-after. These materials, like polymeric matrix composites, are widely used in diverse applications and industries. Examples of these applications are easy to identify, like the automotive industry [1,2], shipbuilding industry [3], wind turbines blades [4], or even sports [5] or furniture [6,7].

One of the advantages of these type of material is their excellent properties, such as high mechanical strength combined with low weight, resulting in a high specific strength. As a result of the way that parts are produced, they also provide a near-net shape, characterized by good surface finishing and no need for large machining processes. However, there are situations in which, due to the need to assemble structural parts, machining, especially drilling, is necessary. Drilling is a complex process that can cause significant damage, delamination being the most relevant, by negatively affecting the mechanical properties of the machined plates. This outcome of the drilling process and its consequences has been extensively studied by a considerable number of authors [8–13] and some significant revisions concerning advances in drilling techniques and delamination evaluation can be found in [14–16]. Delamination, which occurs at the entrance and exit of the hole, is one of the major damages to be studied.

1.1. Composite Plate Drilling

In drilling, material removal occurs due to the contact and rotation of the drill bit on the workpiece. Normally, drills have two cutting edges and two flutes to allow the easy removal of chips from the machining zone. The force that is exerted in the direction perpendicular to the part is called the axial thrust force (F_a), and it is dependent on the plate material, the drill bit geometry and the drilling parameters. This axial thrust force is considered as primarily responsible for the damage caused in the drilling process. In [8], the authors investigated the influence of cutting speed, feed rate and tool diameter on the uncut fiber and delamination damage on composite sandwich structures using DOE (design of experiments), showing the influence of feed rate and the possibility of achieving an optimum point for cutting speed and tool diameter. In [9,10], similar conclusions on the importance of feed rate on damage were confirmed. Rajkumar et al. [11] conducted an experimental investigation using RSM (response surface methodology) to determine the influence of these parameters on the machining of composites, revealing the importance of drill diameter on delamination and thrust force. The influence of different drill geometries was the focus in [12,13], demonstrating the influence of tool geometry.

Figure 1 [15] outlines the main parameters that influence the drilling operation in composite materials. This figure summarizes the inputs that condition and alter the drilling for a composite material. In [14–16], it is possible to find reviews of the issues related to drilling of fiber-reinforced composites, including the minimization of delamination extension and the path to high-quality drilling, presenting the main findings of recent papers. The main parameters that influence conventional drilling are feed rate, cutting speed, and drill bit geometry. Of these parameters, that regarded as the most important is the feed rate as it is directly related to the axial thrust force, thus defining the damage caused by the drilling. Cutting speed is the second most important parameter among those mentioned above [10].

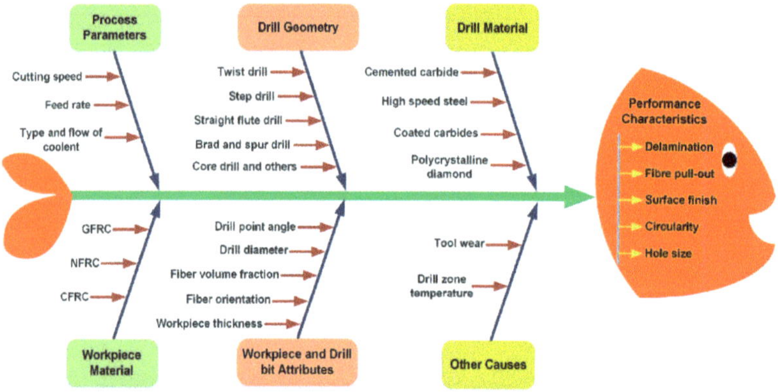

Figure 1. Cause and effect diagram showing the influence of various parameters on performance characteristics of drilled composites [15].

Tool geometry also has some influence on the development of axial thrust force during drilling. Besides the standard twist drill, with variation of the point angle, defined as the angle formed by the two cutting edges of the drill bit, other drill geometries have been used when studying the drilling of composite plates, like Brad and Spur, step, straight-flute or core drill;, see Figure 2, showing some of the drills used to produce holes in composites [15]. The importance of drill point angle on the progress of axial thrust force and damage caused by drilling should be noticed. When using sharp-angled tools, lower cutting forces and less delamination damage is expected, whereas obtuse-angled tools register higher thrust forces, causing more delamination damage [17–20].

Figure 2. Various drill geometries employed in the drilling of composites [15].

1.2. Non-Destructive Testing of Drilled Composite Plates

To achieve the necessary results in order to quantify and qualify the damage extension, it is necessary to analyze the region around the drilled holes with non-destructive techniques to determine the required geometrical parameters of the affected areas and, eventually, the mechanical properties of the materials after machining, ensuring the necessary strength [21–24]. The determination of mechanical properties of the drilled coupons is outside the scope of this study.

Non-destructive testing, or NDT, is particularly important in the evaluation and characterization of materials. Some of the NDT methods most commonly used in composite materials are radiography, ultrasound, eddy-current or tera hertz, among others. All these methods are suitable to detect delamination or other damage in the materials. When using radiography, it is normal to associate the use of liquid penetrant testing, by immersing the plate in a contrasting fluid. In enhanced radiography, a contrasting fluid is used to improve radiographic images. These liquids can help in the detection of small flaws in the interior of the material under analysis. As they are oriented perpendicularly to the radiation beam, delaminations are well detected by this method [25–27].

X-ray computed tomography (CT) has been increasingly used in composite materials as a technique for non-destructive testing. This technique allows for the reconstruction of the part, enabling highly accurate inspection of composite materials, as these materials are characterized by their heterogeneity and damage is sometimes difficult to carry out a proper assessment on. Therefore, 3D imaging can be used to analyze and evaluate the possibility of existing damages and thus ensure structural integrity of the parts. This technique is based on the computational reconstruction of a 3D image, obtained by the beams of radiation made from various angles of a part [28,29]. Examples of the use of this technique can be found in [30,31]

The ultrasound method is based on the incidence of high-frequency sound waves, between 20 kHz and 25 MHz, on the material to be analyzed. Due to the properties of composite materials, frequencies between 1 and 5 MHz are frequently applied. The most common defects detected in composite materials are fiber breakage, inclusions, matrix cracking and delamination [32].

Eddy-current can be adapted to characterize subsurface defects in composites, like delaminations, microcracks, porosity or fiber breakage, providing an effective and economical solution for the non-destructive inspection of CFRP [33,34].

As a final note for the use of Terahertz waves as NDT method, these waves use signal frequencies between 300 GHz and 3 THz, with wavelengths ranging from 1 mm to 100 µm. This method is characterized by using non-ionizing radiation to the detriment of ionizing radiation, such as X-radiation, assuring safety for the operator or avoiding consequences to biological samples. Terahertz demonstrates effectiveness in penetrating non-conductive materials, such as ceramics and plastics. The use of Terahertz as a non-destructive method has deserved the interest shown in recent research [35,36], as some advantages are easily recognized.

1.3. *Image Processing Techniques for Delamination Assessment*

A characteristic that is common to all these non-destructive methods is the recording of images to allow for posterior assessment of the geometrical features of the region around the drilled hole, as delamination criteria apply based on this. The processing and analysis of the images obtained during the process of radiography of the holes is a fundamental step towards the results of this dissertation. Image processing, based on MatLab®, version 23.2 or similar software, is a method frequently used to help on the quantification of the delaminated area around the hole, referred to by previous studies of the team involved in this work, such as Durão et al. [37] or Silva et al. [38]. The ultimate goal in any image processing method is to obtain images in which the pixels that correspond to the damaged area have one color, usually white, and the rest of the pixels in the same image another color, usually black (see Figure 3). The boundaries between the drilled region and the damaged area or between the damaged area and sound plate have various shades, and the same is true for the damaged area, the drilled area or the plate surface. It is therefore necessary to define a strategy that permits a clear definition of the contours and areas. A technique often used for this purpose is the use of threshold. Threshold is an algorithm that calculates an edge value, which divides the grayscale values, turning the image into a binary image. In other words, the algorithm assigns a value to each color in the image, between 0 and 255, calculates an average value of the grayscale of an image and assigns a black pixel for values below the defined threshold and a white pixel for values above it, resulting in a black-to-white image, with sharper contours and areas. Several software applications that allow image processing already have the threshold tool incorporated, as is the case with Matlab®, used in the study here presented.

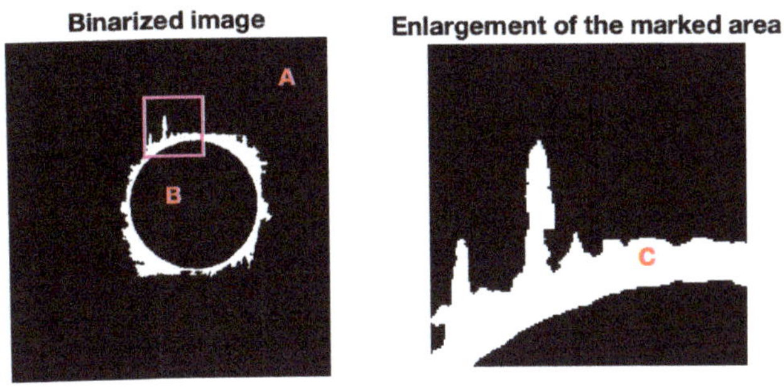

Figure 3. Example of a damaged area region: A (black outer region) is the undamaged coupon; B (black region) is the hole; C (white region) is the damaged area (full line).

Alternative approaches to the assessment of damaged area are the use of ANN (Artificial Neural Networks) or even the use of AI tools, considering the increased computational capacity of modern computers. Both can be used for any process where the starting point is an image obtained by some NDT method, as in those mentioned in 1.2. Concerning ANN, this research team has presented a novel solution based on an artificial neural network in the analysis of radiographic images [39]. In [40], recurrent neural networks (RNNs) were developed and implemented to estimate tool wear during composites drilling.

The study of delamination requires quantifying it in some way. Only in this way is it possible to assess how the various factors already mentioned can affect delamination, to establish comparisons between them and to seek solutions to mitigate this phenomenon. The evaluation of delamination begins with obtaining some kind of image of the drilled hole and its peripherical region [38]. Afterwards, it is fundamental to define some mathematical criteria for this assessment. One of the early criteria presented was the Delamination Factor (F_d) [41], Equation (1), which is obtained by the ratio between the maximum diameter of the damaged area (D_{max}) and the nominal diameter of the hole (D_0).

$$F_d = \frac{D_{max}}{D_0} \tag{1}$$

However, this factor has the limitation of being one-dimensional, not considering the effect of the damaged area, so two holes with different areas of damage can have the same value of F_d (see Figure 4) [42]. Additionally, in cases where the damaged area is more irregular and not circular, it is advisable to use the damaged area to quantify delamination instead of the maximum diameter.

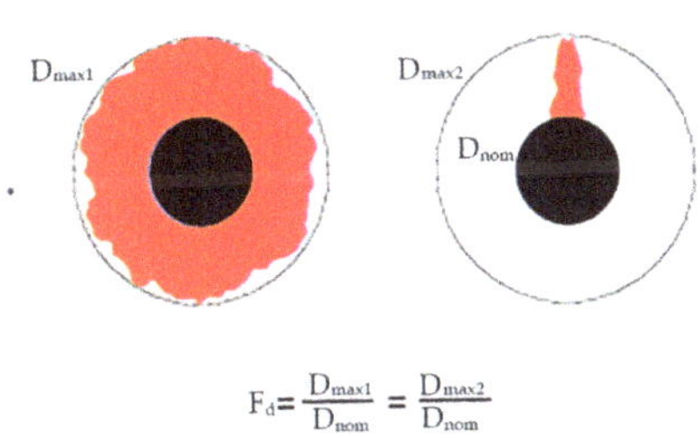

Figure 4. Example of equal delamination factors for different damaged areas (damage area is marked in red) [42].

Several solutions have been proposed to overcome this problem (see Davim et al. [43]), which suggested the Adjusted Delamination Factor, Equation (2):

$$F_{da} = \alpha \frac{D_{max}}{D_0} + \beta \frac{A_{max}}{A_0} \quad (2)$$

where A_{max} is the damaged area and A_0 the nominal hole area.

Recently, Tsao et al. [44] developed the Equivalent Delamination Factor, (F_{ed}), which relates an equivalent diameter (D_e), see Figure 5, to the nominal diameter of the hole (D_0), Equations (3) and (4). This criterion was adopted for the study here presented.

$$F_{ed} = \frac{D_e}{D_0} \quad (3)$$

$$D_e = \left(\frac{4(A_d + A_0)}{\pi} \right)^{0.5} \quad (4)$$

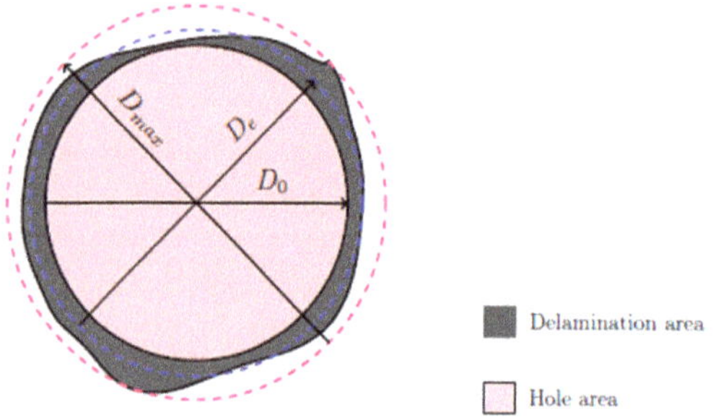

Figure 5. Scheme for Equivalent Delamination Factor.

A tridimensional delamination criterion (F_v) was recommended by Xu et al. [45] and it would be interesting to consider this factor in this work, as it includes the accumulated volume of the various delaminated plies along the plate, Equation (5), where p is the number of layers and k is the number that specifies the delaminated ply. However, for the current state of the radiographic setup, this option was not possible to incorporate, remaining as a challenge to meet in future work.

$$F_v = \frac{1}{p} \sum_{k=1}^{p} \frac{A_d^k}{A_{nom}} \quad (5)$$

Therefore, understanding and predicting how the drilling process can affect the damage extension and, consequently, the mechanical strength of composite plates is the main purpose of this study. The use of the Taguchi method is helpful in designing the experimental sequence and helping with a sound analysis of the results obtained regarding damage evaluation and correlations with the experimental factors defined for this study.

2. Materials and Methods

To evaluate the effect of drilling on composite materials and considering the information collected in the bibliographic review, the main parameters of drilling to include in this study were decided: feed rate, cutting speed and drill bit geometry. Two possible levels have been assigned to each of these characteristics. For feed rate, the values of 0.05 and 0.2 mm/rev were established. The choice of these values was based on bibliographic research and analysis of previously published papers in the same field [13–17,21,22,37,38,46]. These values are within a range of acceptable drilling values for the composite material and from which satisfactory results are predictable. For the cutting speed, the decision was to use spindle speed values of 500 and 2000 rpm, corresponding to 9.4 and 37.7 m/min. For the 6 mm drill geometry, a step drill from INOVA Tools (Kinding, Germany), Ref. 850.037.00, with a first diameter of 3.7 mm and a point angle of 140° was chosen (see Figure 6a). When using a step drill, it is reported that larger feed rates can be used, meaning shorter cycle times, without delamination damage [46]. The other option was the use of a twist drill in a pilot hole drilling sequence, using two twist drills of different diameters, with a ratio of 0.4 from the pilot to the final hole, following a previous published study [47]. The advantage of pilot hole drilling on delamination reduction by cancelling the chisel edge effect during final diameter machining was evidenced in [48,49]. For the twist drills, the point angle was 118°. These tools were also from INOVA Tools, Refs. 701.024.000 for the pilot hole and Ref. 701.060.000 for the final hole (see Figure 6b). The drills selected are capable of drilling composite materials and are within a range that has normally been evaluated in this type of study.

Figure 6. Tools used in experimental work (Source: INOVA Tolls catalogue): (**a**) step drill; (**b**) twist drill.

A summary of the experimental levels and their unfolding for experimental work is provided in Table 1.

Table 1. Experimental plan—parameters levels.

Coupon ID	Drill Geometry	Feed Rate mm/rev	Spindle Speed rpm
1–PD0505	Pilot hole drilling	0.05	500
2-PD0520	Pilot hole drilling	0.05	2000
3-PD2005	Pilot hole drilling	0.20	500
4-PD2020	Pilot hole drilling	0.20	2000
5-ST0505	Step drill	0.05	500
6-ST0520	Step drill	0.05	2000
7-ST2005	Step drill	0.20	500
8-ST2020	Step drill	0.20	2000

With the help of Minitab® software, version 22.1.0, a DOE (design of experiment) was created for the application of the Taguchi method, resulting in two different plans. One is simpler, with only four configurations of the control factors, which results in a sample array of four, which was abandoned. Another DOE was prepared, with all the configurations of the three factors, which resulted in an array of eight, as illustrated in Table 1, together with the identification adopted for the plates drilled in each option. Drill bit wear was considered as the noise factor; therefore, the planning was repeated to check and verify the results.

To run this experimental plan, four laminate plates were produced, each consisting of twelve layers of prepreg, symmetrically stacked and balanced. The laminate was of the cross-ply type, and the stacking sequence was $[0°/90°]_{3S}$. The plates were produced using a HEXCEL prepreg from HEXCEL®, Stamford, CT, USA, composed of AS4 12K carbon fibers (HexTow®) and 8552 epoxy resin (HexPly®). The prepreg has a nominal fiber volume of 57%. Material characteristics of the prepreg material can be seen in Table 2. The plates produced were approximately 300 mm wide and long and 2.2 mm thick. Before drilling, plates were cut in coupons of the appropriate dimension for the testing sequence planned, considering the coupons needed for the complete DOE. A total of 12 holes were drilled at each level, including the necessary repetitions for statistical analysis soundness, and tool wear evaluation, only at two levels, new and worn. Machining was performed in a HAAS VF-2 machining center (Figure 7a,b).

Table 2. Prepreg mechanical properties—HexPly® 8552 UD Carbon Prepregs, in HEXCEL, USA.

Test	Units	Value
0° Tensile Strength	MPa	2207
90° Tensile Strength	MPa	81
0° Tensile Modulus	GPa	141
90° Tensile Modulus	GPa	10
0° Compression Strength	MPa	1531
0° Compression Modulus	GPa	128
0° ILSS	MPa	128
In-plane Shear Strength	MPa	114

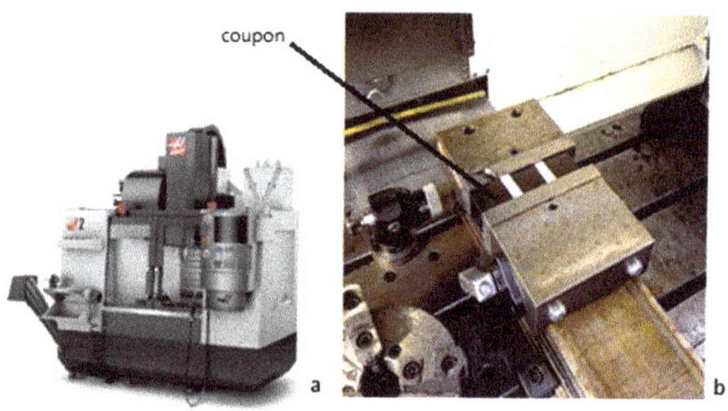

Figure 7. (**a**) HAAS VF-2 CNC machine (HAAS, Oxnard, CA, USA); (**b**) Image of a drilled coupon.

Then, a non-destructive alternative was determined, in this particular case enhanced radiography, since this had already been used in previous works, reducing uncertainty in results [37,38,50]. For enhanced radiography, coupons were immersed in diiodomethane, a contrasting liquid, for 15 min and then radiographed with the help of a digital imaging system consisting of a 60 kV, 300 kHz Kodak 2100 X-ray system (Kodak, Rochester, NY, USA) associated with a Kodak RVG 5100 digital acquisition system. The exposition time was set to 0.25 s [50]. Following the line of this study, the images obtained by enhanced radiography were analyzed and treated using a program developed in MatLab®. This software was used to convert radiographic images into binary images composed of black and white pixels. From this point, the program calculated the values of the geometric parameters that are understood as useful, knowing that each pixel corresponds to a square with a side of 0.0185496 mm.

As already stated, the delamination criteria used in this study was the Equivalent Delamination Factor, (F_{ed}) [45] where the delamination assessment is computed from the measurement of the delaminated area resulting from the use of a computational routine developed on MatLab® for radiographic image treatment, following Equations (3) and (4) (see Section 1) and enabling the values of the delaminated area around the hole. The minimum value of this factor is 1 (one), equal to a situation where no delamination is observed, or greater than 1 (one) if some delamination is observed. Inherently, the greater the value of this factor, the greater the delaminated area.

Finally, the application of the Taguchi method was completed to validate the results and identify the best drilling configuration. With this method, it is also possible to evaluate the interactions that the different factors have with each other and the significance that each of them represents for the result. The drill bit wear was considered a noise factor. Therefore, the values of this factor were inserted for new and worn drills.

To analyze the results obtained by this method, it is necessary to identify the best level for each control factor. Through the S/N ratio (signal/noise), the configurations of the control factor that minimize the variability caused by the noise factor are identified. In the linear analysis of the model, several coefficients are calculated, the most important being the p-factor. This is calculated for each control factor and for the interactions between each of the factors considered. The p-factor determines the statistical significance of each control factor in the response. It is usually assigned a significance level α of 0.05, which indicates a 5% risk of concluding that there is a significance when in fact this does not exist. If the *p*-value is less than or equal to the significance level, it is possible to conclude that the factor in question is statistically significant for the response. If the *p*-value is greater than or equal to the level of significance, it is not possible to conclude that there is significance among the experimental factors involved [51,52].

3. Results and Discussion

3.1. Preliminary Testing

As a starting point, for confirmation of plate properties, a tensile test according to ISO 527-1:2019 [53] was performed. As the relevant properties considered were the elastic modulus and the tensile strength of the material, a test speed of 1 mm/min was adopted until a load of 8 kN was reached and then switched to 2 mm/min. to provide data for both properties. A summary of the results is presented in Table 3, concerning the data from 8 tests.

Table 3. Average mechanical properties of the plates.

Test	Units	Average	Std Dev
Tensile Strength	MPa	971.44	82.09
Tensile Modulus	GPa	66.55	2.34

During the drilling phase, no data were monitored regarding axial thrust force, as there are previous studies confirming the effect of tool geometry, cutting speed or feed rate on this outcome.

3.2. Delamination Measurement

The enhanced radiographic analysis process, described in the previous section, permitted quantification of the damage caused by the drilling operation. For that purpose, the plates were immersed for 15 min in diiodomethane, a contrasting fluid, before image capturing. Through the image processing sequence, it was possible to obtain the following results (see Table 4) that represent the average of 6 coupons under each drilling condition.

Table 4. Measured hole and delamination values.

Coupon ID	Hole Area	Damage Area	F_{ed}
1-PD0505	28.227	1.312	1.023
2-PD0520	28.456	0.759	1.013
3-PD2005	28.566	1.552	1.027
4-PD2020	28.547	1.097	1.019
5-ST0505	28.248	1.694	1.029
6-ST0520	28.365	1.154	1.020
7-ST2005	28.492	1.832	1.032
8-ST2020	28.571	1.121	1.028
Average PD	28.449	1.180	1.022
Average ST	28.419	1.450	1.028

By analyzing the average values calculated, the holes with the pilot hole strategy return lower values for the damaged area, meaning that less damage was caused by drilling.

Taking all the values obtained from the image treatment of the radiographs, it was possible to calculate the equivalent delamination factor for every situation, using the criteria as defined in Section 2 (see [44] and Equations (4) and (5)), where the relevant values in Table 4 are the damaged area and the nominal dimensions of the drilled hole. Resulting F_{ed} average values are also presented in Table 4.

By analyzing these results, it is possible to conclude that a higher cutting speed and a lower feed rate cause less damage, as the higher cutting speed is the common factor in the two best values shown in Table 4.

From the combination of these factors for both drilling strategies, it results that the configuration that has a higher equivalent delamination factor is that with a high feed rate and a low cutting speed, using a step drill geometry. On the other hand, the drilling operation that results in a lower equivalent delamination factor is that which results from using a pilot hole strategy with a low feed rate and a high cutting speed, considering the range of this experimental study. Lower feed rates reduce the thrust force during drilling, keeping these values below the threshold for delamination onset, as demonstrated in [54]. Higher spindle speeds avoid long contact between the cutting edges of the drill and the hole walls, reducing temperature during drilling, which prevents matrix softening and, consequently, delamination or other related damages increase in temperature during dry drilling, reduce the elastic modulus of the CFRP and cause thermal expansion of the drill [13,55].

Based on previous knowledge of the delamination effects on mechanical features of plates [13,22,38,48,56], an increase in the F_{ed} value represents a plausible decrease in the bearing strength resulting from the Bearing test (ASTM D5961 [57]), or another mechanical test with the same objective. This trend is normally expected, meaning that a higher value of F_{ed} represents greater damage around the hole that, in turn, causes a greater loss of the mechanical strength of the plate. The completion of destructive confirmation tests is out of the scope of this study.

3.3. *Taguchi Method Analysis*

Finally, the application of the Taguchi method was completed to validate the results of this experimental design and identify the best drilling configuration among the options considered in this study. With this method, it is also possible to evaluate the interactions that the different factors have with each other and the significance that each of them represents for the result. The data for the equivalent delamination factor (F_{ed}) was then analyzed. For that purpose, drill wear was considered as a noise factor. Therefore, the values of this factor were inserted for new and worn drills. For worn drills, new drills equal to those used in the first experimental step were used to produce 48 consecutive holes in a sacrificial plate identical to those of the experimental sequence.

For the F_{ed} analysis, the ratio chosen for the S/N ratio was "lower is better", because the objective of the experiment was to minimize the response, which in this case is the F_{ed} value. Although the S/N ratio was considered as "lower is better", the choice of optimal levels should be made in such a way that the S/N ratio value is maximum. This is because minimizing the loss-to-function is associated with maximizing the S/N ratio; therefore, the higher the value of this ratio, the better. Tables 5 and 6 represent the linear analysis for the equivalent delamination factor average value and for the correspondent S/N ratios.

Table 5. Analysis of variance for experimental results.

Source	DF	SQ Seq	SQ(aj)	QM(aj)	F	*p*
Drill type	1	0.000094	0.000094	0.000094	214.36	0.043
Feed rate	1	0.000067	0.000067	0.000067	152.60	0.051
Cut speed	1	0.000113	0.000113	0.000113	257.40	0.040
Drill*Feed	1	0.000000	0.000000	0.000000	0.98	0.504
Drill*speed	1	0.000004	0.000004	0.000004	9.09	0.204
Feed*speed	1	0.000005	0.000005	0.000005	10.86	0.188
Residual error	1	0.000000	0.000000	0.000000		
TOTAL	7	0.000284				

Table 6. S/N ratio variance analysis.

Source	DF	SQ Seq	SQ(aj)	QM(aj)	F	*p*
Drill type	1	0.006808	0.006808	0.006808	231.74	0.042
Feed rate	1	0.004790	0.004790	0.004790	163.06	0.050
Cut speed	1	0.008221	0.008221	0.008221	279.83	0.038
Drill*Feed	1	0.000031	0.000031	0.000031	1.04	0.494
Drill*speed	1	0.000302	0.000302	0.000302	10.28	0.192
Feed*speed	1	0.000364	0.000364	0.000364	12.38	0.176
Residual error	1	0.000029	0.000029	0.000029		
TOTAL	7	0.020544				

By analyzing the values in the tables above, it is possible to conclude that for the F_{ed} average, the drill type and the cutting speed are significant as they have a p-factor less than 0.05. However, the lowest value is that of cutting speed, so this control factor is the most significant, followed by the drill type. The feed rate factor has a value of 0.051, which makes it not significant, nonetheless this value is very close to the limit. For the interactions between the control factors, all of them have a p-factor greater than 0.05, which means that they do not have statistical significance for the average calculated for F_{ed}. Regarding the S/N ratios, the p-factors of the control factors are all significant for the result. On the other hand, the interactions between them are not statistically significant.

For a better interpretation of these results, it is possible to elaborate several graphs from the analysis software, as illustrated below, see Figures 8 and 9.

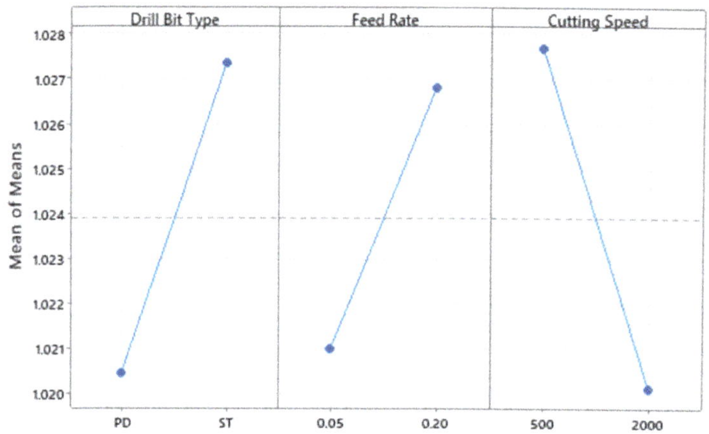

Figure 8. Average values—main effects of experimental factors.

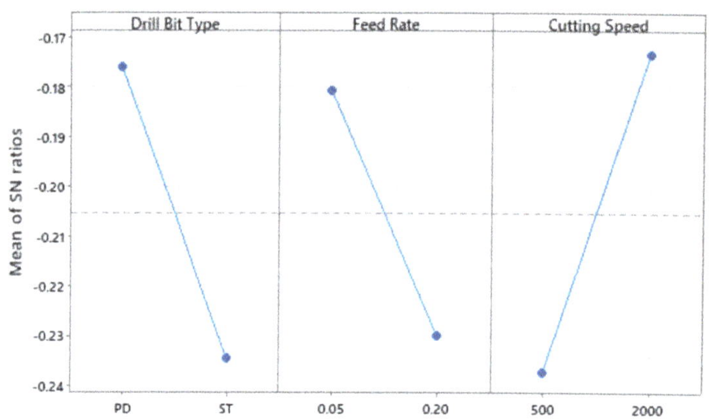

Figure 9. Signal-noise ratio of experimental factors.

As previously stated, the objective is to maximize the value of the S/N ratio. The drilling setup that allows this premise employs a pilot hole strategy using a standard twist drill with a 140° point angle, a feed ratio of 0.05 mm/rev and a spindle speed of 2000 rpm, equal to a cutting speed of approximately 38 m/min when a 6 mm diameter drill is used. Note that different drill diameters can turn into diverse conclusions.

The concern of drilling with a new or with a worn drill bit was also compared for F_{ed}. Therefore, it is possible to say that the values of the equivalent delamination factor are lower in drilling with the new drill than with the worn one. The worn drill is also associated with higher standard deviations of damage extension, characteristic of a more irregular drilling.

Finally, and to validate the results, the Taguchi method was used, and it was possible to identify which control factors were most significant and evaluate the interactions between them. For the F_{ed} data, we can conclude that all factors are significant, with the cutting speed as the most significant factor, and that the ideal drilling configuration is the one performed with a pilot hole strategy, with a feed rate of 0.05 mm/rev and a spindle speed of 2000 rpm.

4. Conclusions

This work aimed to study the effects of drilling on the damage extension of a composite plate by using enhanced radiography, an image processing tool based on MatLab®, and the Taguchi method. With this purpose, several drilling tests were carried out to characterize the damage around the drilled holes under diverse conditions of drill geometry, feed rate and cutting speed.

The analysis by enhanced radiography permitted to quantify the damage extension caused by drilling, calculating the equivalent delamination factor (F_{ed}).

The application of Taguchi method was crucial to validate the results and identify the best drilling configurations. It was concluded that the combination of pilot hole drilling, feed rate of 0.05 mm/rev and a spindle speed of 2000 rpm provided the best results in terms of lower drilling-induced delamination extension. Statistical analysis showed that spindle speed and the drill geometry are significant factors, while feed rate has a lower significance.

This study demonstrated that the drilling parameters can influence damage extension in composite materials. Through a detailed experimental approach and the application of statistical methods, such as the Taguchi method, it was possible to identify the drilling configurations that could improve the quality of the holes.

Future work should include the use of computed tomography (CT-scan), making it possible to apply a three-dimensional delamination factor and, consequently, obtain more accurate results. Another possible improvement would be the use of AI (Artificial Intelligence) regarding the definition of the threshold value in image processing.

Other works could include similar studies with different composite materials, varying the number of layers and the stacking sequence or the fiber direction, thus contributing to an improved understanding of the mechanical behavior of composite materials.

Author Contributions: Conceptualization, L.M.P.D. and J.L.E.; methodology, H.R.C.C.; software, J.E.M.; validation, H.R.C.C. and S.C.F.F.; formal analysis, J.L.E. and S.C.F.F.; investigation, H.R.C.C.; resources, L.M.P.D. and J.L.E.; data curation, J.E.M.; writing—original draft preparation, H.R.C.C. and L.M.P.D.; writing—review and editing, L.M.P.D.; visualization, J.E.M.; supervision, J.L.E.; project administration, L.M.P.D.; funding acquisition, L.M.P.D. and J.L.E. All authors have read and agreed to the published version of the manuscript.

Funding: This work was granted by the Portuguese FCT in the scope of the LAETA project UIDB/50022/2020 (Strategic Funding of R&D Units).

Institutional Review Board Statement: Not applicable.

Informed Consent Statement: Not applicable.

Data Availability Statement: The original contributions presented in this study are included in the article. Further inquiries can be directed to the corresponding author.

Acknowledgments: The authors would like to thank LMC-ISEP for plates production, Victor Moreira from LTM-ISEP for coupons drilling, Rui Silva and Miguel Figueiredo from DEMec–FEUP for supporting the X-ray analysis.

Conflicts of Interest: The authors declare no conflicts of interest.

References

1. Ahmad, H.; Markina, A.A.; Porotnikov, M.V.; Ahmad, F. A review of carbon fiber materials in automotive industry. *IOP Conf. Ser. Mater. Sci. Eng.* **2020**, *971*, 032011. [CrossRef]
2. Continental Reifen Deutschland GmbH. *Off-The-Road Tires—Technical Data Book*; Continental Reifen Deutschland GmbH: Hannover, Germany, 2019.

3. Mouritz, A.P.; Gellert, E.; Burchill, P.; Challis, K. Review of advanced composite structures for naval ships and submarines. *Compos. Struct.* **2001**, *53*, 21–42. [CrossRef]
4. Enercom—Production & Sourcing. Available online: https://www.enercon.de/en/company/production-sourcing (accessed on 24 October 2024).
5. Sreejith, M.; Rajeev, R.S. Fiber reinforced composites for aerospace and sports applications. In *Fiber Reinforced Composites: Constituents, Compatibility, Perspectives and Applications*; Woodhead Publishing: Cambridge, UK, 2021; pp. 821–859.
6. Zimmer, A.; Bachmann, S. Challenges for recycling medium-density fiberboard (MDF). *Results Eng.* **2023**, *19*, 101277. [CrossRef]
7. Composite Panel. Association—Medium Density Fiberboard. Available online: https://www.compositepanel.org/products/medium-density-fiberboard/ (accessed on 21 October 2024).
8. Ghabezi, P.; Khoran, M. Optimization of Drilling Parameters in Composite Sandwich Structures (PVC Core). *Indian J. Sci. Res.* **2014**, *2*, 173–179.
9. Ghabezi, P.; Farahani, M.; Shahmirzaloo, A.; Ghorbani, H.; Harrison, N.M. Defect evaluation of the honeycomb structures formed during the drilling process. *Int. J. Damage Mech.* **2020**, *29*, 454–466.
10. Yasar, N.K.; Korkmaz, M.E.; Günay, M. Investigation on hole quality of cutting conditions in drilling of CFRP composite. *MATEC Web Conf.* **2017**, *112*, 01013. [CrossRef]
11. Rajkumar, D.; Ranjithkumar, P.; Jenarthanan, M.P.; Sathiya Narayanan, C. Experimental investigation and analysis of factors influencing delamination and thrust force during drilling of carbon-fibre reinforced polymer composites. *Pigment. Resin Technol.* **2017**, *46*, 507–524.
12. Xu, J.; Li, L.; Geier, N.; Davim, J.P.; Chen, M. Experimental study of drilling behaviors and damage issues for woven GFRP composites using special drills. *J. Mater. Res. Technol.* **2022**, *21*, 1256–1273.
13. Martins, A.; Carvalho, A.; Bragança, I.M.F.; Barbosa, I.C.J.; Barbosa, J.I.; Loja, M.A.R. A Statistical Assessment of Drilling Effects on Glass Fiber-Reinforced Polymeric Composites. *Materials* **2024**, *17*, 5631. [CrossRef]
14. Geng, D.; Liu, Y.; Shao, Z.; Lu, Z.; Cai, J.; Li, X.; Jiang, X.; Zhang, D. Delamination formation, evaluation and suppression during drilling of composite laminates: A review. *Compos. Struct.* **2019**, *216*, 168–186. [CrossRef]
15. Kumar, G.; Rangappa, S.M.; Siengchin, S.; Zafar, S. A review of recent advancements in drilling of fiber-reinforced polymer composites. *Compos. Part C Open Access* **2022**, *9*, 100312. [CrossRef]
16. Xu, J.; Geier, N.; Shen, J.; Krishnaraj, V.; Samsudeensadham, S. A review on CFRP drilling: Fundamental mechanisms, damage issues, and approaches toward high-quality drilling. *J. Mater. Res. Technol.* **2023**, *24*, 9677–9707.
17. Heisel, U.; Pfeifroth, T. Influence of Point Angle on Drill Hole Quality and Machining Forces when Drilling CFRP. *CFRP Procedia Cirp* **2012**, *1*, 471–476.
18. Feito, N.; Díaz-Álvarez, J.; Díaz-Álvarez, A.; Cantero, J.L.; Miguélez, M.H. Experimental Analysis of the Influence of Drill Point Angle and Wear on the Drilling of Woven CFRPs. *Materials* **2014**, *7*, 4258–4271. [CrossRef]
19. Tejas, N.; Rahul, M.C. Effect of Point Angle in Twist Drill Bit on Delamination in CFRP. *Mater. Today Proc.* **2020**, *21*, 1278–1282.
20. Pardo, A.; Le Gall, J.; Heinemann, R.; Bagshaw, L. The impact of tool point angle and interlayer gap width on interface borehole quality in drilling CFRP/titanium stacks. *Int. J. Adv. Manuf. Technol.* **2021**, *114*, 159–171. [CrossRef]
21. Durão, L.M.P.; Tavares, J.M.R.S.; de Albuquerque, V.H.C.; Marques, J.F.S.; Andrade, O.N.G. Drilling Damage in Composite Material. *Materials* **2014**, *7*, 3802–3819. [CrossRef] [PubMed]
22. Silva, P.; Matos, J.E.; Durão, L.M.P. Analysis of damage outcome in the strength of polymer composite materials. *J. Compos. Mater.* **2019**, *53*, 547–560.
23. Camanho, P.P.; Erçin, G.H.; Catalanotti, G.; Mahdi, S.; Linde, P. A finite fracture mechanics model for the prediction of the open-hole strength of composite laminates. *Compos. Part A* **2012**, *43*, 1219–1225. [CrossRef]
24. Bao, H.; Liu, G. Progressive failure analysis on scaled open-hole tensile composite laminates. *Compos. Struct.* **2016**, *150*, 173–180.
25. Waqar, M.; Memon, A.M.; Sabih, M.; Alhems, L.M. Composite pipelines: Analyzing defects and advancements in non-destructive testing techniques. *Eng. Fail. Anal.* **2024**, *157*, 107914. [CrossRef]
26. Kapadia, A. Best Practice Guide Non Destructive Testing of Composite Materials. National Composites Network. Available online: https://avaloncsl.com/wp-content/uploads/2013/01/ncn-best-practice-ndt.pdf (accessed on 12 November 2024).
27. Gholizadeh, S. A review of non-destructive testing methods of composite materials. *Procedia Struct. Integr.* **2016**, *1*, 50–57. [CrossRef]
28. Garcea, S.C.; Wang, Y.; Withers, P.J. X-ray computed tomography of polymer composites. *Compos. Sci. Technol.* **2018**, *156*, 305–319. [CrossRef]
29. Song, X.; Zhou, J.; Zhang, D.; Zhang, S.; Li, P.; Bai, L.; Yang, X.; Du, F.; Wang, J.; Chen, X.; et al. Trans-scale analysis of 3D braided composites with voids based on micro-CT imaging and unsupervised machine learning. *Compos. Sci. Technol.* **2024**, *249*, 110494. [CrossRef]

30. Tsao, C.C.; Hocheng, H. Computerized tomography and C-Scan for measuring delamination in the drilling of composite materials using various drills. *Int. J. Mach. Tools Manuf.* **2005**, *45*, 1282–1287.
31. Wang, Y.; Burnett, T.L.; Chai, Y.; Soutis, C.; Hogg, P.J.; Withers, P.J. X-ray computed tomography study of kink bands in unidirectional composites. *Compos. Struct.* **2017**, *160*, 917–924.
32. Coelho, B. Deteção de defeitos em materiais compósitos usando tecnologia de ultrassons. Universidade do Minho, 2020. Available online: https://repositorium.sdum.uminho.pt/handle/1822/65770 (accessed on 22 November 2024). (In Portuguese).
33. Xu, C.; Zhang, W.; Wu, C.; Xie, J.; Yin, X.; Chen, G. An improved method of Eddy current pulsed thermography to detect subsurface defects in glass fiber reinforced polymer composites. *Compos. Struct.* **2020**, *242*, 112145.
34. Versaci, M.; Laganà, F.; Morabito, F.C.; Palumbo, A.; Angiulli, G. Adaptation of an Eddy Current Model for Characterizing Subsurface Defects in CFRP Plates Using FEM Analysis Based on Energy Functional. *Mathematics* **2024**, *12*, 2854. [CrossRef]
35. Strąg, M.; Świderski, W. Non-destructive inspection of military-designated composite materials with the use of Terahertz imaging. *Compos. Struct.* **2023**, *306*, 116588. [CrossRef]
36. Besson, A.; Minasyan, A. Terahertz Imaging for Composite Non Destructive Testing. In Proceedings of the 15th Asia Pacific Conference for Non-Destructive Testing, Singapore, 13–17 November 2017; Volume 23. Available online: https://www.ndt.net/?id=22114 (accessed on 15 November 2024).
37. Durão, L.M.P.; Tavares, J.M.R.S.; de Albuquerque, V.H.C.; Goncalves, D.J.S. Damage evaluation of drilled carbon/epoxy laminates based on area assessment methods. *Compos. Struct.* **2013**, *96*, 576–583. [CrossRef]
38. Silva, D.; Teixeira, J.P.; Machado, C.M. Methodology analysis for evaluation of drilling-induced damage composites. *Int. J. Adv. Manuf. Technol.* **2014**, *71*, 1919–1928. [CrossRef]
39. De Albuquerque, V.H.C.; Tavares, J.M.R.S.; Durão, L.M.P. Evaluation of Delamination Damage on Composite Plates using an Artificial Neural Network for the Radiographic Image Analysis. *J. Compos. Mater.* **2009**, *44*, 1139–1159. [CrossRef]
40. Caggiano, A.; Mattera, G.; Nele, L. Smart Tool Wear Monitoring of CFRP/CFRP Stack Drilling Using Autoencoders and Memory-Based Neural Networks. *Appl. Sci.* **2023**, *13*, 3307. [CrossRef]
41. Chen, W.-C. Some experimental investigations in the drilling of carbon fiber-reinforced plastic (CFRP) composite laminates. *Int. J. Mach. Tools Manuf.* **1997**, *37*, 1097–1108. [CrossRef]
42. Lissek, F.; Tegas, J.; Kaufeld, M. Damage quantification for machining of CFRP: An introduction about characteristic values considering shape and orientation of drilling-induced delamination. *Procedia Eng.* **2016**, *149*, 2–16. [CrossRef]
43. Davim, J.P.; Campos Rubio, J.; Abrao, A.M. A novel approach based on digital image analysis to evaluate the delamination factor after drilling composite laminates. *Compos. Sci. Technol.* **2007**, *67*, 1939–1945. [CrossRef]
44. Tsao, C.C.; Kuo, K.L.; Hsu, I.C. Evaluation of a novel approach to a delamination factor after drilling composite laminates using a core-saw drill. *Int. J. Adv. Manuf. Technol.* **2012**, *59*, 617–622. [CrossRef]
45. Xu, J.; Li, C.; Mi, S.; An, Q.; Chen, M. Study of drilling-induced defects for CFRP composites using new criteria. *Compos. Struct.* **2018**, *201*, 1076–1087. [CrossRef]
46. Durão, L.M.P.; Gonçalves, D.J.; Tavares, J.M.R.; de Albuquerque, V.H.C.; Marques, A.T. Comparative analysis of drills for composite laminates. *J. Compos. Mater.* **2011**, *46*, 1649–1659. [CrossRef]
47. Wang, G.D.; Kirwa, M.S. Comparisons of the use of twist, pilot-hole and step-drill on influence of carbon fiber-reinforced polymer drilling hole quality. *J. Compos. Mater.* **2018**, *52*, 1465–1480. [CrossRef]
48. Hocheng, H.; Tsao, C.C. Effects of special drill bits on drilling-induced delamination of composite materials. *Int. J. Mach. Tools Manuf.* **2006**, *46*, 1403–1416. [CrossRef]
49. Tsao, C.C. The effect of pilot hole on delamination when core drilling composite materials. *Int. J. Mach. Tools Manuf.* **2006**, *46*, 1653–1661.
50. Durão, L.M.P.; Matos, J.E.; Loureiro, N.C.; Esteves, J.L.; Fernandes, S.C.F. Damage Propagation by Cyclic Loading in Drilled Carbon/Epoxy Plates. *Materials* **2023**, *16*, 2688. [CrossRef] [PubMed]
51. Roy, R.K. A Primer on the Taguchi Method—Google Books. Available online: https://books.google.pt/books?id=OUI54mrYdqIC&printsec=frontcover&hl=pt-PT#v=onepage&q&f=false (accessed on 13 December 2024).
52. Taguchi Robust Design and Loss Function. Available online: https://sixsigmastudyguide.com/taguchi-robust-design/ (accessed on 16 December 2024).
53. *ISO 527-1*; Plastics—Determination of tensile properties. Part 1: General principles. International Organization for Standardization: Geneva, Switzerland, 2019.
54. Hocheng, H.; Dharan, C.K.H. Delamination during drilling in composite laminates. *J. Eng. Ind.* **1990**, *112*, 236–239.
55. Karpat, Y.; Karagüzel, U.; Bahtiyar, O. A thermo-mechanical model of drill margin-borehole surface interface contact conditions in dry drilling of thick CFRP laminates. *Int. J. Mach. Tools Manuf.* **2020**, *154*, 103565.

56. Wysmulski, P. Failure Mechanism of Tensile CFRP Composite Plates with Variable Hole Diameter. *Materials* **2023**, *16*, 4714. [CrossRef]
57. *ASTM D5961*; Standard Test Method for Bearing Response of Polymer Matrix Composite Laminates. ASTM International: West Conshohocken, PA, USA, 2017.

MDPI AG
Grosspeteranlage 5
4052 Basel
Switzerland
Tel.: +41 61 683 77 34

Materials Editorial Office
E-mail: materials@mdpi.com
www.mdpi.com/journal/materials

www.ingramcontent.com/pod-product-compliance
Lightning Source LLC
Chambersburg PA
CBHW061522210926
51850CB00008B/2005